Hydrocarbon Reservoir Characterization

Geologic Framework and Flow Unit Modeling

Organized and Edited

by

Emily L. Stoudt

and

Paul M. Harris

SEPM Short Course No. 34
Houston, March 4 - 5, 1995

Copyright 1995 by SEPM (Society for Sedimentary Geology)

SEPM gratefully acknowledges financial support from the Harlton Publication Fund of the SEPM Foundation, Texaco, and Chevron Petroleum Technology Company for the publication of this volume.

These SEPM Short Course Notes have received independent peer review. In order to facilitate rapid publication, these notes have not been subjected to the editorial review required for SEPM Special Publications.

ISBN #1-56576-019-0

Additional copies of this publication may be ordered from SEPM. Send your order to:

SEPM
P. O. Box 4756
Tulsa, Oklahoma 74159-0756
U.S.A.

© Copyright 1995 by

SEPM (Society for Sedimentary Geology)
Printed in the United States of America

PREFACE

SEPM Short Course #34 - **Hydrocarbon Reservoir Characterization - Geologic Framework and Flow Unit Modeling** - and the associated compilation of papers were designed to demonstrate to development geologists and engineers a variety of effective methods for creating valid conceptual and numerical geologic models of producing hydrocarbon reservoirs for use in engineering flow simulation. We feel that the continuing shift in emphasis in the United States from exploration to maximizing recovery of hydrocarbons from existing fields makes the topic extremely timely.

The eight papers included within the course notes summarize up-to-date studies of reservoirs of varied ages, geographic locations, geologic frameworks, and stages of reservoir development. These reservoir examples most definitely illustrate a broad spectrum of reservoir problems that face development geologists and reservoir engineers and varied approaches that are being utilized to attack the problems. There is a strong emphasis in the papers on the need for detailed rock studies in a reservoir analysis to provide facies heterogeneity and layering data, and suggestions are given on how to combine the geologic data with other field data, i.e., core analysis data and engineering performance analysis, to create the most realistic reservoir models.

- The first two papers discuss in detail the geologic and engineering data for Aneth Field in southeast Utah, currently under waterflood and producing oil from a Middle Pennsylvanian-age carbonate.

- The large volume of residual, mobile oil in Permian-age carbonate and mixed carbonate-siliciclastic San Andres/Grayburg reservoirs of the Permian Basin in west Texas and southeastern New Mexico has been an impetus for enhanced oil recovery projects. Four reservoir studies are included as the next grouping of papers: Yates and Seminole fields that are under waterflood, Mabee Field currently under a CO_2 miscible flood, and McElroy Field, under waterflood but with a pilot CO_2 project under way.

- The Hatter's Pond Field of Alabama produces wet gas from both Upper Jurassic-age siliciclastics of the Norphlet Formation and carbonates of the Smackover Formation. The geological and reservoir engineering studies involved in building an integrated model for this reservoir, which is now under gas injection, are discussed in the next paper.

- The final paper concerns the Kern River Field of California, which is under steamflood and producing oil from late Miocene- to Pleistocene-age siliciclastics.

This collection of papers presents documentation for (1) approaches to be taken in developing a geologic framework for explaining layering, heterogeneity, and compartmentalization of a reservoir; (2) the value of outcrop data in improving understanding of reservoir performance; (3) methods for integrating, analyzing, and displaying geologic, petrophysical rock property, and engineering data to be used during field evaluation, management, and simulation; (4) geostatistical approaches that are being used to characterize the spatial distribution of reservoir properties and augment geologic descriptions; and (5) methods of displaying quantitative models of reservoir properties and reservoir simulation in three dimensions.

The workshop and notes were made possible with the help of many people. We especially thank the authors and their respective companies for their contributions at the workshop and to the notes. The SEPM staff and Continuing Education Committee supported the concept of the workshop and handled the logistical preparations. We thank Texaco E&P Technology Department and Chevron Petroleum Technology Company for supporting our efforts in organizing the workshop and editing the notes. We especially thank Julie T. Law of Chevron Petroleum Technology Company for her help in editing and typing of the manuscripts.

Emily L. Stoudt and Paul M. (Mitch) Harris

HYDROCARBON RESERVOIR CHARACTERIZATION - GEOLOGIC FRAMEWORK AND FLOW UNIT MODELING

TABLE OF CONTENTS

RESERVOIR DELINEATION AND PERFORMANCE: APPLICATION OF SEQUENCE STRATIGRAPHY AND INTEGRATION OF PETROPHYSICS AND ENGINEERING DATA, ANETH FIELD, SOUTHEAST UTAH, U.S.A.
By L. James Weber, Frank M. Wright, J. F. (Rick) Sarg, Ed Shaw, Leslie P. Harman, Jim B. Vanderhill, and Don A. Best 1

CONTRIBUTION OF OUTCROP DATA TO IMPROVE UNDERSTANDING OF FIELD PERFORMANCE: ROCK EXPOSURES AT EIGHT FOOT RAPIDS TIED TO THE ANETH FIELD
By Donald A. Best, Frank M. Wright, III, Rajiv Sagar, and L. James Weber 31

RESERVOIR CHARACTERIZATION OF A PERMIAN GIANT: YATES FIELD, WEST TEXAS
By Scott W. Tinker and Denise H. Mruk 51

FLUID-FLOW CHARACTERIZATION OF DOLOMITIZED CARBONATE RAMP RESERVOIRS: SAN ANDRES FORMATION (PERMIAN) OF SEMINOLE FIELD, AND ALGERITA ESCARPMENT, PERMIAN BASIN, TEXAS AND NEW MEXICO
By F. Jerry Lucia, Charles Kerans, and Fred P. Wang 129

RESERVOIR CHARACTERIZATION AND THE APPLICATION OF GEOSTATISTICS TO THREE-DIMENSIONAL MODELING OF A SHALLOW-RAMP CARBONATE, MABEE SAN ANDRES FIELD, ANDREWS AND MARTIN COUNTIES, TEXAS
By Dennis W. Dull 155

GEOSTATISTICAL INTEGRATION OF CROSSWELL DATA FOR CARBONATE RESERVOIR MODELING, MCELROY FIELD, TEXAS
By William M. Bashore, Robert T. Langan, Karla E. Tucker, and Paul J. Griffith 199

RESERVOIR CHARACTERIZATION AND MODELING OF THE JURASSIC SMACKOVER AND NORPHLET FORMATIONS, HATTER'S POND UNIT, MOBILE COUNTY, ALABAMA
By Elliott P. Ginger, Andrew R. Thomas, W. David George, and Emily L. Stoudt 227

THE IMPACT OF GEOLOGIC RESERVOIR CHARACTERIZATION ON THE FLOW UNIT MODELING AT THE KERN RIVER FIELD, CALIFORNIA, U.S.A.
By Elliott P. Ginger, William R. Almon, Susan A. Longacre, and Cynthia A. Huggins .. 317

RESERVOIR DELINEATION AND PERFORMANCE: APPLICATION OF SEQUENCE STRATIGRAPHY AND INTEGRATION OF PETROPHYSICS AND ENGINEERING DATA, ANETH FIELD, SOUTHEAST UTAH, U.S.A.

L. JAMES WEBER,[1] FRANK M. WRIGHT,[2] J. F. (RICK) SARG,[1] ED SHAW,[2]
LESLIE P. HARMAN,[3] JIM B. VANDERHILL,[3] AND DON A. BEST[2]

[1]Mobil Exploration and Producing Technical Center,
P. O. Box 650232, Dallas, Texas 75265-0232;
[2]Mobil Exploration and Producing Technical Center,
P. O. Box 819047, Farmers Branch, Texas 75381-9047;
[3]and Mobil Exploration and Producing U.S., Inc.,
P. O. Box 633, Midland, Texas 79702

ABSTRACT

Rock petrophysical data and production/performance (i.e., engineering) data from the McElmo Creek Unit of the Giant Aneth Field, southeastern Utah, were integrated into a reservoir architecture or stratigraphic layer model that is based on high-resolution sequence stratigraphy. The layer model describes the architecture of high-frequency depositional cyclicity. Nineteen layers (i.e., parasequences or depositional cycles) are described within the Middle Pennsylvanian Desert Creek and lower Ismay section at McElmo Creek. Time-slice mapping of these synchronous layers, and of the facies contained within the layers, provides the basis for predicting the distribution and continuity of reservoirs. Geologic maps and cross sections were constructed to illustrate facies distribution and predict reservoir quality and continuity. Facies and layering data, coupled with core analysis data and engineering performance analysis, contribute to the understanding of fluid pathways. Several examples are selected that relate reservoir performance to changes in facies.

At the field scale, porous and permeable facies stack to form a thick and areally extensive reservoir. Reservoir performance is related to the position of shallow-water facies along a platform-to-basin transect. High-resolution sequence stratigraphy provides architectural detail that permits mapping of successive stages of platform development. Field performance is improved in areas where injection and production wells are completed in platform algal buildup facies.

Facies heterogeneity and reservoir compartmentalization occur within a synchronous, genetically related cycle of deposition. Production/performance anomalies are observed at the interwell scale in laterally discrete reservoirs. Geologic maps show the distribution of facies and are used to predict the occurrence of reservoir quality rock. Engineering maps are compared with geologic maps to identify wells or areas of the field that require remedial action.

Areas of improved reservoir performance are tied to diagenetic processes that crosscut depositional fabric. Basinally restricted fluids are interpreted to have flowed through porous and permeable rock along localized areas of the northern or windward margin of the carbonate platform, causing pervasive dolomitization. Cumulative oil production in these areas is much higher than for other areas of the field.

A reservoir characterization study of the McElmo Creek Unit was conducted by technologists with Mobil Exploration and Producing U.S., Inc. A synergistic approach led to an improved geologic model, but, more importantly, aspects of this study have been used to increase reserves, increase production, and decrease production costs on a $/barrel basis.

INTRODUCTION

Enhanced recovery of hydrocarbons requires an understanding of reservoir anisotropy by both geoscientists and engineers. Heterogeneity in rock properties affects fluid flow from the microscopic to reservoir scale of investigation. The ability to describe rock properties, produce oil, and predict recovery on the microscopic scale are well documented, but the ability to describe the reservoir lags behind. The challenge is to describe reservoirs in sufficient detail to identify remaining hydrocarbons, and then to produce these reserves efficiently. The focus of this paper is on the description of the reservoir and the role that the description has on understanding reservoir performance. Sequence stratigraphy is used to develop a predictive facies model. Mapping of facies distribution and integration of petrophysical and engineering data allow delineation of reservoirs and a better understanding of interwell and field-wide production and performance anomalies. As a result of this work, an action plan was implemented to maximize economic recovery of hydrocarbons.

Location and Stratigraphy

The Paradox Basin is a late Paleozoic basin that is Pennsylvanian to early Permian in age. The basin is located in the southwestern United States and covers parts of Colorado, Utah, Arizona, and New Mexico (Fig. 1). The largest oil field in the Paradox Basin is the Greater Aneth Field, located in southeastern Utah (Fig. 2). The producing area covers 75 square miles, with an original oil in place that is estimated to exceed 1.3 billion barrels. Anticipated ultimate recoverable reserves will approach 600 million barrels through Tertiary recovery. Since discovery of the Aneth Field in 1956, 385 million barrels of oil have been produced. Current daily production is approximately 17,000 barrels of oil per day.

In the Paradox Basin, oil and gas fields are developed in the Middle Pennsylvanian (Desmoinesian) Paradox Formation (Fig. 3). The Paradox Formation is divided into five intervals that are, from oldest to youngest, Alkali Gulch, Barker Creek, Akah, Desert Creek, and Ismay (Fig. 3). Areally the Paradox Formation is separated into three broadly distributed facies zones. Proximal to the Uncompahgre Uplift and extending to the south and west to the central portion of the Paradox Basin is a thick, siliciclastic wedge of quartz arenite and arkose. In the central basin area, siliciclastics interfinger with thick evaporites that are punctuated by thin, organic-rich dolostone and shale. Estimates of original depositional thickness of bedded halite, anhydrite, dolostone, and shale range from approximately 4000 to 7000 ft (1220-2135 m) (Hite, 1960; Peterson and Hite, 1969). Laterally equivalent to the clastic and evaporite units in the southern and southwestern portion of the Paradox Basin is the carbonate shelf facies (Wengerd and Strickland, 1954; Wengerd and Matheny, 1958; Peterson and Hite, 1969). According to Wengerd and Matheny (1958), the carbonate shelf facies of the Paradox Formation is as much as 1000 ft (305 m) thick. However, near local clastic source areas (e.g., Monument Upwarp, Zuni-Defiance Uplift, etc.), the carbonate shelf facies thins drastically and is clastic-rich. Conspicuous carbonate-mounded buildups are observed in the south-central and southwestern portion of the Paradox Basin and form important petroleum reservoirs. More detailed descriptions of rocks in

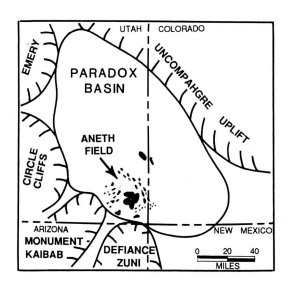

Figure 1. Paradox basin bounded by major uplifts (after Herrod and others, 1985).

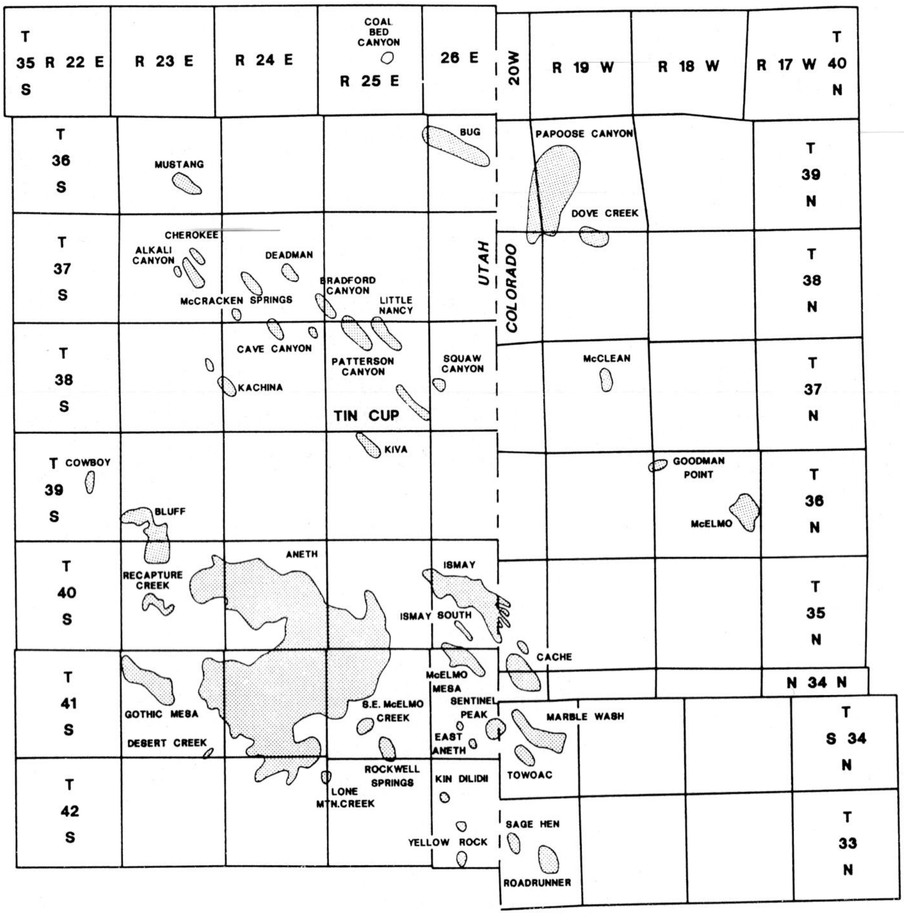

Figure 2. Selected oil and gas fields in the Paradox basin (after Herrod and Gardner, 1988).

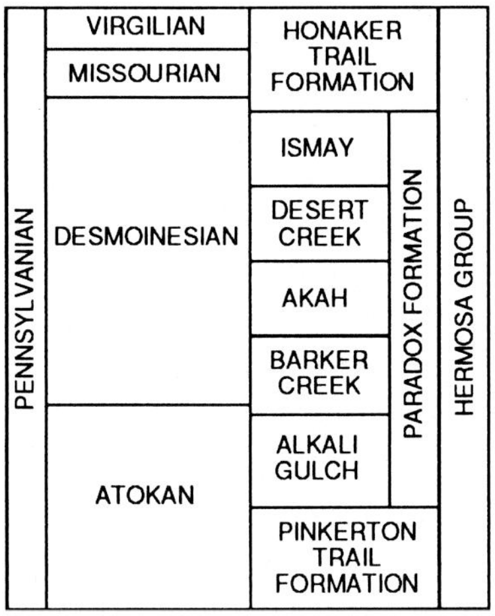

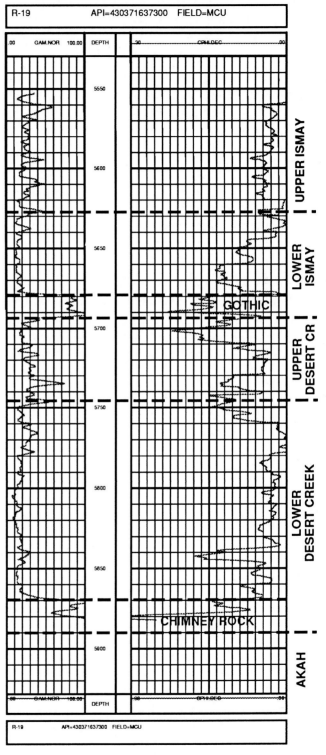

Figure 3. General stratigraphy of the Paradox basin.

Figure 4. Typical log of the McElmo Creek Unit of the Aneth field.

the Paradox Formation are given by Wengerd (1951), Choquette and Traut (1963), Peterson and Hite (1969), Hite (1970), Wilson (1975), Choquette (1983), Goldhammer and others (1991), and Weber and others (in prep.).

The Aneth Platform was deposited downdip from the carbonate shelf facies as an isolated carbonate buildup with steep flanks. A steepened platform margin is compatible with the probable north-northeast wind and south-southeast current directions existing in the basin at the time of deposition (Peterson and Ohlen, 1963). In addition, the porous algal mound intervals thin and the overall carbonate thickness decreases rapidly toward the flanks of the Aneth Platform. Net thickness of producing zones range from 200 ft (61 m) to near 0 ft at the field boundary (Peterson, 1992).

Desert Creek and Ismay reservoirs produce oil and gas in the Aneth Field. The Desert Creek and Ismay are each subdivided into lower and upper units (Fig. 4). Approximately ten percent of the total hydrocarbon pore volume resides in the lower Ismay. The remaining hydrocarbon pore volume is split equally between the lower and upper Desert Creek. The lower Desert Creek and the lower Ismay are composed of meter-scale depositional cycles that are dominated by phylloid algae. Phylloid algal communities form individual mounds that, with time, stack and coalesce into algal mound buildups, characterized by well-developed primary porosity. Reservoir quality is enhanced by solution-formed moldic and vuggy porosity. Algal buildups develop on laterally extensive tidal-flat dolostones. In the lower Ismay, dolostones have intercrystalline and moldic porosity. Much of the oil that is produced from the Ismay is derived from dolomitized tidal-flat accumulations. Oolitic and peloidal limestone reservoirs occur in the upper Desert Creek. Moldic porosity accounts for the high storage capacity of these reservoirs, but low permeability makes these reservoirs difficult to produce.

Background Information

The Greater Aneth Field is divisible into four operating units: Aneth, McElmo Creek, Ratherford, and White Mesa (Fig. 5). Currently Mobil operates McElmo Creek and Ratherford. In March 1990 Mobil Exploration and Producing U.S., Inc., Midland, initiated a reservoir characterization study of the McElmo Creek Unit. During this three-year study, geoscientists, engineers, and technical support staff worked toward a better characterization of reservoir continuity and quality. Data acquisition and quality control, database management, reservoir architecture, geologic mapping, volumetric and performance calculations and mapping, and operations support comprise the major components of this study. Eleven person years were required to complete the initial phase of work at a cost that exceeded $2MM. The emphasis of this paper is on the reservoir characterization work that was conducted within the McElmo Creek Unit. Knowledge gained from this study is applied to Mobil's operations in the Ratherford Unit. In addition, methods and concepts are applicable to producing ventures worldwide.

Prior to the McElmo Creek reservoir study, stratigraphic and reservoir layering of the Aneth Field was based on lithostratigraphic correlation (Fig. 6). Interwell correlations were determined in the early 1960s by geologists and engineers representing the major operators of the Aneth Field. Correlations were based on well-to-well linkage of similar log response and

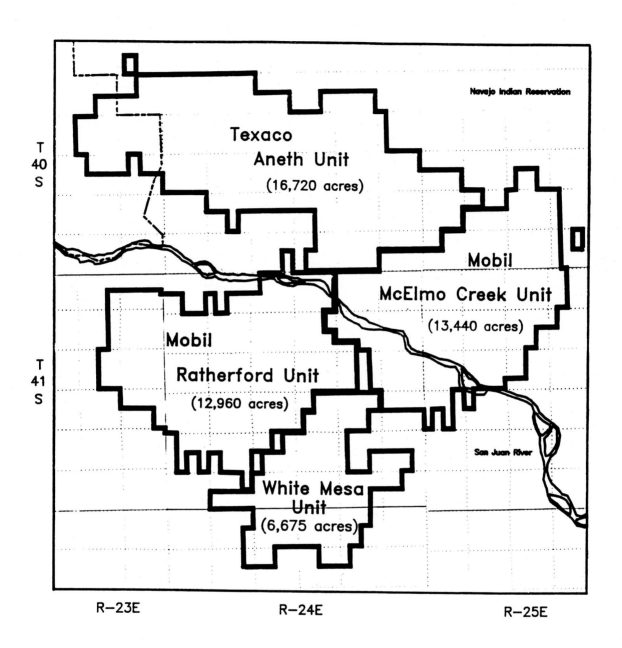

Figure 5. Location of the four operational units of the Greater Aneth field. The focus of this study is on the McElmo Creek Unit.

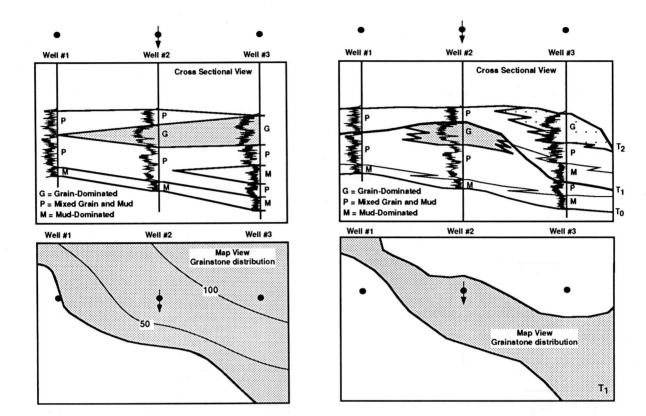

Figure 6. A hypothetical example is used to illustrate lithostratigraphic correlation. Three wells (injector with two producers) are indicated in cross-sectional view, each with a hypothetical porosity curve. Curve inflection to the left indicates higher porosity. Lithologies are annotated as G = grain-dominated (stippled pattern = main reservoir) lithology, P = mixed grain and mud lithology, and M = mud-dominated lithology. Distribution of grain-dominated reservoir rock is indicated in map view. Numbers associated with contour lines reflect thickness of grain-dominated lithology. If we assume that porosity and permeability are developed only in the grain-dominated lithology, the majority of injected fluid from Well #2 would support production in Well #3.

Figure 7. Raw data in this hypothetical example is identical to Figure 6. Rock data suggest that shallowing- and coarsening-upward depositional cycles are identified. Bounding surfaces between cycles are indicated as time lines: T_0, T_1, and T_2. Rock contained within the interval T_0 to T_1 for example, represents a genetically-related succession of rock types in time and space. Correlation of lithology, porosity, etc. can not cross-cut time surfaces. The corresponding map view for the interval T_0 to T_1 shows the distribution of reservoir (stippled pattern) and non-reservoir rock. Fluid injection into Well #2 would not support production in Well #1 or #3. In this example, natural and drilling induced fractures are not important to reservoir performance.

lithology. During the 1970s and 1980s, results of engineering performance analysis (i.e., production history, waterflood optimization studies, pressure test analysis, black oil reservoir simulation, etc.) revealed discrepancies between predicted and actual performance. Production anomalies on both the interwell and reservoir scale were not easily explained geologically.

The method of correlation used in the McElmo Creek reservoir study involves correlation of time-synchronous, genetically related units of deposition (Fig. 7). Using this sequence stratigraphic approach, interwell and field-wide correlations are improved (Fig. 8). This has led to a more realistic delineation of stratified reservoirs and to a better prediction of reservoir quality and performance. With the integration of engineering data, the geologic model is being tested and refined.

STRATIGRAPHIC FRAMEWORK, RESERVOIR FACIES, AND PORE TYPES

A sequence stratigraphic framework is established for Middle Pennsylvanian (Desmoinesian) strata within the Paradox Basin using (1) surface exposures at Honaker Trail, Raplee Anticline, and Eight Foot Rapids located 25 to 40 miles (40-64 km) west of the Aneth Field; (2) core and well logs in S.E. Utah, S.W. Colorado, N.W. New Mexico, and N.E. Arizona; and (3) seismic data (Weber, 1992; Weber and others, 1994; Weber and others, in prep.). Bounding discontinuities (major exposure and flooding surfaces) serve as regional time or chronostratigraphic surfaces that are correlative over several thousand square miles (km) in the Four Corners region. Systems tracts of 3rd-order composite sequences (1.2-1.5 Ma) comprise 4th-order sequences (~400 ka) and 5th-order depositional cycles or parasequences (<100 ka)(Weber and others, in prep.). Three 4th-order depositional sequences are recognized through the gross reservoir section in the Aneth Field. Each depositional sequence is composed of facies that were deposited during lowstand, transgressive, and highstand depositional systems. During highstands of sea level, sediment accumulation occurred on the carbonate platform; topographic lows adjacent to the platform received little or no sediment. During lowstands of relative sea level, the platform was emergent and sedimentation took place in topographic lows adjacent to the platform. Facies analysis of 15,000 ft (4575 m) of core is tied into the chronostratigraphic framework to constrain correlation of high-frequency depositional cyclicity. Mapping of facies within parasequences permits the prediction of porous and permeable facies and the characterization of variability in reservoir pore systems (Weber, 1992).

Within the McElmo Creek Unit of the Aneth Field, stratified reservoirs occur within lowstand, transgressive, and highstand systems tracts of the Desert Creek and lower Ismay intervals (Figs. 9 and 10). Siltstone, dolostone, and evaporites form lowstand wedges that were deposited 150 ft below the crest of the Aneth Platform. Porous dolomudstone and dolowackestone are productive where they onlap and pinch out against the Aneth carbonate platform and are isolated from reservoirs on the platform. Within transgressive systems tracts, lagoonal/ tidal-flat dolomudstone/wackestone comprise parasequences and display intercrystalline and solution-enhanced secondary porosity. Core analysis and production/performance data indicate that significant fluid pathways are developed in dolomudstone deposited on the carbonate platform on paleodepositional highs. In the lower Desert Creek, initial parasequences of the highstand systems tract represent a time of mound building and platform development as a result

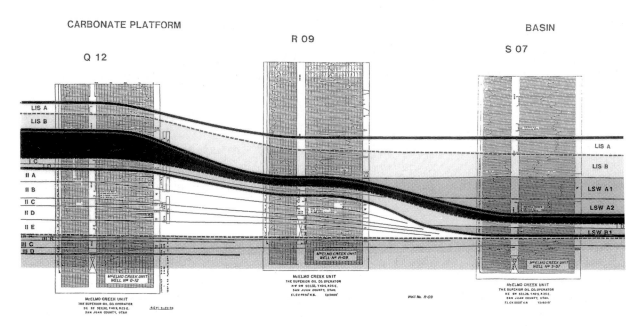

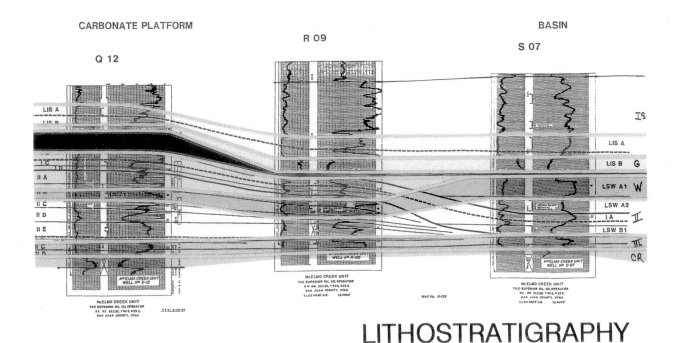

Figure 8. Comparison of the sequence stratigraphic versus the lithostratigraphic correlation. This example is a three well cross-section from the McElmo Creek Unit.

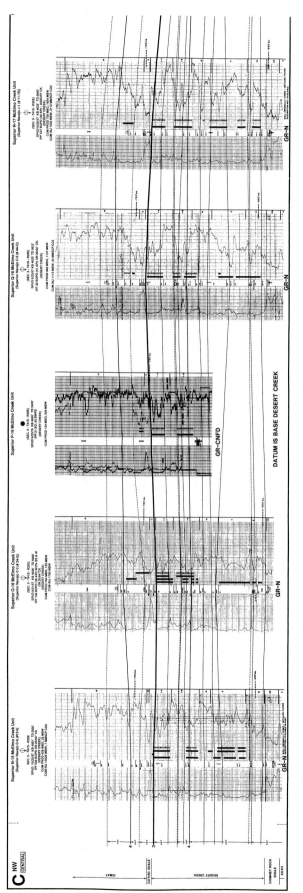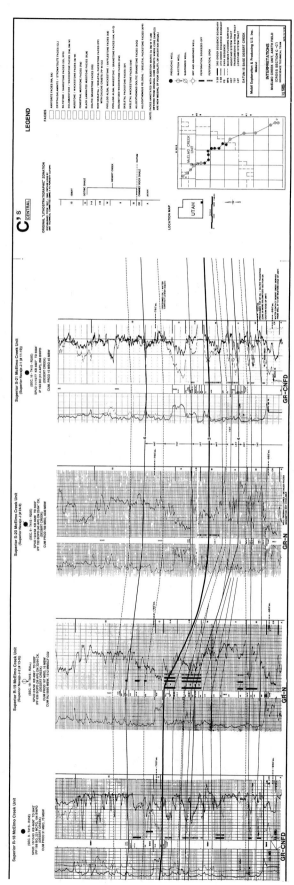

Figure 9. McElmo Creek C-C' cross-section.

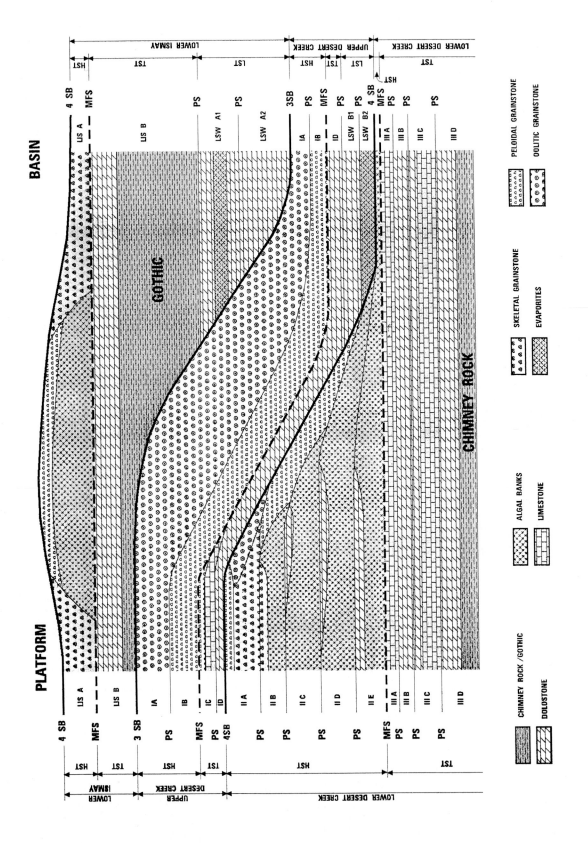

Figure 10. Generalized cross-section of the platform and basin at McElmo Creek.

of coalescing biologic communities of phylloid algae. Interparticle and shelter porosity dominate. Subsequent parasequences within the lower Desert Creek highstand systems tract are composed of skeletal and nonskeletal wackestone to grainstone. Porosity is developed on paleodepositional highs at the top of parasequences, where shoal-water facies have preserved primary pore systems that are secondarily enhanced by leaching of less stable carbonate minerals by meteoric water. Reservoirs dominated by primary pore systems provide the best long-term production and account for the majority of oil produced in McElmo Creek. In the upper Desert Creek highstand systems tract, ooid/peloid grainstones aggrade and prograde to fill available depositional space. Hydrocarbons are produced on the platform and along the platform-to-basin margin from carbonate sand sheets and allochthonous debris aprons. Grainstone debris aprons may also be deposited during early lowstand conditions of the lower Ismay sequence. On the platform, meteoric diagenesis resulted in the formation of moldic porosity in ooid grainstone deposits beneath the upper Desert Creek sequence boundary. Within moldic pore systems, storage capacity is favorable, but permeability is low, generally less than 1 or 2 md. Facies composed of moldic porosity and lacking significant primary porosity are poor reservoirs.

The stratigraphic layer model describes the architecture of high-frequency depositional cyclicity (Fig. 10). Nineteen parasequences (i.e., stratigraphic layers or depositional cycles) were described within the Desert Creek and lower Ismay section at McElmo Creek. Time-slice mapping of these synchronous layers, and of the facies contained within the layers, provides the basis for predicting the distribution and continuity of reservoirs (i.e., Predictive Geology). A regionally extensive, low-relief tidal flat serves as the stratigraphic datum. Successively younger layers are added to the datum. Each layer top mimics the paleodepositional topography, and the consequent depositional geometry is used to predict facies within each layer. Geologic maps and cross sections were constructed to illustrate facies distribution and predict reservoir quality and continuity within each depositional layer. In addition, these layers are used to allocate pore-ft of net reservoir and, ultimately, the original oil in place. Facies and layering data, coupled with core analysis data and engineering performance analysis (i.e., pattern analysis), contribute to the understanding of fluid pathways. Rock petrophysical data (i.e., porosity and permeability) and production/performance data are integrated into the reservoir architecture or stratigraphic layer model of the McElmo Creek Unit. Three examples are discussed below. This approach has led to a better understanding of (1) layer stacking, geometry, and associated production/performance anomalies at the field scale; (2) facies heterogeneity and compartmentalization within individual time-slice layers at the interwell scale; and (3) early diagenesis, including dolomitization, and the effect on porosity, permeability, and hydrocarbon recovery at the field to interwell scale.

DISCUSSION

Geometry, Facies, and Performance in Stacked Cycles

Lower Desert Creek Highstand Systems Tract (LDC-HST).—
Five shallowing-upward depositional cycles (layers IIA-IIE) characterize the LDC-HST (Fig. 10). Only the lowermost three are discussed here. These high-frequency cycles are composed primarily of algal facies that stack to form the Aneth Platform. The platform margin contracts as each successive layer is added (Fig. 11). In the basin these layers are composed primarily of

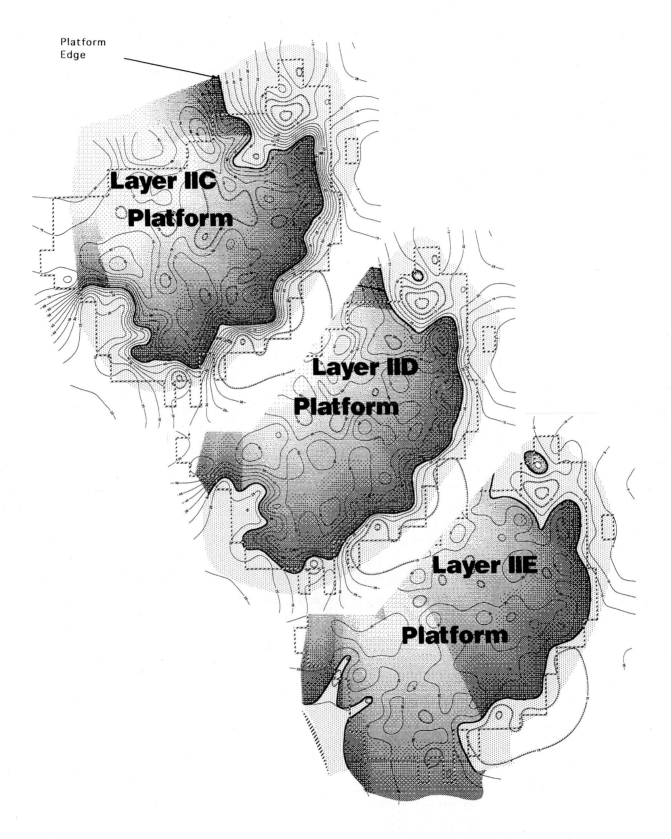

Figure 11. Paleodepositional maps for three successive layers (IIE, IID, and IIC) showing platform margin evolution through time.

skeletal wackestone and/or packstone. Basin deposits are thin and generally less than 10 to 20 ft (3-6 m) thick; whereas, on the platform, time-equivalent rocks exceed 100 ft (31 m) in thickness. Shelf-to-basin depositional relief progressively increases with each successive cycle.

Cycle caps display subaerial exposure, but duration and areal extent of exposure were limited. Subaerial exposure is indicated by the development of localized, partially dolomitized grainstone shoals and tidal flats that cap parasequences and display fenestral fabric, subvertical syndepositional fracturing, and very fine-grained dolomite crystal growth. Significantly, alveolar fabric, calcretes, silcretes, and pedolithic nodules and pisoids are absent at the tops of high-frequency cycles (layers IIE-IIC). In general, porosity and permeability are developed best in areas of subaerial exposure, where meteoric diagenesis resulted in dissolution of unstable carbonate allochems. In these areas the storage capacity of the rock was increased and permeability was enhanced within thin stringers. High-permeability thief zones occur toward the top of depositional cycles, especially in layer IIC. They direct injection fluids along conduits and reduce injection into, and production out of, adjacent strata.

Operational Significance.—
Gross changes in reservoir geometry, facies, and performance are observed in stacked, shallowing-upward depositional cycles of the lower Desert Creek (layers IIE, IID, and IIC). Porosity and permeability are generally well developed in platform facies dominated by phylloid algae. Pressure test analysis indicates that effective porosity in these layers is in excellent pressure communication. Low porosity and permeability are observed in skeletal wackestone, located downdip from time-equivalent platform lithologies. Associated with the facies change from platform to basin is a corresponding change in reservoir performance. The spatial location of platform facies evolves with time (Fig. 11). Individual well and pattern performance is adversely affected by the location of production and/or injection wells in nonreservoir facies. Thus platform development and areal distribution of reservoir layers are important to understand. The application of sequence stratigraphy provides architectural detail that permits mapping of successive stages of platform development. This understanding of the architecture allows better alignment of completion intervals.

Layers IIE, IID, and IIC account for 30 percent of the total hydrocarbon pore volume in the McElmo Creek Unit. Gross changes in reservoir geometry, facies, and performance are observed along the margin of the Aneth Platform as these depositional cycles thin, facies become mud-dominated, and production/injection rates decline. Throughout much of the McElmo Creek area, nonreservoir rock composed of skeletal wackestone and packstone is located above layer IIC. Salient features of each depositional cycle or layer are discussed below.

Layer IIE.—
Layer IIE ranges from 2 to 32 ft (1-10 m) in thickness and is composed of facies dominated by phylloid algae. An ideal shoaling cycle from base to top consists of phylloid algal bafflestone, phylloid algal packstone/grainstone, and phylloid algal wackestone (Fig. 12). This cycle may be capped by thin dolomudstone or dolomitized peloidal grainstone. Phylloid algal bafflestone was deposited in shallow water. Algal fronds are well preserved. Millimeter- and centimeter-scale

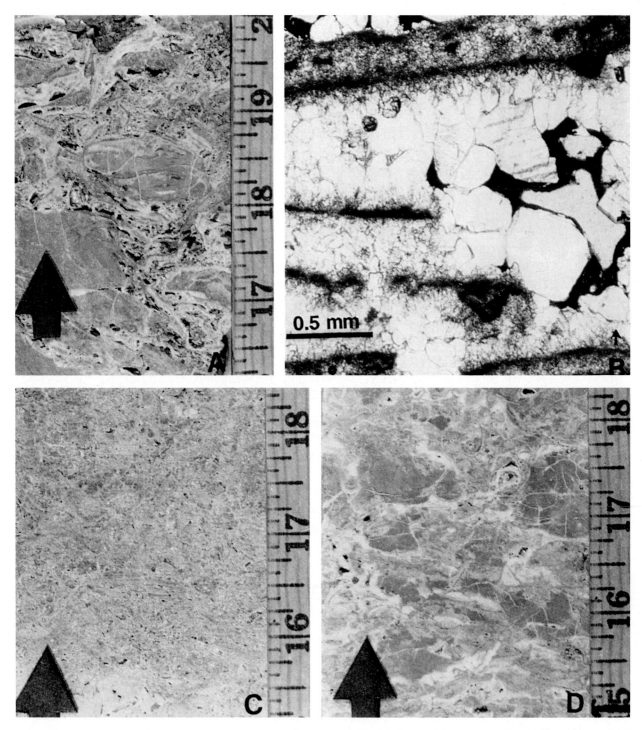

Figure 12. Selected features of lithologies dominated by phylloid algae. A. General character of phylloid algal bafflestone showing abundant primary (shelter) porosity. Slab photo, MCU K-231, 5778', scale in inches. B. Fibrous marine cement rimming phylloid algal plates. Equant calcite cement fills most remaining pore space. Note oil stained pores. Photomicrograph, MCU T-17, 5730', cross-polarized light, scale in mm. C. General character of phylloid algal packstone/grainstone. Interparticle porosity is formed by irregular packing of abraded and rounded algal plates. Slab photo, Aneth Unit K-231, 5752', scale in inches. D. General character of phylloid algal wackestone. White areas are blocky calcite cement. Molds and vugs are more abundant than interparticle porosity and result from leaching by meteoric water. Light gray areas are dolomitized. Slab photo, MCU Q-16, 5586', scale in inches.

shelter porosity is recognized as the most significant pore type. Most pores are partially filled with marine fibrous cement. Permeability is generally less than 50 md. Phylloid algal packstone/grainstone was deposited on a beach or nearshore high-energy environment. Algal fronds are fragmented, abraded, and well sorted. Porosity and permeability are generally favorable (6-10 percent and 1-1000 md). Leaching of algal plates is prevalent along paleodepositional highs (Fig. 13). Pattern and well performance data indicate that storage and flow capacity are improved in these areas. The parasequence is capped by thin phylloid algal wackestone, dolomudstone, and/or peloidal grainstone. Dolomudstone and dolomitized peloidal grainstone are observed along paleodepositional highs as isolated island complexes. On average, permeability is much lower in the capping facies.

Layer IID.—
The compositional makeup of layer IID is very similar to layer IIE, with one exception: phylloid algal bafflestone is not observed in layer IID. Layer IID ranges from 0 to 30 ft (0-9 m) in thickness. Layer IID is composed of algal grainstone on the platform overlain by partially dolomitized algal wackestone, peloidal grainstone, and/or dolomudstone. Meteoric diagenesis is responsible for the development of algal moldic and isolated vugular porosity. Blocky calcite cement partially fills primary and secondary pores toward the top of the cycle. On the platform, downdip areas are characterized by increased pore filling by blocky calcite cement. The best permeability is developed in phylloid algal grainstone intervals that are dominated by interparticle porosity. Moldic porosity does not contribute significantly to flow, but does add to storage capacity. As with layer IIE, the best performance occurs along subtle paleodepositional highs (i.e., grainstone thicks) at the top of layer IID. The meteoric diagenetic overprint adds to storage capacity in these areas.

Layer IIC.—
Layer IIC is composed of algal packstone/grainstone overlain by algal wackestone. Dolomudstone and/or dolomitized peloidal grainstone caps this depositional cycle. Layer IIC is similar to layer IID in composition and average thickness, but a big difference exists in the stratigraphic partitioning of the facies. Less depositional space was created at the base of this cycle, and, as a result, algal grainstone accounts for a much smaller proportion of the gross thickness. Algal wackestone is volumetrically the most important facies. Relative to layers IID and IIE, permeability in layer IIC is more variable and ranges from 0.01 to >1000 md. Significant thief zones identified in the algal wackestone are evident on injectivity profiles. Thief zones may be related to subaerial exposure and associated vadose and meteoric leaching that occurred on paleodepositional highs after deposition of layer IIC.

Facies and Performance in Time-Equivalent Reservoirs

Analysis of engineering data is constrained by input of geologically significant surfaces that define stratigraphic layers. Measurement of rock fabric petrophysical properties of facies is integrated into the stratigraphic layer model. Geologic and engineering data are examined for spurious outliers. Geologic concepts are refined and validated. A reservoir management plan

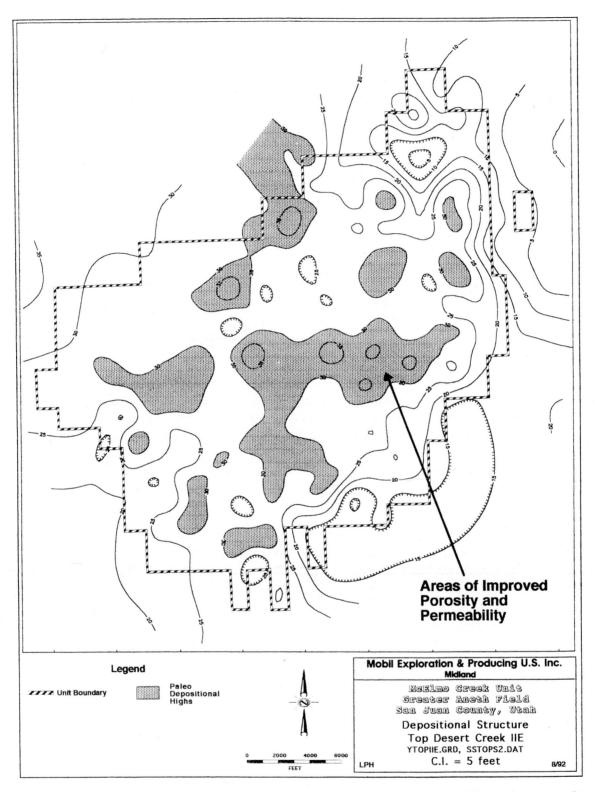

Figure 13. Depositional structure map on top of Layer IIE. Highlighted areas show paleodepositional highs where diagenesis improved storage capacity of the rock. Core analysis data suggests that permeability is improved in many of the paleohighs. Contour numbers correspond to number of feet above stratigraphic datum.

is developed that may include development and infill drilling, well recompletions, conversions, workovers, stimulation, etc., to maximize economic recovery of hydrocarbons. This approach was used in the McElmo Creek study and is presented below.

Approximately 35 percent of the total hydrocarbon pore volume within the McElmo Creek Unit occurs in the uppermost parasequence of the Desert Creek (layer IA; see Fig. 10). Performance anomalies are commonly related to vertical and lateral changes in facies within this time-synchronous, genetically related package. Variability is observed in reservoir performance because depositional processes were modified significantly by early diagenesis.

Layer 1A occurs in the highstand systems tract of the upper Desert Creek 4th-order depositional sequence. Layer 1A is bounded by nonreservoir rock. Layer 1B underlies layer 1A and is composed of mud-rich skeletal wackestone and packstone that in some areas is capped by a thin, fine-grained peloid and skeletal packstone/grainstone. The Gothic, an organic-rich, black, laminated carbonate mudstone overlies layer 1A.

Layer 1A is from 12 to 30 ft (4-9 m) thick on the platform and is made up of ooid and peloid grainstone that is subdivided into a stabilized sand flat and a mobile sand belt (Fig. 14A and B). Carbonate grains formed along the crest of the platform and were redistributed to other parts of the platform, slope, and basin. On the platform, carbonate production ceased temporarily, midway through layer 1A. Stabilization and bioturbation homogenized the ooid-rich sediment (Fig. 14A). Shortly after stabilization, carbonate production resumed and an active sand belt became reestablished. Cross-stratified bed forms, coarser grain size, and active marine cementation characterize the upper portion of layer 1A (Fig. 14B). Along the margin of the platform, ooids and peloids accumulated as prograding sand sheets or as carbonate sand debris aprons at the basin margin. Allochthonous grainstone exceeds 80 ft (24 m) in thickness at the basin margin and extends into the basin for several miles as a thin sand sheet (<10 ft in thickness).

Examination of slabbed core and thin sections, integration of porosity and permeability data, transformation of core-derived facies to log character, and the correlation of log character to noncored wells form the basis for predictive geologic assessment of reservoir continuity and quality within layer 1A. Favorable reservoir quality is developed on the platform on subtle paleodepositional highs (Figs. 15 and 16). Decreased circulation of sea water resulted in reduced marine cementation. Interparticle porosity is only partially filled by blocky calcite cement. Porosity varies from 10 to 30 percent, and permeability is generally in the tens of millidarcys. Platform areas adjacent to paleodepositional highs reveal high porosity, often exceeding 25 percent, but low permeability is typically less than 1 or 2 md. Interparticle porosity is mostly filled by marine fibrous and early meteoric bladed to equant cement. Moldic porosity is abundant but poorly connected. Within isochore thicks along the platform margin and extending into localized areas along the basin margin (Figs. 15 and 16), carbonate sand sheets fill available accommodation space. Preserved interparticle porosity is coupled with meteoric dissolution of ooid and peloid grains, resulting in rocks with high flow and storage capacity. Further downdip in the basin, allochthonous ooid grainstone is dominated by interparticle porosity. A close relationship exists between grain size and permeability. Coarsest grains onlap and abut the carbonate platform. With increasing distance from the platform, permeability decreases.

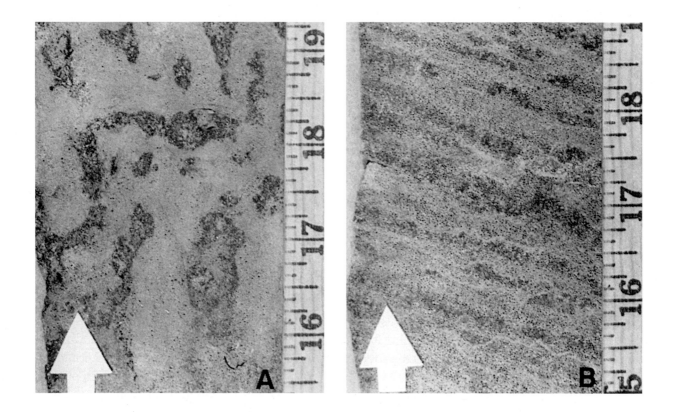

Figure 14. Selected features of lithologies dominated by ooid and peloid grainstone. A. Stabilized sand flat. Note distinctive burrows. Oomoldic porosity is visible. Slab photo, Aneth Unit K-231, 5644', scale in inches. B. Well-sorted, current stratified, and mud-free mobile sand belt. Oomoldic porosity is readily apparent. Laminae are inclined and result from changes in cementation and grain size. Aneth Unit K-231, 5628', scale in inches.

Figure 16 shows the relationship between facies, porosity, permeability, and pore type in cross-sectional view along the platform-to-basin transect. Four rock fabric, petrophysically defined flow facies are recognized within ooid and peloid grainstone in layer 1A. Flow facies are defined on the basis of geologic characteristics, position in the vertical sequence, and on petrophysical properties, especially porosity and permeability. These flow units are distinctive rock types with petrophysical characteristics significantly different from adjoining units.

Analysis of geologic data has defined perspective areas of enhanced reservoir quality within layer IA of the upper Desert Creek (Fig. 15). Engineering data are analyzed and compared to geologic interpretations to test, refine, and validate geologic concepts. As an example, a maximum water injection rate map is shown in Figure 17. This map shows the rate at which water enters layer IA. Flow rates are not related to flow into hydraulically fractured zones. Rock compressibility tests indicate that the parting pressure of the rock has not been exceeded in most areas of the field. Thus large-diameter circles represent areas of higher rates of injection corresponding to higher reservoir permeability relative to smaller circles. Areas of good reservoir quality, as interpreted from Figure 15, compare favorably to areas of high permeability in Figure 17. In general, a close correspondence exists between engineering and geologic data. However, some areas that are predicted to have good reservoir quality facies do not appear to be

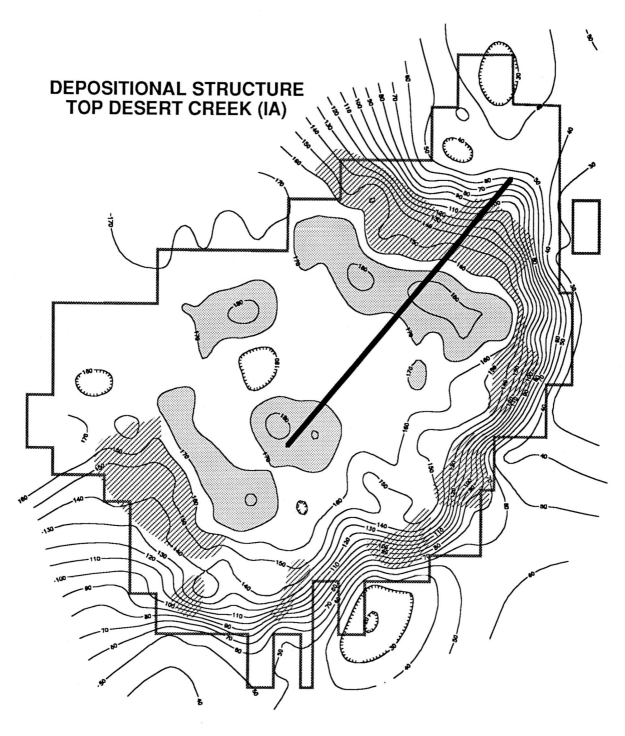

Figure 15. Depositional structure map on top of Layer IA. Dot pattern indicates subtle paleodepositional highs superimposed on a relatively flat platform surface. Fine diagonal lines (isochore thicks) show where grainstone accumulated in down dip areas along the platform margin or in the basin. The deeper basin extends to the north, east, and south. Contour numbers correspond to the number of feet above the stratigraphic datum. Heavy line indicates the location of the cross-section seen in Figure 16.

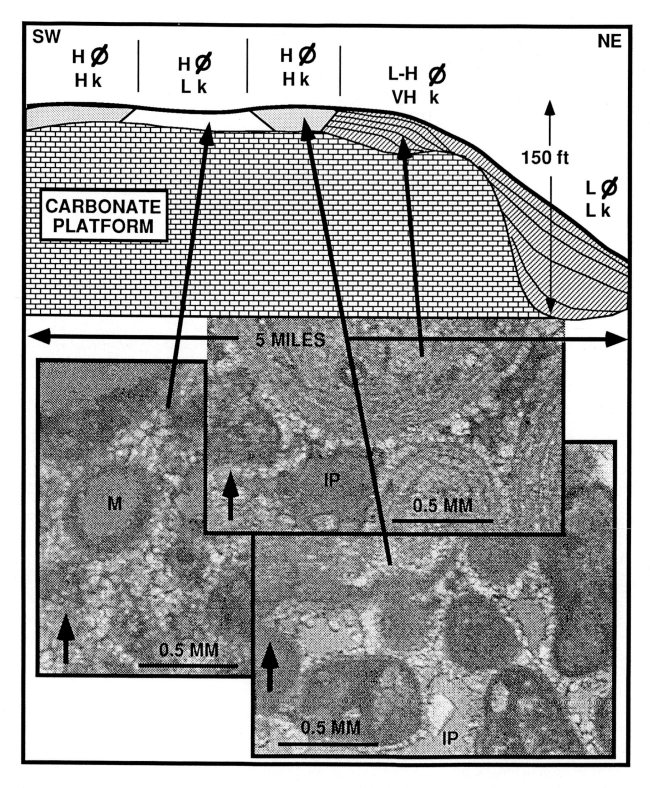

Figure 16. Cross-sectional view of Layer IA showing lateral position of rock fabric and petrophysically-defined flow facies. Photomicrographs are tied to cross-sectional view. Patterns on cross-section tie to patterns on Figure 15. Interparticle porosity (IP) is volumetrically important on the platform associated with paleohighs and along the platform-basin margin. Moldic porosity (M) is most apparent on the platform.

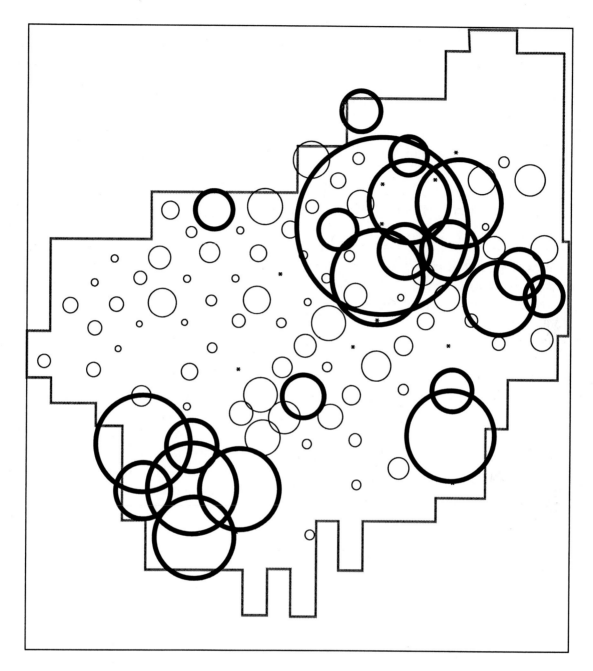

Desert Creek 1a – McElmo Creek Unit.
Maximum Water Injection Rate

Figure 17. Maximum water injection rate map showing daily injected volumes into Desert Creek Layer IA. Diameter of circle shown at the bottom of the page reflects water injection rate equal to 1000 barrels per day. Larger circles reflect higher rates, smaller circles reflect lower rates.

receiving injection support. These areas are found along the basin margin in the southeastern portion of McElmo Creek. Through this comparative approach predictive geologic maps and engineering performance maps are used to identify wells or areas of the field that require remedial action. An inventory of future work is identified to access underperforming reservoirs. A reservoir management plan was implemented that included well recompletions, conversions, workovers, and acid stimulation to maximize economic recovery of hydrocarbons. The McElmo Creek Unit has been producing for nearly 40 years, yet valuable information was gained from this work to improve well, pattern, and field-wide performance.

Predictive Model, Dolomitization, and Production Anomalies

Prior to the McElmo Creek Unit reservoir study, many areas of unusual well or pattern performance could not be explained from a geologic perspective. One area of concern involved the northern portion of the field (Fig. 18). Two specific areas are identified where individual wells have cumulative oil production in excess of 2.5 million barrels, much higher than in any other area of the field.

Log, standard core, and pressure test analyses indicate high storage and flow capacity within the lower Desert Creek section in some wells along the northern margin of McElmo Creek. Separation of neutron and density curves through the lower Desert Creek section suggests that pervasive dolomitization may contribute to better-than-expected reservoir performance. No direct rock data is available from this area (i.e., no core is available); however, to the northwest along trend, core is available in the Aneth Unit, and the lower Desert Creek section is dolomitized. A paleodepositional map drawn near the top of the lower Desert Creek depositional sequence shows two extensions of the carbonate platform (Fig. 19). These terraces are located in areas of high cumulative production.

During the lower Ismay, after deposition of the Desert Creek, relative fall of sea level exposed the carbonate platform to meteoric diagenesis. Carbonate grains were cemented and leached soon after exposure. Porous platform margin deposits (layers IA-IIE) are interpreted to have been dolomitized during this regional sea-level fall. Restricted basinal fluids interacted with porous rock along the margin of McElmo Creek and are interpreted to have caused pervasive dolomitization (Fig. 20). Dolomitization resulted in enhanced porosity and permeability. This is especially so along the northern, windward margin of the platform, where storms and other higher-energy events pushed dolomitizing fluids onto terraces, where refluxing brines percolated down through the platform margin deposits.

Production and injection wells situated on or near the terraces cycle large volumes of water. Opportunities for increased production in this area are minimal. However, opportunities may exist along trend to the northwest. Also, perhaps in some of the small isolated platforms that produce to the north and east, wells should be positioned along the northern margin to access dolomitized lithofacies with improved storage and flow capacity.

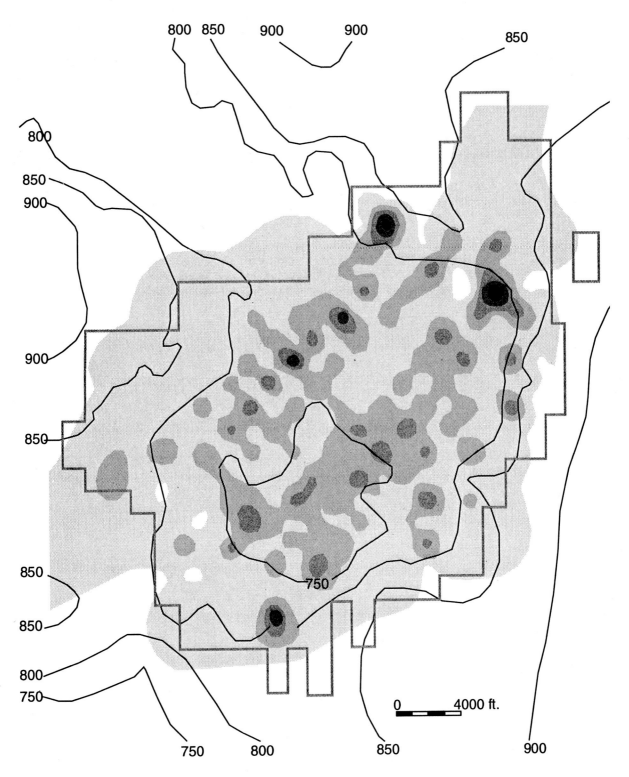

Figure 18. Map showing cumulative oil production superimposed on the present-day structural top of the Desert Creek (Layer IA). Black "blobs" along the northern part of McElmo Creek indicate production in excess of 2.5 million barrels. Map is based on co-mingled production from all producing zones from discovery to 1991. As a result, this map tends to be biased by development history, however, the primary reason for showing this map is to point out anomalous production.

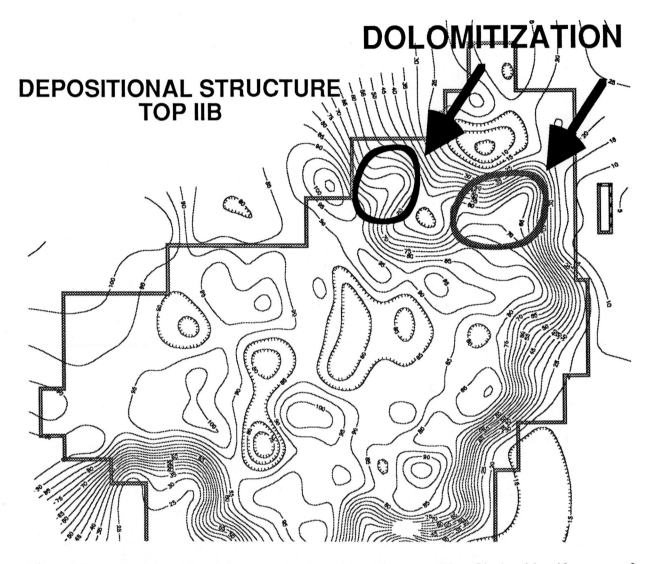

Figure 19. Paleodepositional structure map on top of Layer IIB. Circles identify areas of pervasive dolomitization. Contour numbers correspond to number of feet above the stratigraphic datum.

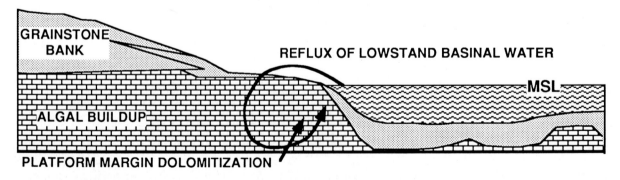

Figure 20. Platform to basin cross-section (northern McElmo Creek Unit) showing relationship between basin fluids associated with the lower Ismay lowstand systems tract and dolomitization of platform margin deposits.

CONCLUSIONS

High-resolution sequence stratigraphy is used to develop a stratigraphic layer model. Each layer is composed of a shallowing-upward succession of facies that form a depositional cycle or parasequence. Core descriptions are used to define facies and facies stacking patterns. Facies are tied to log character and are correlated from well to well.

A regionally extensive, low-relief tidal-flat dolostone serves as the stratigraphic datum. The datum occurs beneath the platform buildup. Each layer is added to the datum, and layer tops mimic paleodepositional topography. Consequent depositional geometry is used to predict facies within each layer. Core descriptions are used to calibrate facies predictions. Standard core analysis data and engineering data are integrated into the stratigraphic layering to better understand reservoir continuity and the distribution of fluid pathways. Three examples are selected that integrate geology, petrophysics, and engineering:

1. Porous and permeable facies stack into three depositional layers on the Aneth Platform (layers IIE, IID, and IIC) that form a large reservoir. Reservoir performance is related to the position of shallow-water facies along a platform-to-basin transect. Platform facies exhibit good reservoir performance, but platform margin and basinal facies that were deposited contemporaneously with platform facies are nonproductive. High-resolution sequence stratigraphy provides architectural detail that permits mapping of successive stages of platform development. Field performance is improved in areas where injection and production wells are completed in platform facies.

2. Within a synchronous, genetically related cycle of deposition (i.e., layer IA), production/performance anomalies are observed in laterally discrete reservoirs. Geologic maps show the distribution of facies within layer IA and are used to predict the occurrence of reservoir quality rock. Engineering maps are compared with geologic maps to identify wells or areas of the field that require remedial action.

3. Areas of improved reservoir performance are tied to diagenetic processes that crosscut depositional fabric. In the lower Ismay a fall in relative sea level caused exposure of the Desert Creek section on the Aneth Platform. Shallow subaqueous evaporites and dolostones accumulated in the basin surrounding the Aneth Platform. Basinally restricted fluids were pumped through porous and permeable rock along the northern or windward margin of the platform. Pervasive dolomitization enhanced porosity and permeability. Cumulative oil production in this area is much higher than for other areas of the field.

Oil and gas have been produced from the McElmo Creek Unit for almost 40 years, yet valuable information is gained from the reservoir characterization study that improves well, pattern, and field-wide performance. A synergistic approach that involved the integration of disciplines led to an improved geologic model, but, more importantly, this approach increased reserves, increased production, and decreased production costs on a $/barrel basis. Increased production and decreased operating expense paid for this study.

ACKNOWLEDGMENTS

We would like to thank the following people for their significant contribution to this project: R. Bernhart, B. Evans, G. Evans, W. McPherson, and J. Olmos. Ideas found in this paper have been improved through reviews and discussions with the following people: N. Humphreys, J. Markello, B. Goldhammer, and B. Loucks. V. Jenkins handled and prepared core and made thin sections. S. Sims constructed and plotted log cross sections. Data input into the mainframe database was coordinated by L. Pool. We are indebted to Mobil's Midland and Dallas drafting/reproduction departments for their effort.

REFERENCES

CHOQUETTE, P. W., 1983, Platy Algal Reef Mounds, Paradox Basin, *in* Scholle, P. A., Bebout, D. G., and Moore, C. H., eds., Carbonate Depositional Environments: American Association of Petroleum Geologists Memoir no. 33, p. 454-462.

CHOQUETTE, P. W., AND TRAUT, J. D., 1963, Pennsylvanian Carbonate Reservoirs, Ismay Field, Utah and Colorado, *in* Bass, R. O., ed., Shelf Carbonates of the Paradox Basin: Four Corners Geological Society 4th Field Conference Guidebook, p. 157-184.

GOLDHAMMER, R. K., OSWALD, E. J., AND DUNN, P. A., 1991, The Hierarchy of Stratigraphic Forcing: an Example from Middle Pennsylvanian Shelf Carbonates of the Paradox Basin, *in* Franseen, E. K., and others, eds., Sedimentary Modelling: Computer Simulations and Methods for Improved Parameter Definition: Kansas Geological Survey, Bulletin 233, p. 361-413.

HERROD, W. H., AND GARDNER, P. S., 1988, Upper Ismay Reservoir at Tin Cup Mesa Field, *in* Goolsby, S. M., and Longman, M. W., eds., Occurrence and Petrophysical Properties of Carbonate Reservoirs in the Rocky Mountain Region: Rocky Mountain Association of Geologists, p. 175-192.

HERROD, W. H., ROYLANCE, M. H., AND STRATHOUSE, E. C., 1985, Pennsylvanian Phylloid-Algal Mound Production at Tin Cup Mesa Field, Paradox Basin, Utah, *in* Longman, M. W., and others, eds., Rocky Mountain Carbonate Reservoirs: A Core Workshop: Society of Economic Paleontologists and Mineralogists Core Workshop No. 7, Golden, Colorado, p. 409-445.

HITE, R. J., 1960, Stratigraphy of the Saline Facies of the Paradox Member of the Hermosa Formation of Southeastern Utah and Southwestern Colorado, *in* Smith, K. G., ed., Geology of the Paradox Basin Fold and Fault Belt: Four Corners Geological Society 3rd Field Conference Guidebook, p. 86-89.

HITE, R. J., 1970, Shelf Carbonate Sedimentation Controlled by Salinity in the Paradox Basin, Southeast Utah, *in* Ron, J. L., and Dellwig, L. F., eds., Third Symposium on Salt: Northern Ohio Geologic Society, v. 1, p. 48-66.

PETERSON, J. A., 1992, Aneth Field, U.S.A., Paradox Basin, Utah, *in* Foster, N. H., and Beaumont, E. A., eds., Stratigraphic Traps III, Treatise of Petroleum Geology, Atlas of Oil and Gas Fields: American Association of Petroleum Geologists Special Publication, p. 41-82.

PETERSON, J. A., AND HITE, R. J., 1969, Pennsylvanian Evaporate Carbonate Cycles and Their Relation to Petroleum Occurrence: American Association of Petroleum Geologists Bulletin, v. 53, p. 884-908.

PETERSON, J. A., AND OHLEN, H. R., 1963, Pennsylvanian Shelf Carbonates, Paradox Basin, *in* Bass, R. O., ed., Shelf Carbonates of the Paradox Basin: Four Corners Geological Society 4th Field Conference Symposium, p. 65-79.

WEBER, L. J., 1992, Geologic Description and Aspects of Reservoir Description (Desmoinesian) Carbonates, Aneth Field, McElmo Creek Unit, Southeastern Utah: Mobil Internal Report, 290 p.

WEBER, L. J., SARG, J. F., AND WRIGHT, F. M., in prep., Sequence Stratigraphy and Reservoir Delineation of Middle Pennsylvanian (Desmoinesian) Carbonates, Aneth Field, Southeast Utah.

WEBER, L. J., SARG, J. F., WRIGHT, F. M., AND HUFFMAN, A. C., 1994, High-Resolution Sequence Stratigraphy: Reservoir Description and Geologic Setting of the Giant Aneth Oil Field, Southeast Utah: Field Trip Guidebook, Pre-Meeting Field Trip for the American Association of Petroleum Geologists Annual Convention in Denver, Colorado, sponsored by the Rocky Mountain Association of Geologists, Field Trip #2, June 8-12, 1994, 71 p.

WENGERD, S. A., 1951, Reef Limestones of Hermosa Formation, San Juan Canyon, Utah: American Association of Petroleum Geologists Bulletin, v. 35, p. 1038-1051.

WENGERD, S. A., AND MATHENY, M. L., 1958, Pennsylvanian System of the Four Corners Region: American Association of Petroleum Geologists Bulletin, v. 42, no. 9, p. 2048-2106.

WENGERD, S. A., AND STRICKLAND, J. W., 1954, Pennsylvanian Stratigraphy of the Paradox Salt Basin, Four Corners Region, Colorado and Utah: American Association of Petroleum Geologists Bulletin, v. 38, no. 10, p. 2157-2199.

WILSON, J. A., 1975, Carbonate Facies in Geologic History: Springer-Verlag, New York, 471 p.

CONTRIBUTION OF OUTCROP DATA TO IMPROVE UNDERSTANDING OF FIELD PERFORMANCE: ROCK EXPOSURES AT EIGHT FOOT RAPIDS TIED TO THE ANETH FIELD

DONALD A. BEST,[1] FRANK M. WRIGHT, III,[1] RAJIV SAGAR,[2] AND L. JAMES WEBER[3]

[1]*Mobil Exploration and Producing Technical Center, P. O. Box 819047, Farmers Branch, Texas 75381-9047;*
[2]*Kelkar & Associates, Inc., 3528 E. 104th St., Tulsa, OK 74137;*
and [3]*Mobil Exploration and Producing Technical Center, P. O. Box 650232, Dallas, Texas 75265-0232*

ABSTRACT

Building an accurate geologic model of a reservoir from well data is a challenge. Subsurface geologic data exhibit considerable variability, and interpretations are dependent on well data that are widely spaced. Outcrop analogs provide an additional source of data. Outcrops are used by geoscientists to gain a better understanding of subsurface geology, particularly reservoir architecture. Reservoir engineers would do well to follow the geoscience lead and use outcrop depiction to investigate how reservoir architecture might influence fluid flow. Rock exposures at Eight Foot Rapids exhibit an excellent Aneth Field reservoir analog, in all its detail. This locality is used to construct a model suitable for investigating, in cross section, the interwell behavior of a waterflood of the Aneth Field algal buildup reservoir. The work is particularly useful for demonstrating the impact of heterogeneity on unsteady state waterflood performance. Results illustrate the consequences of alternative well-to-well correlations.

INTRODUCTION

Commonly geoscientists study outcrops to aid in the understanding of subsurface geology, but it is less common for engineers to use outcrops as analogs for understanding subsurface fluid flow in hydrocarbon reservoirs. Yet outcrops allow us to investigate fluid flow at the intrawell, interwell, and reservoir scale, and may provide insight into areal and vertical sweep efficiency of subsurface reservoir fluids. An outcrop is selected at Eight Foot Rapids, along the San Juan River canyon in southeast Utah, to demonstrate the utility of outcrop analogs in reservoir engineering studies. The purpose of this paper is to investigate the impact of heterogeneity of rock properties on the direction of waterflood and on the reservoir performance of an unsteady state system. By design, this study focuses on two-dimensional fluid flow.

Geologists have several ways to reconstruct reservoir architecture from limited well data. Two common methods are straight line correlation of facies between wells and geostatistical re-creation. Outcrop data are used to evaluate each of these methods.

Location

The Paradox Basin is a northwest-southeast trending, paleotectonic depression of Pennsylvanian and Early Permian age that is located in the southwestern United States. The basin covers parts of Utah, Colorado, New Mexico, and Arizona (see Fig. 1 of Weber and others, this volume). In the Paradox Basin oil is produced from Middle Pennsylvanian (Desmoinesian) limestone, dolostone, and siltstone/sandstone. The best development of reservoir and seal facies occurs in the southern and southwestern part of the Paradox Basin. These facies are available for study in outcrop at Eight Foot Rapids and in the subsurface within the Greater Aneth Field. Eight Foot Rapids is located 26 miles (42 km) west of the Aneth Field.

Geologic Setting

In the Aneth Field oil is produced primarily from the Desert Creek and lower Ismay (Fig. 3 of Weber and others, this volume). Lithostratigraphically the Desert Creek extends from the base of the underlying black laminated dolomudstone, informally named the Chimney Rock, to the base of the overlying black laminated dolomudstone, informally called the Gothic. Above the Desert Creek is the lower Ismay, which is capped by a silty, dark-colored dolomudstone. A typical 1950s GNT neutron and gamma-ray log is used to show the stratigraphy (Fig. 4 of Weber and others, this volume).

In the lower Desert Creek and lower Ismay, carbonate mounds may form and are comprised of biologic communities, dominated by phylloid algae. Mounds may be isolated from one another by intermound fossil debris and carbonate mud lenses, or they may aggrade and coalesce to form laterally extensive algal buildups. Algal buildups are observed in the lower Ismay in the Aneth Field, where they produce oil and gas. Identical buildups are observed in the lower Ismay along the San Juan River canyon at Eight Foot Rapids. In this study a 400 ft (122 m) long by 70 ft (21 m) thick section of outcrop is selected for fluid-flow simulation (Fig. 1).

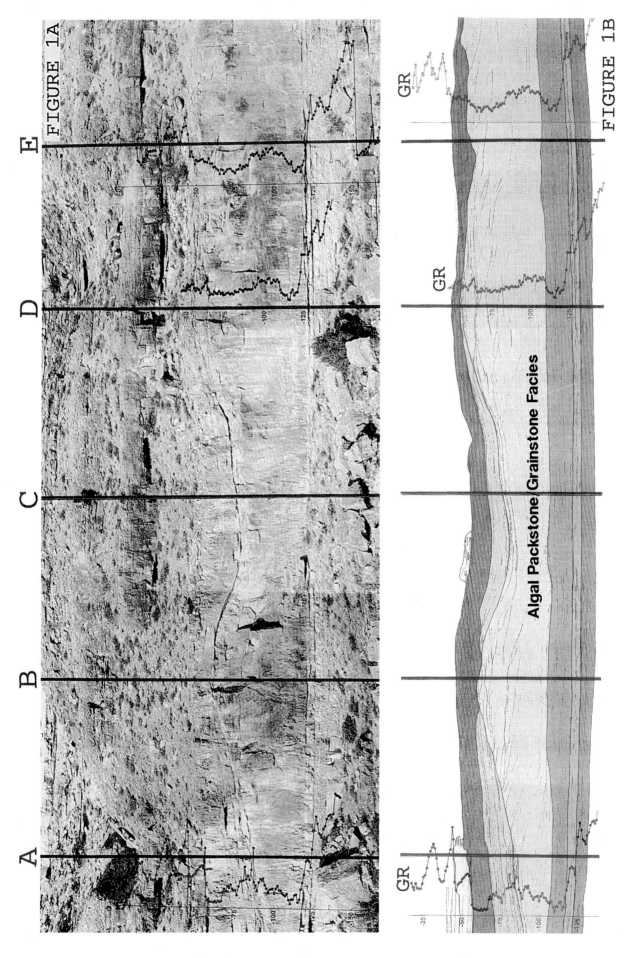

Figure 1 A & B. 1A. Shows outcrop face at Eight Foot Rapids that is selected for simulation. Five hypothetical wells are positioned along the outcrop 100 ft (31 m) apart. 1B. Depicts the distribution of facies across the rock face. Vertical thickness of the interpreted section is about 70 ft (21 m). Rock unit that displays a flat lower surface and undulating upper surface is the Algal packstone/grainstone facies. Porosity and permeability are well developed in this facies. Bounding facies are tight.

Surface exposures of the lower Ismay at Eight Foot Rapids are directly comparable to the subsurface in the Aneth Field. Facies successions in both areas are identical. The following facies are recognized from base to top in the lower Ismay: (1) black laminated carbonate mudstone and dolomudstone, (2) dolomudstone/dolowackestone, (3) skeletal-algal wackestone, (4) phylloid algal packstone/grainstone, (5) skeletal/peloid wackestone to grainstone, and (6) siltstone/sandstone. In this study the upper part of the lower Ismay is of primary interest. This interval represents the highstand systems tract of the lower Ismay, fourth-order depositional sequence (see Fig. 10 of Weber and others, this volume). The maximum flooding surface separates dolostone below from skeletal limestone above. Algal mound development was less extensive in the lower Ismay than in the lower Desert Creek. Generally algal buildups in the lower Ismay are less than 40 ft (12 m) thick and occur in areas where sufficient depositional space was available for optimum growth of phylloid algae. In the lower Ismay highstand systems tract, only the phylloid algal packstone/grainstone facies displays good reservoir quality (6-10 percent porosity, 10s of millidarcys of permeability). Generally other facies exhibit low porosity (<4 percent) and low permeability (<0.1 md). The phylloid algal packstone/grainstone facies ranges from 0 to 40 ft (0-12 m) thick and, where present, is characterized by excellent development of primary shelter porosity. Typically reservoir productivity is highest where algal mound buildup facies are encountered.

The phylloid algal packstone/grainstone facies in the lower Ismay was chosen for the engineering simulation study because it represents a confined reservoir. The phylloid algal facies is bounded by tight facies. A discussion of the phylloid algal packstone/grainstone facies is found in Weber and others (this volume).

Field History

Production history of the Paradox Basin began in 1908 with the development of an oil field at Mexican Hat, Utah. Renewed interest in the hydrocarbon potential of the Paradox Basin was stimulated by a discovery in 1954 by the Shell Oil Company. In February 1956 Texaco, when drilling an expiring lease on a low-relief, surface-mapped closure, completed the discovery well in the Aneth Field. Fifty feet (15 m) of the Desert Creek was perforated at a drilling depth of approximately 5800 ft (1770 m). Initial open-flow production of the Aneth discovery was 1702 barrels of oil per day through 2-inch tubing.

With these indications of important oil potential, exploration activity in this part of the Paradox Basin increased greatly, resulting in the discovery of a number of other oil and gas fields within a radius of 30 miles (48 km)(Fig. 2 of Weber and others, this volume). The largest and most important are the Aneth, McElmo Creek, Ratherford, and White Mesa. The development history of the area began in 1957 to 1958. Approximately 500 wells were producing by May 1960. Average daily production was on the order of 90,000 barrels of oil per day (Harman and Vanderhill, 1992). Engineering analysis revealed that the four fields were connected. They were grouped into the Greater Aneth Field.

Secondary recovery was initiated in 1961 to 1962. By 1970 daily oil production averaged less than 25,000 barrels per day. In the mid 1970s an infill drilling program was implemented and well spacing was reduced from 80 to 40 acres. In the late 1970s oil production stabilized at approximately 20,000 barrels a day. Tertiary recovery began in 1985 at McElmo Creek. In recent years oil production at McElmo Creek has increased (Harman and Vanderhill, 1992).

BASE CASE MODEL

To evaluate the effect of heterogeneity on reservoir performance, a two-dimensional model of the outcrop section at Eight Foot Rapids was created for a numerical reservoir simulation study. Facies correlations comprise the architectural detail that was observed in the outcrop (Fig. 1). Five hypothetical wells were placed along the outcrop 100 ft apart. Input data for simulation were derived from subsurface core within the McElmo Creek Unit of the Aneth Field. Standard core analysis data (permeability and porosity) from equivalent facies within the same stratigraphic position in the lower Ismay of the Aneth Reservoir were assigned on a foot-by-foot basis to each of the hypothetical wells. Spatial distribution of permeability and porosity between wells was generated geostatistically using a method known as sequential Gaussian simulation. This method retains the available well data, honors spatial structure and correlation, and preserves the facies distribution of the modeled region. The result is a 70 (vertical or "z" direction) by 60 (horizontal or "x" direction) grid block showing the distribution of permeability and porosity values (Figs. 2 and 3). The dimensions of each grid cell are 1 ft (0.3 m) in the "z" direction and 6.7 ft (2 m) in the "x" direction. A porosity-to-permeability transform was not used.

Although five wells were used to construct the cross-sectional model, only two wells, A and E, were used in the study of waterflood efficiency reported here. Both wells were assumed to be perforated along the entire interval, and both were assumed to be on primary production for one year prior to converting one of them to a water injector. Initial reservoir pressure was fixed at 5000 pounds per square inch (psi); bubble point pressure at 3500 psi. The rock properties, fluid properties, and pertinent well constraints used in this simulation are summarized in Table 1. Both wells are produced at a bottomhole pressure limit of 3500 psi, and water injection rates are limited to 50 stock tank barrels per day (STB/D). These are realistic constraints, consistent with the desire to keep production drawdown pressures above bubble point. In absolute terms, the water injection rate is low. This is a result of the size of the reservoir model used in the simulation. Fifty STB/D is significant, however, as a percentage of the oil in place within the model and is consistent with oil field practices. In the analysis that follows, the consequence of flood direction is considered between end member wells A and E.

ANALYSIS OF WATERFLOOD PERFORMANCE

Simulator-generated primary performance is shown in Figure 4. The pressure profile at 70 days of production is illustrated in Figure 5. The pressure is shown to deplete faster at well E.

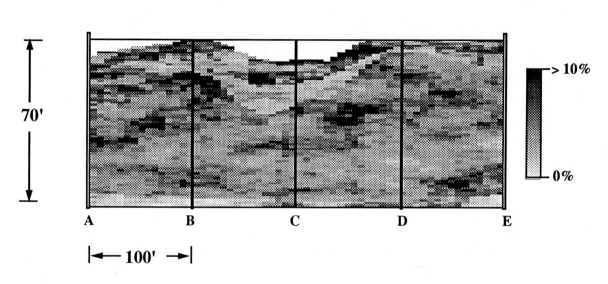

Figure 2. Eight Foot Rapids porosity map.

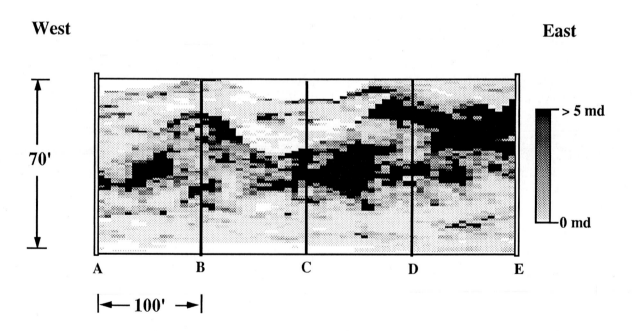

Figure 3. Eight Foot Rapids permeability map.

(Best et al., 1995) (Middle Pennsylvanian, Paradox Basin, Utah)

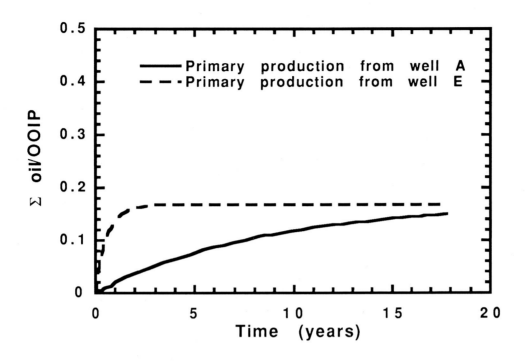

Figure 4. Primary production performance.

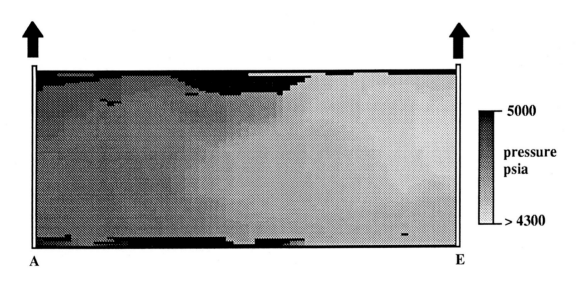

Figure 5. Primary production pressure map at 70 days.

(Best et al., 1995) (Middle Pennsylvanian, Paradox Basin, Utah)

TABLE 1. OUTCROP ROCK AND FLUID PROPERTIES AND PERFORMANCE CONSTRAINTS.

Initial pressure	5000 psia
Oil Gravity	45 °API
Residual water saturation	0.3
Formation compressibility	2×10^{-6} psi^{-1}
Oil compressibility	9×10^{-5} psi^{-1}
Water compressibility	2×10^{-5} psi^{-1}
Bo @ 5000 psi	1.8 RB/STB
Bw @ 5000 psi	1.0 RB/STB
μ_o	2.5 cp
μ_w	1 cp
Well flowing pressure	3500 psia
Water injection rate	50 STB/D
Dead oil, No free gas	

Two different water injection scenarios are considered at the end of year one: (1) inject at well A and produce at well E, and (2) inject at E and produce at A. In both cases the injection rate is 50 STB/D and the producing wells are drawn down to 3500 psi (bottomhole pressure limit).

Water saturation profiles at year 2 and 8 for the case "A to E" are shown in Figure 6a and 6b. The comparable saturation profiles for the "E to A" case are illustrated in Figure 7a and 7b. The saturation distributions clearly indicate swept regions and connected pore volume of this outcrop analog.

Watercut for each of the two injection strategies is shown as a function of time in Figure 8 and pore volumes injected in Figure 9. Breakthrough is defined as the point when the watercut curve changes from zero slope. Oil recovery is shown in Figure 10.

The watercut results (Figs. 8 and 9) show that breakthrough occurs later when injecting from well E to A, suggesting a better sweep efficiency in this direction than experienced when injecting from well A to E. However, the oil recovery results for the two scenarios (Fig. 10) indicate that a higher recovery is obtained if water is injected from well A to E. Although breakthrough time is usually considered a reasonable measure of sweep efficiency, this work shows that it may not be a direct measure of optimal strategy. The reason in this instance is due to the initial unsteady-state behavior of fluids within this reservoir section, induced by various geologic heterogeneities. This is illustrated graphically in Figure 11. The ordinate in this figure

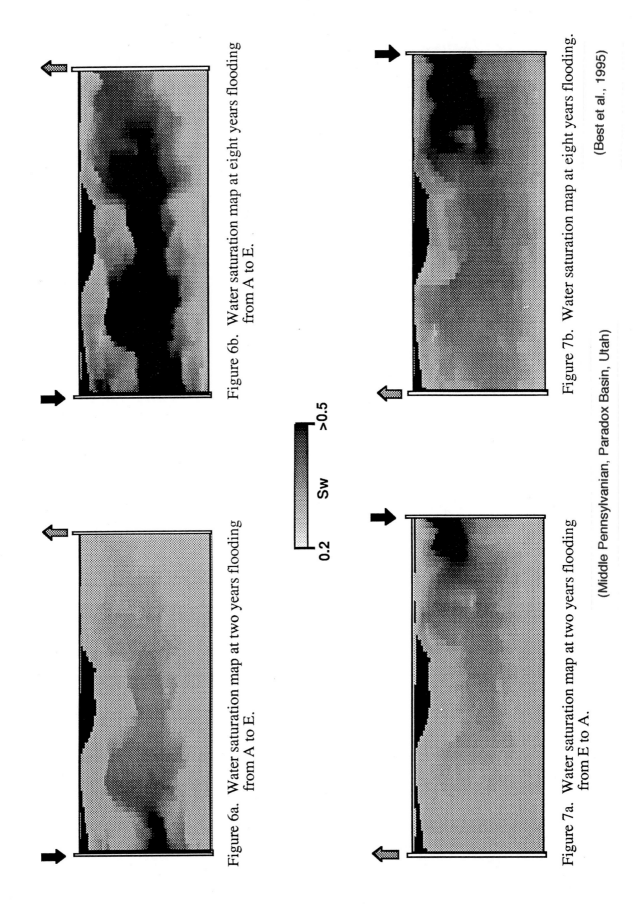

Figure 6a. Water saturation map at two years flooding from A to E.

Figure 6b. Water saturation map at eight years flooding from A to E.

Figure 7a. Water saturation map at two years flooding from E to A.

Figure 7b. Water saturation map at eight years flooding.

(Best et al., 1995)

(Middle Pennsylvanian, Paradox Basin, Utah)

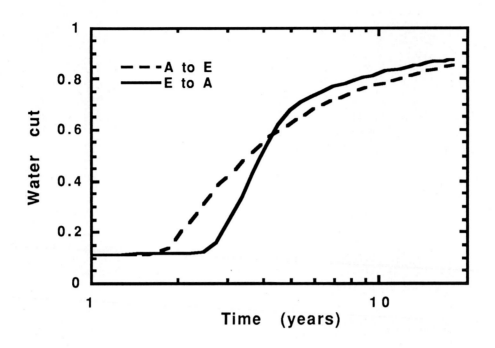

Figure 8. Water cut as a function of time.

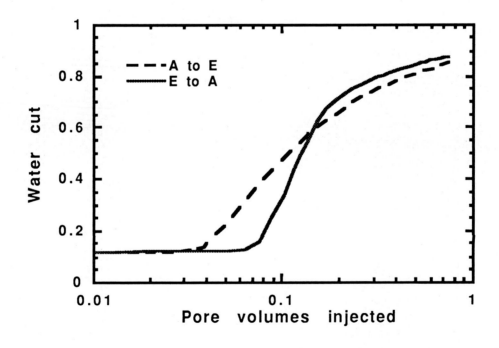

Figure 9. Water cut for pore volumes injected.

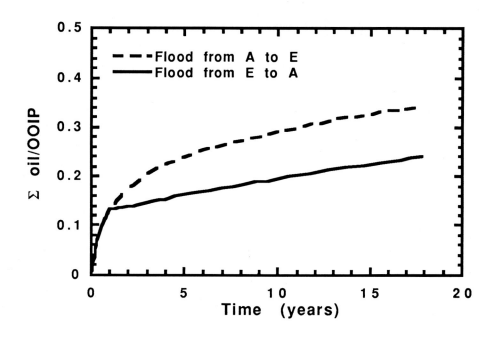

Figure 10. Cumulative oil recovery for different flood directions.

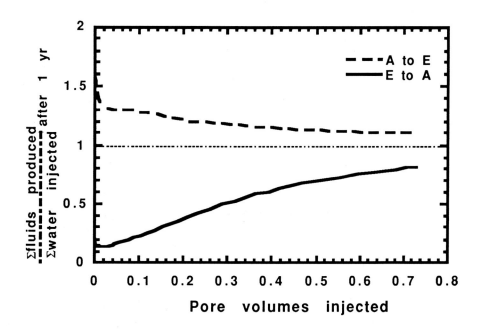

Figure 11. Cumulative recovery after the beginning of the waterflood as a function of pore volumes injected illustrating non-steady state effect.

is the ratio of cumulative fluids produced to cumulative fluids injected. If injection and production were balanced, then the ratio would be one. The ratio would be less than one for overinjection and greater than one for underinjection.

Well A has a lower injectivity and productivity than well E, due to the choking effect caused by an overall reduction of gross reservoir section and associated permeability pinchouts near well A. Injection exceeds production if 50 bwpd is injected into well E. Since there is overinjection, reservoir pressure increases. Since pressure in the reservoir increases, the production rate at A continues to increase, and, hence, the rise in the "E to A" curve in Figure 11. Production at A will continue to increase until steady state is established; that is, injection into E equals production at A.

The alternate situation exists with injection of 50 bwpd at well A. Initial production from well E is higher than 50 bpd because of the high productivity at this well related to reservoir quality and initial reservoir pressure. The reservoir is being overproduced relative to volume of fluids injected. However, the production rate in this scenario declines as reservoir pressure declines, and finally stabilizes at 50 bpd, equivalent to the injection rate.

In this study the productivity index of well A is 0.0024 STB/D/psi and that at well E is a factor of 10 higher at 0.03 STB/D/psi. Differences in productivity indices for these two wells are due to permeability variations associated with the facies. Figure 12 shows a series of permeability cross sections colored white, where permeability is greater than a given value. The lack of high-permeability continuity becomes apparent in the progress from the 0.32 md figure through to the 50 md figure.

On the basis of the production profiles illustrated in Figure 13, one would conclude that if a waterflood was carried out in this two-dimensional reservoir, then A to E would be the preferred direction. In this instance both absolute and incremental performance is better. This may have important implications to field performance. Perhaps water injection wells should be positioned along the margin of the algal buildup. Under this scenario production wells would be located updip where buildup facies are thicker. Of course, this assumes that the permeability structure of the rock does not change significantly.

Additional questions are raised as a result of this work. Would well spacing have influenced the conclusion? What about the effect of the third dimension? Would it have been sensible to use well A as a waterflood candidate in the first place? If not, what would you have used the well for? These are useful reservoir engineering questions to explore. For now, however, we will focus our discussion on the effect of geologic model-building methods on performance predictions.

ALTERNATIVE GEOLOGIC MODEL CONSTRUCTION

In this section the consequences of simplifying the reservoir architecture and the effect on prediction of performance are examined. Two alternatives have been studied and the results reported here. The first alternative is referred to as a "snap line" model of facies architecture,

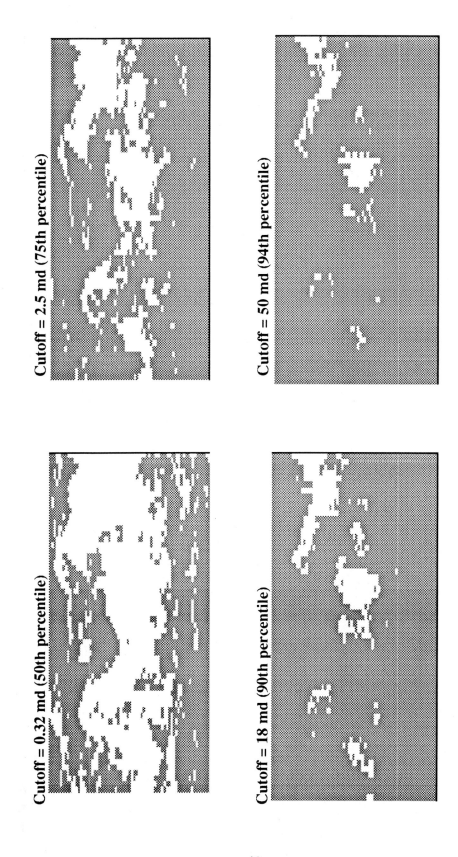

Figure 12. Permeability threshold cutoffs for Eight Foot Rapids (white is greater than cutoff value).

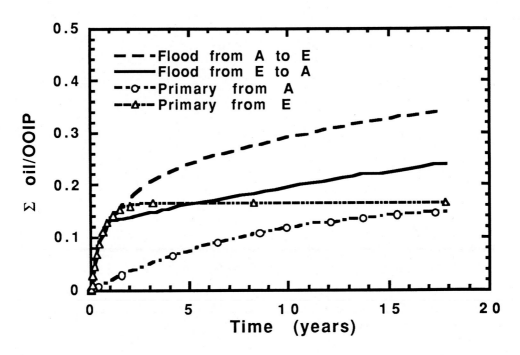

Figure 13. Cumulative oil recovery under primary and secondary recovery for different scenarios.

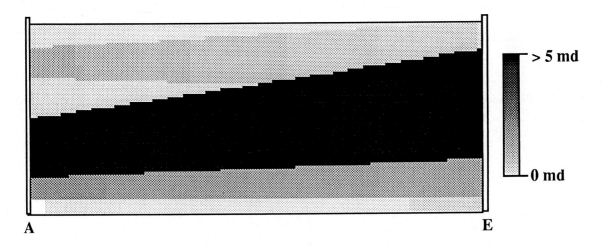

Figure 14. Snap line model of inverse distance averaged permeability.

with permeability and porosity averaged between wells. In this model, facies architecture is established by drawing straight lines between correlated features identified at wells A and E of the outcrop. In the second approach the same straight line facies architecture is used, but interwell values for permeability and porosity are generated geostatistically.

The Snap Line Model

Where a particular facies is not correlated across the section, it is terminated half way between the wells. The interwell values for porosity and permeability are calculated as the weighted average of the respective values at the wells. The weighted average used here is determined using the inverse distance method (for discussion of the method, see Isaaks and Srivastava, 1989). According to this method, the value of permeability or porosity at some intermediate distance between wells is proportional to the distance from the point being estimated to the value at the wells according to:

$$\hat{v} = \frac{\left(\dfrac{v_a}{d_a} + \dfrac{v_e}{d_e}\right)}{\left(\dfrac{1}{d_a} + \dfrac{1}{d_e}\right)}$$

where $\hat{v}$ is the value of permeability or porosity being estimated; v_a the value at well A; v_e the value at well E, d_a the distance to well A, and d_e the distance to well E. The resulting permeability model is illustrated in Figure 14.

The Geostatistical Model

The interwell values for porosity and permeability are calculated geostatistically using a method called sequential Gaussian simulation. This is a common geostatistics technique discussed in detail by Deutsch and Journel (1992). Since it is a geostatistical simulation method, a number of possible models of permeability and porosity have been generated. The porosity and permeability cross sections for one of these possibilities is illustrated in Figures 15 and 16.

Effect on Predicted Performance

A simulated waterflood using Table 1 parameters was carried out using well A as the injection well and E as the production well. The water saturation distribution after approximately 0.25 pore volumes (pv) of water injected is illustrated in Figure 17. The distribution is consistent with the cross sections for each of the models shown in Figures 14-16.

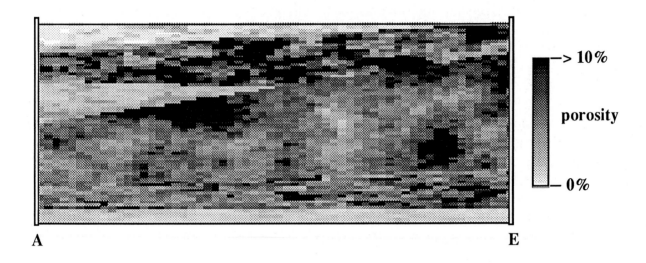

Figure 15. Snap line model of stochastic porosity (realization 1).

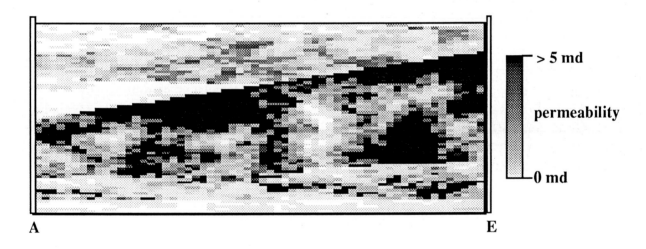

Figure 16. Snap line model of stochastic permeability (realization 1).

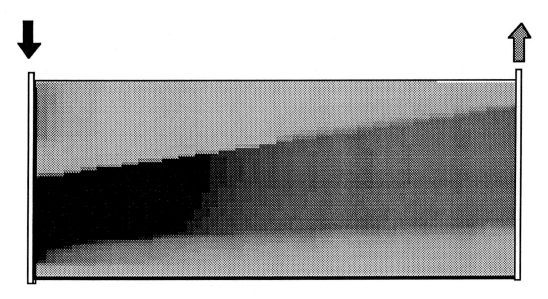

Figure 17a. Water saturation at 0.25 injected pore volumes for snap line averaged model.

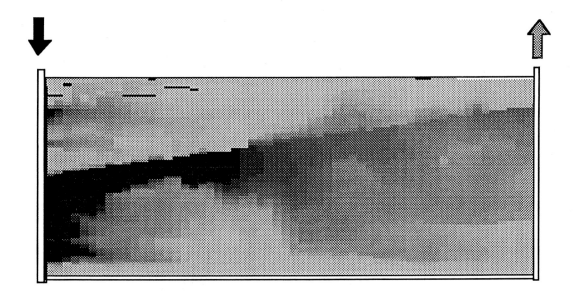

Figure 17b. Water saturation at 0.25 injected pore volumes for snap line stochastic model.

The watercut at well E is illustrated in Figure 18 as a function of pore volumes of water injected. The solid line in this figure is obtained from the detailed model study that was shown earlier in Figure 9. This earlier work is referred to here as the base case. Also shown are the results from a number of the geostatistical realizations and the results obtained with average permeability and porosity.

The snap line with averaged permeability gives an optimistic forecast of water breakthrough in that breakthrough was predicted to occur later than that predicted using the detailed model. As a consequence, sweep efficiency would be higher than that observed with the base case. More oil would be produced prior to water breakthrough (Fig. 19). The geostatistical models tended to bracket the results observed with the base case. Some realizations had permeability distributions that exacerbated water channeling; others mitigated it. Since both the base case and the geostatistical models had heterogeneous distributions of porosity and permeability, we have illustrated what should have been evident: that channeling of water is more significant in the heterogeneous model than in the averaged model, resulting in earlier breakthrough of water in the heterogeneous model with less efficient sweep.

A similar conclusion is reached using production data. Different geostatistical models have different porosity maps and therefore different estimates of oil in place. This leads to variations in cumulative production for the models. The difference in oil in place at the end of primary production for the different models is eliminated by reporting the production performance as cumulative production divided by the oil in place at the start of waterflooding. These normalized recovery results are shown in Figure 19. Oil production for the averaged permeability case and the geostatistical models are consistently higher than the base case. That difference is due to a better sweep efficiency after breakthrough of the facies models constructed by straight line interpolation between wells. The stochastic realizations bracket the averaged permeability case. Perhaps, in hindsight, this was to be expected. A geostatistical realization represents one possible permeability and porosity distribution. The results from enough realizations would generate a distribution of performance curves that bracket an average performance.

Speculation on the significance of using a straight line approximation for facies and different averaging methods for permeability or porosity is interesting. The following postulates are based only on Figure 19 and should not yet be construed to be generally the case.

1. Using a straight line approximation to correlate the facies gives a conservative result which means oil production performance is better than with the heterogeneous model.

2. For the same facies architecture, the geostatistical generation of permeability and porosity will bracket an average performance that is similar to the performance determined using the inverse distance-averaged permeability and porosity. This results because the difference between the two averages is small.

3. The difference between the average performance and the base case at 0.5 pv injected is approximately (0.24 - 0.21) normalized oil recovery after waterflood, or approximately 14 percent of the base case. In comparison, the difference in strategy had a 52 percent

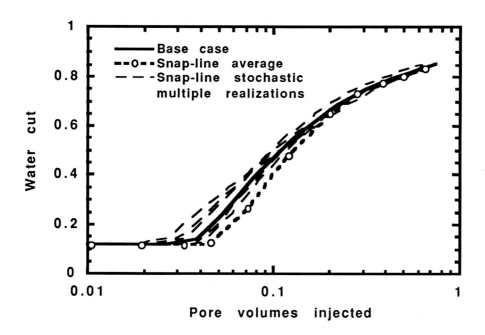

Figure 18. Water cut as a function of injected pore volumes for alternate geological models.

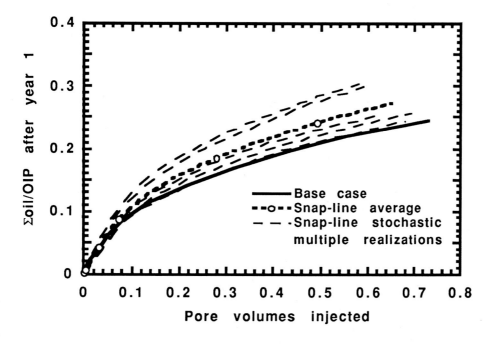

Figure 19. Normalized recovery from the start of the waterflood as a function of injected pore volumes.

impact on the performance.* In this instance, strategy had a more significant performance impact than architectural definition, and the reason strategy was so significant was due to permeability heterogeneity.

*(Fifteen years is approximately 0.5 pv injected. For consistency, the results in Figure 13 converted to incremental oil produced after the start of waterflood, which, for the case A->E is 0.34 - 0.13 = 0.21, and for the case E->A is 0.23 - 0.13 = 0.10. The impact is determined as [0.21-0.1]/0.21.)

SUMMARY

The objective of this work is to illustrate the value of outcrop analogs for reservoir engineering. The Eight Foot Rapids outcrop is used to examine interwell-scale issues that may exist at the Aneth Field. Averaged data for permeability, porosity, or facies architecture result in an estimate of waterflood breakthrough time which is later than that predicted from a detailed heterogeneous model. Production performance for the averaged model (even with heterogeneous permeability but averaged facies architecture) is greater than that predicted from the detailed heterogeneous model. Also, complex heterogeneity needs to be considered when developing an optimized field development strategy. For this outcrop example, flood direction affects breakthrough time and ultimate recovery. Overall, one can conclude that studying applicable outcrop analogs does provide insight into reservoir engineering issues that could occur within adjacent subsurface reservoirs.

REFERENCES

DEUTSCH, C. V., AND JOURNEL, A. G., 1992, GSLIB: Geostatistical Software Library and User's Guide: New York, Oxford University Press, 340 p.

HARMAN, L. P., AND VANDERHILL, J. B., 1992, Greater Aneth Field, San Juan County, Utah: Reservoir Evaluation Scoping Study: Unpublished Mobil Internal Report, 42 p.

ISAAKS, E. H., AND SRIVASTAVA, R. M., 1989, Applied Geostatistics: New York, Oxford University Press, 561 p.

RESERVOIR CHARACTERIZATION OF A PERMIAN GIANT: YATES FIELD, WEST TEXAS

SCOTT W. TINKER AND DENISE H. MRUK
Marathon Oil Company, Petroleum Technology Center, Littleton, Colorado; and Marathon Oil Company, Mid Continent Region, Midland, Texas

ABSTRACT

The Yates Field reservoir characterization project provided the geologic framework, data, and tools that support the ongoing reservoir management of Yates Field. Geologic and engineering data, including 1800 wells with digital log data; 23,000 ft of quantified core analysis and description; and six decades of production data were integrated, analyzed, and output in a format that could be used for field evaluation, management, and simulation.

The Yates Field reservoir characterization products include quantified, standardized, digital core descriptions for 118 cores in the field; a digital tops database for every well in the field; a digital cave location database; 2-D digital cross sections through every well in the field; 2-D structure and isochore maps for major and internal marker horizons, net and gross reservoir maps, net and gross shale maps, secondary calcite distribution maps, cave distribution maps, and fracture distribution maps; and a 6.8 million cell, 3-D geologic model of the complete reservoir that includes log, core, and production data.

The reservoir characterization project resulted in a quantified description of the heterogeneous matrix and fracture network in Yates Field. The efficient, ongoing management of this classic dual-porosity system has stabilized production from this 68-year-old, 4.2 billion barrel field.

FOREWORD

In 1989 a comprehensive reservoir characterization project involving geoscientists and engineers from Marathon's Mid Continent Region (MCR) and Petroleum Technology Center (PTC) was initiated. At that time production in Yates Field was on a steep decline. Geologic reservoir characterization was identified as an integral step in quantifying the significant oil remaining for additional recovery beyond that producible from the current field operation practices.

The primary goal of the Yates Field reservoir characterization project was to quantify rock matrix and fracture reservoir parameters that are essential to the understanding of dual-porosity reservoir behavior and its effect on oil recovery. Dual-porosity systems are characterized by a large pore volume matrix that is effectively communicated by a low pore volume, high-permeability fracture system. The Yates Field dual-porosity system is comprised of typically high-quality carbonate matrix connected by a highly transmissible fracture and cave network. Matrix fluid movement in Yates Field is dominated by gravity and capillary forces that support sustained oil migration toward the existing fracture "free oil" column. The fracture network provides the flow conduit for oil transfer from the matrix to the fracture "free oil" column, to the producing completions. Shales on the west side of the field may create local "baffles" to this vertical matrix oil migration; however, pressure, fluid flow, and fluid contact data all indicate long-term vertical communication between stratigraphic units throughout most of the reservoir system.

The Yates Field reservoir characterization project provides the geologic data that support the fundamental changes in reservoir management being implemented by the MCR engineering team. The core descriptions, correlations, maps, and 3-D reservoir model are the geologic foundation on which dual-porosity flow concepts can be applied, reservoir flow behavior can be accurately modeled, and fracture EOR technology evaluated. Implementation of the dual-porosity reservoir management plan and effective management of the oil column have resulted in stabilized production in Yates Field.

INTRODUCTION

Objectives

The objectives of the Yates Field reservoir characterization project were to describe the producing San Andres, Grayburg, Queen, and Seven Rivers formations in a format that could be evaluated using log and production data; to map reservoir and nonreservoir facies distribution in 2-D and 3-D; and to integrate the reservoir description into operations projects and field management strategies.

Project Components and Database

Core Description—
Yates Field contains 118 cores (23,000 ft), and each foot of recovered core has whole-core porosity, permeability, and saturation measurements. The objective of the core project was to describe each foot of core in a numerical format that would allow for entry into a digital database, and direct comparison and statistical analysis with log and production data. The Geologic WorkSheet (GWS) description format was developed to meet this requirement. The following parameters were quantified or coded for every foot of core: core-log calibration, recovery, oil staining, core analysis data, fracture type and aperture, pore type, Dunham texture (Dunham, 1962), lithology, cements, stylolites, grain types, sedimentary structures, environments, and general remarks. These data were all entered into an SAS (Statistical Analysis Systems) digital database and merged with log digits and production data. A core report, including the description, color photographs, logs, and a summary, was made for every core. In addition, definitions and photographic examples of most features described in core are documented in a Core Atlas. This atlas serves as a descriptive glossary of terms referenced in core description and is organized in sections mirroring the core description format. Features are documented by color core photographs, thin-section photomicrographs, and scanning electron micrographs.

The core descriptions were fundamental to the interpretation of depositional environments and development of a facies distribution model. The core data were also used to examine the distribution of lithology, depositional texture, and pore type in the producing formations; define core/log porosity mapping parameters and porosity distribution; and evaluate formation porosity-permeability relationships.

Correlation—
Log and core correlation provides a three-dimensional framework of stratigraphic units. The objectives of the correlation project were to define the basic lithologic units and depositional cycles in the Yates Field, to correlate the lithologic and time-stratigraphic packages across the field, and to define the vertical distribution of reservoir and nonreservoir facies. The correlation project included every well ($\approx$1800) in Yates Field in a series of dip and strike lines. Correlation sections, which incorporated graphic logs from the core description project, were constructed digitally in GEOLOG. Tops picked from the correlation work were entered into a digital SAS database, and maps were constructed using Landmark/ZYCOR Z-Map Plus software.

Lithofacies and Porosity—
Maps were made to illustrate the stratigraphic and structural framework of the producing formations and the areal distribution of the reservoir and nonreservoir facies. The maps, representatives of which appear in this paper, include structure, formation/internal unit thickness, net shale thickness, gross sand thickness and porosity, gross dolomite porosity, and porosity "elevation slice" maps (maps of porosity distribution within horizontal elevation "slices").

Calcite—
Calcite is a secondary pore-filling cement that generally decreases matrix and fracture permeability in the Yates Field. The objectives of the calcite analysis were (1) to determine the relationship between the volume percent calcite and matrix permeability, and (2) to map the areal and vertical distribution of calcite in the reservoir. The calcite map set includes a structure map on the first continuous occurrence of secondary calcite and elevation slice maps of the mean percent calcite.

Fractures and Caves—
A critical link between the reservoir facies (matrix) distribution and reservoir performance is the distribution of fractures and open caves in Yates Field. A primary objective of the fracture analysis was the construction of a set of elevation slice maps that show the areal and vertical fracture distribution. Regional structure, Formation MicroScanner (FMS) logs, Fracture Identification Logs (FIL), depositional trends, San Andres cave lineaments, sinkholes and karst depression lineaments, San Andres flexure lineaments, and fluid flow trends were all integrated in an attempt to determine the fracture distribution. These maps may indicate variations in transmissibility that greatly impact individual well performance. In addition to fractures, every well in the field was examined for the occurrence of open caves, and these cave data were entered into an SAS digital database.

3-D Integration and Modeling—
All of the core and log data were integrated into a 3-D geologic model using SGMs StrataModel software. The result was a 6.8 million cell StrataModel of the entire San Andres through Seven Rivers stratigraphic interval in Yates Field. The model, which includes log, core, and production data, was built to mimic the geometries of aggrading and prograding carbonates in the San Andres, erosional truncation at the end of San Andres time, and onlap of clastics and carbonates of the Grayburg, Queen, and Seven Rivers. Integrating the data into an accurate 3-D stratigraphic framework results in a tool that can be used to examine (1) reservoir, nonreservoir, and cave distribution; (2) reservoir connectivity based on imposed pay cutoffs; and (3) volumetrics. The results can be "grossed-up" for input into reservoir simulation models and used to address reservoir management issues.

History and Reservoir Management

Yates Field is located in Pecos and Crockett counties, approximately 90 miles south of Midland, Texas. Yates Field was discovered in October 1926 with the drilling of the I. G. Yates No. 1 (YFU No. 4901). This well was drilled to a remarkably shallow depth of 1032 ft and flowed at a rate of 2220 BOPD. By 1929 over 200 wells had been drilled and tested at rates in excess of 10,000 BOPD. During these early years of field development, production was predominantly from the east side of the field and was driven by solution gas and secondary gas cap growth. The total cumulative production from the Yates Field Unit is in excess of 1.3 billion barrels of oil. Daily production is currently above 50,000 BOPD.

Yates Field was unitized in July 1976, with Marathon Oil Company designated as Unit Operator. The unit encompasses 26,400 acres (~41 square miles) and is divided into 122 tracts. A reservoir management strategy was developed at unitization to maintain reservoir pressure through the reinjection of all produced gas along with supplemental volumes of inert gas.

In 1985 an east-west "engineering operations" line was established based on the geology, bottom hole pressure data, well performance histories, and drive mechanisms of the field. The reservoir was operated as two "separate" parts, primarily to facilitate increased recovery from the west side of the field, which was developed with a 20-acre, 5-spot waterflood. A polymer-augmented waterflood was initiated in December 1983 and completed in February 1989. The east side of Yates Field is developed on 10-acre spacing. The recovery mechanism for the east side is gravity drainage supplemented by produced gas and carbon dioxide injection for pressure maintenance.

Yates Field is a classic, naturally fractured, dual-porosity reservoir, and gravity drainage is the dominant producing mechanism for the entire field. Since completion of this project, the fluid contacts in the field are now managed on a field-wide basis. The west side waterflood has been shut in, gas cap pressure is now maintained through gas injection, and aquifer support achieved through deep injection of produced water. This reservoir management strategy will allow the gas-oil contact to move down to expose more matrix to gravity drainage while the water-oil contact is maintained.

PHYSIOGRAPHIC AND STRUCTURAL SETTING

Yates Field is positioned at the southeastern tip of the Central Basin Platform, with the Midland Basin to the east and the Sheffield Channel to the south (Fig. 1). Post-Permian uplift of the southern portion of the platform (Wessel, 1988) and/or eastward regional tilting (Hills, 1970) have placed the field at the highest structural position on the Central Basin Platform within the Upper Permian. The tilting and resultant structural high coincident with the Yates Field may help to explain the extensive karst system that developed in the Yates Field as relative sea-level fall caused more extensive exposure of the shelf margin. The structurally high position of Yates Field also helps to account for charging of the reservoir, overpressuring to approximately +750 psi (+1150 elevation), and the prolific production rates in comparison to other Late Permian reservoirs on the Central Basin Platform (Craig, 1988).

The Late Permian was a quiet tectonic time in the Permian Basin, with anticlinal structures often resulting from sedimentary features being draped over pre-Permian anticlines or shelf/platform margins. The major structural trends in the subsurface were largely developed between the last half of the Mississippian and the early part of the Permian. Late Paleozoic, pre-San Andres lateral faulting was probably responsible for setting up the dominant northwest-southeast platform margin trends (Hills, 1970).

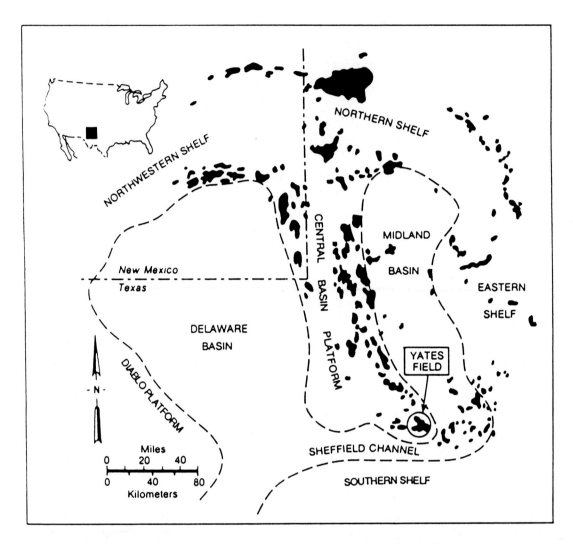

Figure 1. Regional paleogeographic elements of the Permian Basin during the late Permian, with the distribution of oil and gas fields that produce from San Andres and Grayburg reservoirs and their age-equivalents shown in black. Yates Field is located on the southeastern tip of the Central Basin Platform (modified from Craig, 1990). Reprinted by permission.

The Yates Field structure is an asymmetric, horseshoe-shaped anticline that dips more steeply on the basinward and channel sides to the north, east, and south than on the landward (lagoonal) side to the west (Fig. 2). At the San Andres stratigraphic level, the structure has greater than 400 ft of closure over a 26,400-acre area. The structure is broad and gentle with average flank dips of 1 to 4 degrees.

The formation of the Yates Field structure can be attributed to a number of processes. As with most Late Permian structures on the Central Basin Platform, the primary components of structure in Yates Field are interpreted to be drape folding over pre-Permian structure and paleotopographic relief (Craig, 1990; Wessel, 1988). Craig (1990) emphasized that differential compaction between relatively rigid grainstone shoal facies and the highly compactible lagoonal and intertidal facies contribute greatly to the Grayburg and San Andres structure in Yates Field.

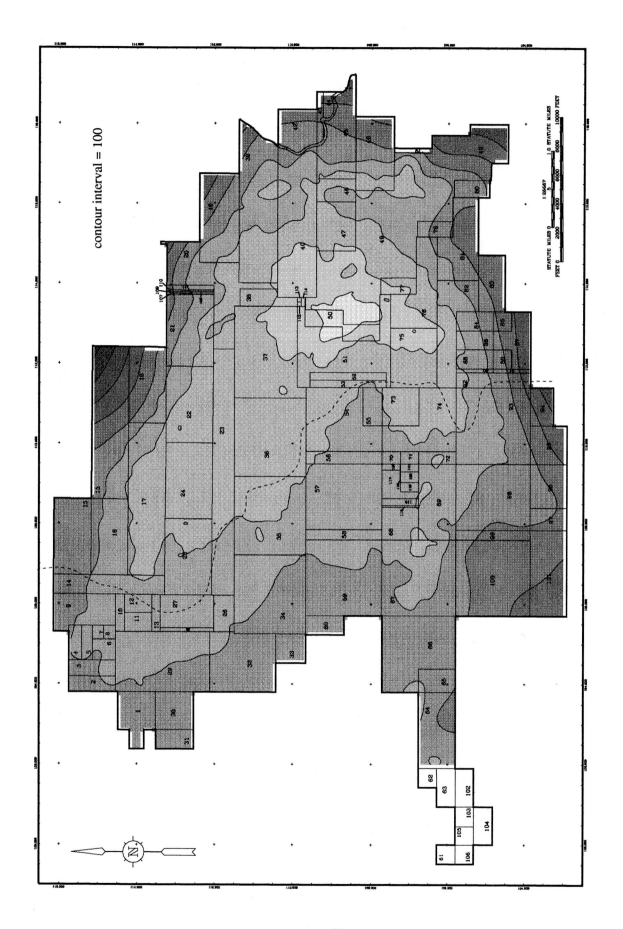

Figure 2. Structure on top of the San Andres Formation within the Yates Field Unit. Note the gentle structural dips to the west and the steeper dips to the south, east, and northeast.

The gentle folding that is responsible for the Yates anticlinal structure can be dated on the basis of local and regional events to be Late Permian (post-Salado and pre-Rustler) in age. Many of the San Andres and Grayburg anticlinal structures in the Permian Basin were formed over buried hills, anticlines, or fault scarps, indicating a second-order relation to older tectonic structures (Hills, 1970).

The top of the San Andres Formation topography has been modified significantly by unconformity-related erosion, karst-related solution collapse, and differential compaction. Internal San Andres structural characteristics are strongly influenced by clinoformal geometries of the carbonate/shale cycles. Internal-unit structural axes shift eastward, higher in the section, and reflect the influence of progradational geometries on the structural framework.

STRATIGRAPHIC SETTING

Regional Stratigraphy

Subsurface—
A north-south cross section along the eastern margin of the Central Basin Platform (Fig. 3) documents the depositional thinning of the Grayburg, Queen, and Seven Rivers toward the southern end of the Central Basin Platform. This thinning reflects the paleostructural high at the southern end of the Central Basin Platform as a result of pre-Grayburg structural tilting (Hills, 1970; Craig, 1990). The southern end of the Central Basin Platform remained high from latest San Andres through Seven Rivers time.

Although Garrett and Kerans (1990) were able to make consistent log correlations along the Central Basin Platform to the Yates Field, the correlations are not entirely in agreement with the stratigraphy as defined in Yates Field. The differences in the definition of the Queen and the Seven Rivers sands, and differentiation of the Grayburg and San Andres carbonate section, account for the major correlation discrepancies. The Queen, as defined by Garrett and Kerans (1990), includes the Seven Rivers "J", "M", and "P" intervals defined in Yates Field (Fig. 3). On the regional cross section, the Grayburg Formation includes both the Grayburg and the Queen of Yates Field.

The definition of the top of the San Andres and its correlation along the Central Basin Platform is problematic. The San Andres, as defined in Yates Field and by Garrett and Kerans (1990), included the top of the predominantly carbonate interval, which includes restricted subtidal/lagoonal shales of the west side (Fig. 4). This is in contrast to the work of Ward and others (1986), who consider the Grayburg the major producing formation of the Yates Field and thus include the restricted subtidal/lagoonal interval within the Grayburg.

Outcrop—
San Andres stratigraphy has been studied in outcrop by Sarg and Lehmann (1986) and Kerans and others (1994) on the Algerita Escarpment, and Sonnenfeld and Cross (1993) in Last Chance Canyon. Kerans and others (1994) divide the San Andres Formation into eight high-frequency

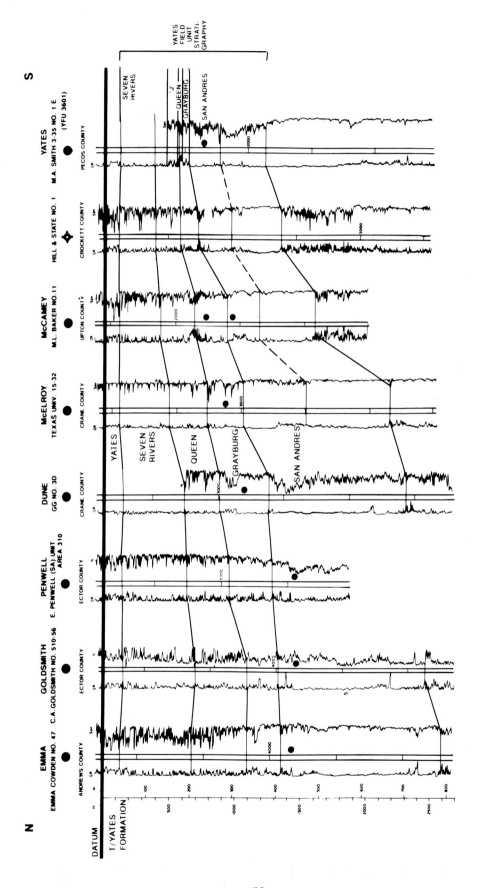

Figure 3. North-south cross section along the eastern edge of the Central Basin Platform. Regional correlations are form Garrett and Kerans (1990). Yates Field stratigraphy is shown on the right. (modified from Garrett and Kerans, 1990) Reprinted by permission.

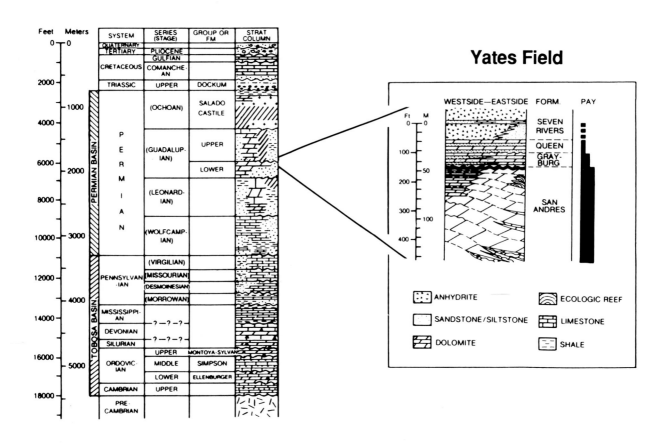

Figure 4. Stratigraphic nomenclature for Yates Field, illustrating the position of the major unconformity at the top of the San Andres Formation. Depth scale shown is measured from the Seven Rivers 'M' marker datum (after Craig, 1990). Reprinted by permission.

sequences (Fig. 5). Leonardian 7 and 8 and part of Guadalupian 1 (lower San Andres) are retrogradational and represent a major transgression and flooding of the underlying Yeso Platform. Guadalupian 1-4 (middle San Andres) highstand, high-frequency sequences are comprised of subtidal-ramp facies at the base that prograde and shallow overall to ramp-crest deposits at the top. The top of Guadalupian 4 represents a major bypass surface over which Brushy Canyon sands were transported and deposited in the basin as Guadalupian sequences 5-11. This sequence boundary is interpreted to be correlative to the unconformity separating the Grayburg and San Andres, as defined in Yates Field. A major floodback followed and established carbonate production in Guadalupian 12-13 (upper San Andres), as seen in the prograding clinoforms of Last Chance Canyon (Sonnenfeld and Cross, 1993) and Plowman Ridge.

Yates Field Sequence Stratigraphic Framework

Recognition and application of cycle-stacking patterns and clinoform geometry within the San Andres are fundamental to understanding and mapping of reservoir continuity and

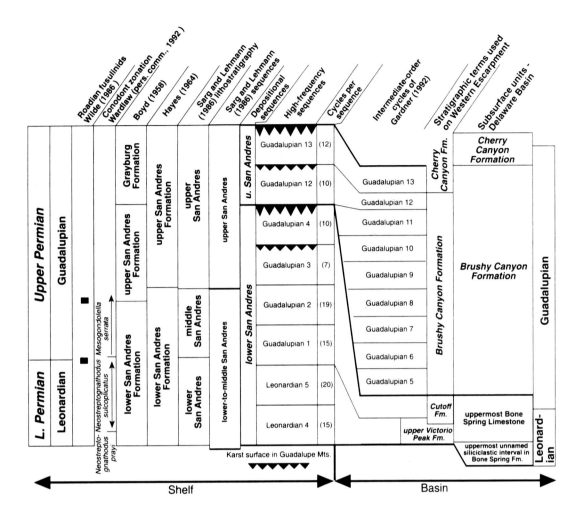

Figure 5. History of stratigraphic nomenclature for the San Andres Formation and basinal equivalents. (from Kerans, et. al. 1994). Reprinted by permission.

nonreservoir barriers. "Layer cake" correlations in the Yates Field result in miscorrelated carbonate/shale packages and erroneous interpretation of reservoir continuity and distribution. Models for San Andres internal stratigraphic architecture in Yates Field are based in part on outcrop work from the Algerita Escarpment (Kerans and others, 1994) and Last Chance Canyon (Sonnenfeld and Cross, 1993). The basic mapping unit from these studies and in Yates Field is the shallowing-upward hemicycle or cycle.

Only two wells penetrated and logged the entire San Andres section in Yates Field. These wells indicate the development of two gross depositional "packages." The lower "package" is interpreted as a low-energy, open-ramp facies. The upper "package" is represented by three to four overall-shoaling, high-frequency sequences (HFS) consisting of restricted subtidal mudstone and wackestone on the west side of the field, and shallow, subtidal packstone-grainstone shoals on the east side of Yates Field. These high-frequency sequences show a strong component of aggradation evidenced by shale-carbonate cycles stacking vertically on the west side of the field, and nearly 300 vertical feet of fusulinid packstone and grainstone preserved on the east side. In contrast, the late San Andres Guadalupian 12-13, high-frequency sequences observed in outcrop and on seismic along the Central Basin Platform show strongly progradational clinoform geometries. The Guadalupian 12-13 sequences likely were either not deposited in the Yates Field area, were deposited east and south of Yates Field, or have been removed by post-San Andres erosion. The seismic lines over Yates Field do not have the resolution to verify these hypotheses.

The four high-frequency sequences and their component cycles are illustrated on a log from the west side of the field (Fig. 6). A typical minor cycle began with a dolomitic shale with a sharp basal contact. Clays in the shales are predominantly illite (90 percent) and chlorite (10 percent) that were transported and trapped in lagoons behind shoal islands during times of

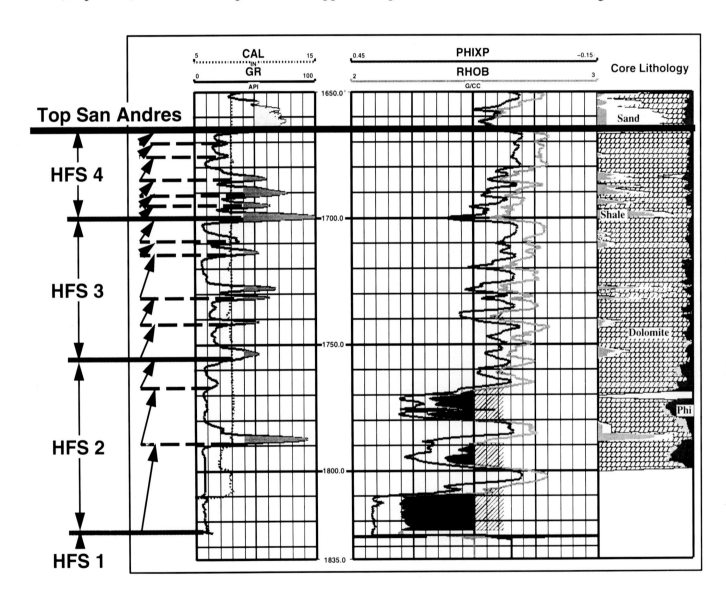

Figure 6. Cycle stratigraphy in Yates Field illustrated on a log from the west side of the field. High frequency sequences (HFS) and component cycles are illustrated for HFS 2, 3 and 4. The gamma ray log is shaded for shales and silts, and core lithology is shown on the far right.

relative sea-level fall. Clays were later reworked during subsequent relative sea-level rise and grade upward into subtidal mudstone-wackestone, overlain by shallow shelf, peloid/fusulinid packstone/grainstone, and are capped locally by peritidal facies. The thickest and most areally extensive shales and shale-carbonate cycles are present at the base of a high-frequency sequence; thinner, laterally discontinuous shales and minor cycles are present near the top. This overall thinning- and shallowing-upward stacking pattern within each high-frequency sequence indicates a decrease in accommodation upward as all of the available space was filled with sediment.

High-frequency sequence boundaries are projected from cycles that could be correlated in the west into stacked fusulinid shoals and inner shoals that could not be correlated to the east. Relative sea-level fall resulted in subaerial exposure of the shelf-crest, shoal-island complex and associated joint dissolution and cave formation at the end of each high-frequency sequence (Tinker and others, in press). When the San Andres lithostratigraphy is superimposed on the sequence stratigraphy (Fig. 7), the relationship between depositional time lines and major lithofacies is apparent. The ramp complex at the base of the section could be equivalent to the retrogradational Leonardian 1-2 and Guadalupian 1 on the Algerita Escarpment. The subsequent aggradational, high-frequency sequences (HFS) could be equivalent to Guadalupian 1-4, as defined by Kerans and others (1994). A major erosional unconformity, marked by significant karst-related features and cave formation, defines the top of the San Andres in Yates Field. The San Andres unconformity surface is onlapped by sediments of the Grayburg Formation that fill most karst-related topographic irregularities.

LITHOFACIES AND DEPOSITIONAL ENVIRONMENTS

Yates Field produces from four Guadalupian-age formations: Seven Rivers, Queen, Grayburg, and San Andres. The producing formations are believed to correlate with the Seven Rivers, Queen, Grayburg, and San Andres formations that have their type sections along the Northwest Shelf of the deep Delaware Basin. In contrast, Yates Field is bordered by the shallower Midland Basin to the east and Sheffield Channel to the south. The differences in the shelf-edge morphology, depositional environment, and tectonics on the Central Basin Platform help to explain lithofacies and thickness variations between the Guadalupian formations in the Yates Field area and the outcrop sections.

There is a strong relationship between lithofacies type and reservoir quality in Yates Field, especially with regard to matrix porosity. Primary matrix porosity influences the flow of later diagenetic fluids that may enhance porosity through solution or reduce porosity and permeability through secondary cementation. Primary matrix porosity and permeability also effect storage capacity and matrix drainage rates. The carbonate shoal packstone and grainstone in Yates Field have maintained the highest matrix porosity of any lithofacies. This is a critical factor, as the bulk of remaining mobile oil in the reservoir, although primarily produced through fractures, is contained in the matrix.

Core observations of lithologic, textural, and porosity characteristics form the basis for the delineation of depositional environments in the field. Data from 118 cores (23,000 ft) were used. A qualitative overview of core data indicates that a continuum of 12 depositional environments

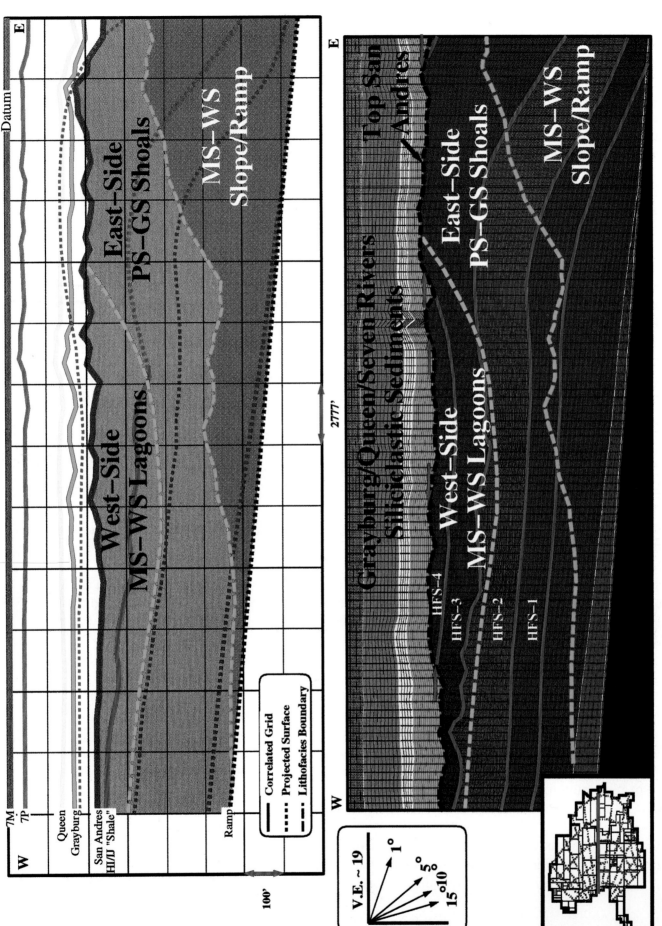

Figure 7. Relationship between sequence stratigraphy and lithostratigraphy. Upper figure is a **ZYCOR** stratigraphic dip section, datumed on the Seven Rivers "M", showing correlated grids and major lithofacies. Note the aggrading and prograding nature of the major lithofacies. Lower figure is cut from **Stratamodel** in approximately the same W-E position, and illustrates the stratigraphic framework used in the 3-D model. Note the proportional fill in the San Andres, erosional truncation surface at the top of the San Andres, and post-San Andres onlap of the Grayburg.

represents the entire stratigraphic section in Yates Field. These environments are listed in Table 1 according to relative stratigraphic order, beginning at the top with the Seven Rivers and extending downward into the San Andres. Three characteristic lithofacies are also listed for each environment, although these lithofacies are broadly overlapping.

Carbonate Environments

Stratigraphic and areal distribution of cores provides an excellent data base for the evaluation of carbonate environments. Although cores in stratigraphically lower sections of the San Andres are scarce, it can be reasonably inferred that the depositional setting in the early San Andres was subtidal mid to outer ramp. Carbonate fabrics and textures in these environments range from mudstone to packstone enriched in algal peloids and skeletal debris, to laminated, organic-rich mudstone and wackestone (Figs. 8 and 9). These textures represent relatively low-energy, possibly oxygen-depleted and stressed depositional conditions. Water depths may have ranged from a few tens of feet to greater than 100 ft.

As deposition continued, carbonate shoal complexes were established in higher-energy, shallower portions of the shelf or ramp setting, building in size and relief throughout San Andres time. These shoals were dominated by packstone and grainstone textures that were partly to completely obliterated by subsequent pervasive dolomitization. Resulting fabrics include "sucrosic" packstone, probably consisting of algal peloids with varying amounts of skeletal debris, to tabular and cross-bedded fusulinid grainstone (Figs. 10 and 11). Peloids are ubiquitous within shoal and shoal-margin lithofacies, and represent the principal constituent of the carbonate "sand" that comprised these shoals. Although recognition of individual peloids in fine-to-coarse crystalline sucrosic dolomite is difficult, petrographic observations suggest that much of the sucrosic fabric observed in the San Andres was originally packstone or grainstone. Some of the relief preserved at the top of these shelf-crest island shoals possibly represents dunes formed by wind deposition of carbonate grains on exposed islands.

Accommodation on the platform decreased toward the top of each major high-frequency sequence as space was filled, probably causing increased bypass of terrigenous materials and carbonate shedding into the Midland Basin to the east and Sheffield Channel to the south. Behind these shoals a variety of shallow, subtidal environments was established. Fabrics and textures in these environments range from peloid and fine skeletal wackestone to burrowed and laminated mudstone (Fig. 12). Under exceptionally low-energy conditions established in protected back-shoal areas or during widespread restriction caused by sea-level fall, clay- and silt-rich mudstone was deposited, forming the lithofacies represented by the San Andres shale marker beds (Fig. 13).

Toward the end of San Andres time and during deposition of the Grayburg, Queen, and Seven Rivers formations, broad tidal-flat environments developed in response to ever-shallowing conditions (Fig. 14). This spectrum of peritidal environments is characterized by variably dense and fenestral mudstone and stromatolitic boundstone. These fabrics are locally mudcracked, brecciated, and bleached as a result of periodic to sustained subaerial exposure.

Figure 8.

A. YFU# 2212 1260.2 ft. Slightly argillaceous, silty dolomite; peloid-intraclast-lithoclast (?) packstone. Tiny grains dominantly composed of partly lithified mudstone are scattered throughout a clay- and silty-rich matrix. Clay content imparts a greenish-gray tint to rock in upper half of photograph. Compaction of ductile clay contributes to poor reservoir quality.

Grayburg Fm
Restricted Subtidal

$\emptyset = 6.6\%$
kmax = 0.1 md
k90 = 0.1 md

B. YFU# 64101 1657.4 ft. Dolomitic limestone; skeletal wackestone. Skeletal fragments consisting principally of molluscs (near B6) bryozoans (most fragments in area of D-E, 9-10) and crinoids (small, circular grain near F8) form a skeletal-rich wackestone texture. Patches of calcite cement (near E2 and G7) locally fill irregular vugs. Irregular nodular fabric results from differential dolomitization of limestone matrix and subsequent compaction, which is greatly exaggerated along stylolite at C11-G9. Matrix in lower right portion of photograph is relatively organic rich.

San Andres Fm
Subtidal

$\emptyset = 13.1\%$
kmax = 2.6 md
k90 = N/A

C. YFU# 3566 1833.8 ft. Slightly argillaceous dolomite; crinoid-skeletal wackestone. Numerous crinoid fragments (light colored grains throughout) and bivalve molluscs (now as molds D-E, 6-7) are distributed within a dense dolomite matrix. An oblique microfracture is visible crossing the rocker slab at A9-G7. Calcite cement (stained pink) fills a lens-shaped vug at B-D, 10-11.

San Andres Fm
Subtidal

$\emptyset = 6.2\%$
kmax = 0.93 md
k90 = 0.16 md

D. YFU# 3566 1806.0 ft. Dolomite, skeletal wackestone. Molds of fusulinids, some filled by white calcite cement (near E5) are visible within a patchy dense dolomite matrix which also hosts numerous crinoid fragments (B-D7). A brachiopod or mollusc appears at F6. Calcite cement (stained pink) also fills microfractures and vugs in the left half of photograph. Brown colored matrix represents dolomicrospar which bears some mBC/BC porosity.

San Andres Fm
Subtidal

$\emptyset = 9.7\%$
kmax = 2.0 md
k90 = 1.4 md

Figure 9

A. YFU# 64101 1604.8 ft. Slightly bituminous, dolomitic limestone; fine skeletal wackestone. A cross section of a crinoid ossicle (D-E, 2-3) is the only identifiable fragment within this sparsely fossiliferous wackestone. Most other skeletals (white, round to platy grains) have been highly fragmented and altered. Orange kerogen particles and wisps are visible throughout the matrix. A portion of an irregular stylolite seam cuts the sample at A7-G11. Bedding inclination is largely due to post-depositional compaction.

San Andres Fm $\emptyset = 9.7\%$
Subtidal $k_{max} = 0.4$ md
45X (3.4mm) PPL $k_{90} = 0.3$ md

B. YFU# 3535 1353.9 ft. Dolomite; fine skeletal-peloid wackestone. Numerous mollusc fragments (B-D, 3-4 and D2-E6) and probable ostracods (near B6) can be identified along with dark, amorphous peloids (D9 and D-E1) within a dense, micritic matrix. Very fine quartz silt and organic material is concentrated as a wispy microstylolite at D4-E11. This sample is devoid of visible porosity.

San Andres Fm $\emptyset = 3.3\%$
Shallow Subtidal-Lagoonal $k_{max} = 0.2$ md
22X (6.8 mm) PPL $k_{90} = 0.1$ md

C. YFU# 9807 1722.0 ft. Dolomite; peloid-skeletal packstone-grainstone. A relatively well preserved fusulinid (C-G, 1-6) is clearly defined by internal microstructure within otherwise obscure, highly altered matrix. Close examination reveals that much of the matrix is probably packstone, as suggested by grain ghosts and fabrics near B5 and G8-9. Porosity in this view includes isolated WP pores within fusulinid, as well as connected BC/mBC pores within matrix. Microfractures are visible along A-G, 10-11.

San Andres Fm $\emptyset = 12.3\%$
Shoal Margin-Restricted Subtidal $k_{max} = 489.$ md
22X (6.8 mm) PPL $k_{90} = 340.$ md

D. YFU# 9807 1735.1 ft. Dolomite; peloid wackestone-packstone. Original texture is obscured by fine dolomicrospar fabric, although micritized grain ghosts visible at C-F, 3-5 and D11 suggests a probable grain-rich texture. Abundant microporosity (mBC pores) in this view appears well connected, although limited pore throat size limits permeability as indicated by core analyses. Larger BC pores (near C4 and E3) are only locally developed.

San Andres Fm $\emptyset = 12.6\%$
Shoal Margin $k_{max} = 3.7$ md
22X (6.8 mm) PPL $k_{90} = 1.7$ md

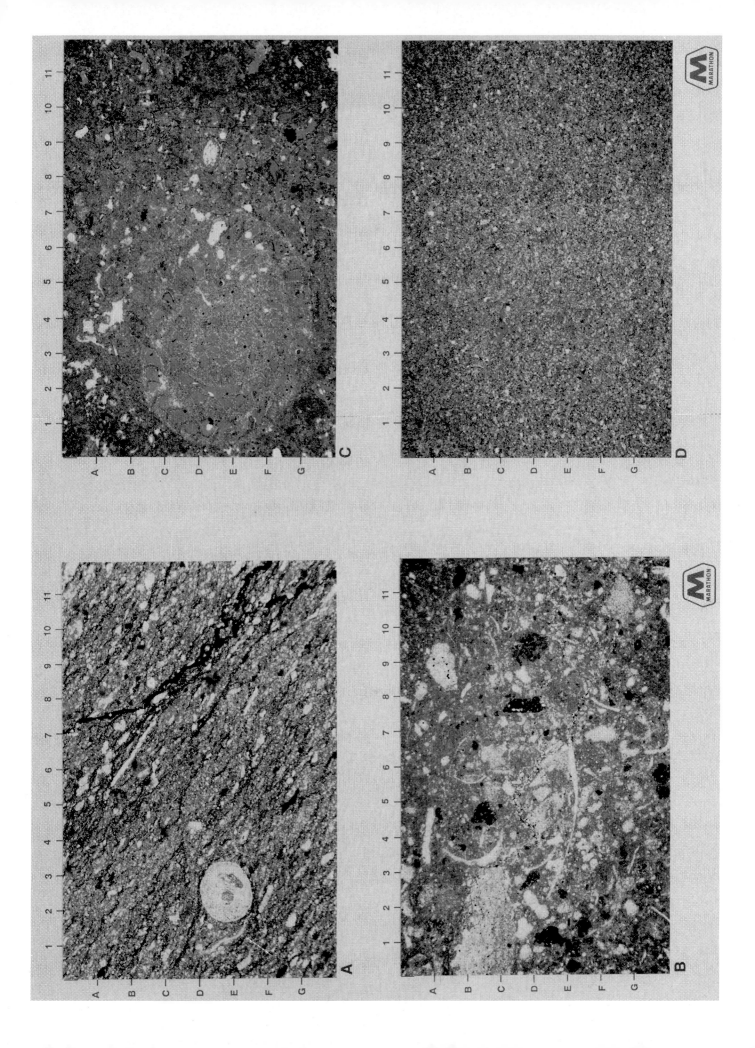

Figure 10.

A. YFU# 3566 1748.3 ft. Dolomite; fusulinid-peloid packstone. Relatively good preservation of packstone texture is illustrated by this sample which contains roughly equal amounts of fusulinids (many as molds; C9) and finer peloids (many as "pinpoint microvugs"; A-G, 1-2). Faint inclined bedding or low-angle crossbedding is somewhat vaguely indicated by alignment of pelmolds within thin, dense bed at A-G, 3-4.

San Andres Fm $\emptyset$ = 15.7%
Shoal kmax = 653. md
 k90 = 507. md

B. YFU# 2433 1391.2 ft. Dolomite.; peloid-skeletal wackestone, grading to packstone. The distribution of molds (fusulinids and molluscs) within dense, upper portion of core appears to represent a wackestone texture, while porous matrix below appears to contain abundant peloids which define a probable packstone texture. It is likely that the apparent wackestone texture is a diagenetic relict resulting from dolomite cementation and obliteration of fine grained packstone.

San Andres Fm $\emptyset$ = 23.3%
Shoal kmax = 347. md
 k90 = 212. md

C. YFU# 3525 1419.6 ft. Dolomite. fusulinid-peloid grainstone. Fusulinids in various stages of preservation are visible throughout this grainstone, and are especially well preserved in the area of C-D, 9-10 where circular to oval cross sections are evident. Some specimens retain intraskeletal (WP) porosity. Inclined bedding or low-angle cross bedding is defined by a thin bed of fine grained material at A-G, 3-4. Small, dark grains and "pinpoint" microvugs represent peloids and their molds, respectively.

San Andres fm $\emptyset$ = 21.4%
Shoal kmax = 87.0 md
 k90 = 87.0 md

D. YFU# 3923 1180.4 ft. Dolomite; fusulinid-peloid grainstone, Highly altered grainstone has evolved to a coarse sucrosic fabric, although fusulinids (large white grains) and peloids (small, sand-size grains throughout matrix) are still evident. Leaching has produced crude vertical channels of connected BC pores and vugs (D-E, 6-9). Short fractures are visible in the upper portion of the photograph, along which partial solution enhancement has occurred (G4-5).

San Andres Fm $\emptyset$ = 24.6%
Shoal kmax = 1127. md
 k90 = 1054. md

Figure 11

A. YFU# 3923 1096.7 ft. Dolomite; peloid packstone-grainstone. Variable cementation of highly altered peloids results in contrasting diagenetic textures in this grain-supported rock. Limited cementation clearly illustrated in area of B-D, 8-10 preserves original grainstone texture and BG porosity; whereas, nearly complete cementation such as in the area of E-G, 1-5 has filled most BG pores, imparting a fabric which appears as packstone with mud-filled pores.

San Andres Fm $\qquad\qquad\qquad\qquad\qquad\qquad\qquad\qquad\qquad\qquad$ Ø = 11.5%
Shoal Margin $\qquad\qquad\qquad\qquad\qquad\qquad\qquad\qquad\qquad\qquad$ kmax = 59.0 md
45X (3,4 mm) PPL $\qquad\qquad\qquad\qquad\qquad\qquad\qquad\qquad\qquad\qquad$ k90 = 6.8 md

B. YFU# 5428 1690.5 ft. Dolomite; peloid-intraclast packstone. Micritized peloids and larger intraclasts form a packstone texture along with clotted mud lumps. BP/mBP pores and a network of microfractures (near D8) are dominantly filled by dolomite cement (white), although some solution -enlarged fenestrae or microvugs remain open (area directly above A2). Microbrecciated fabric suggests possible partial desiccation during early diagenesis.

San Andres Fm $\qquad\qquad\qquad\qquad\qquad\qquad\qquad\qquad\qquad\qquad$ Ø = 8.1%
Shoal Margin $\qquad\qquad\qquad\qquad\qquad\qquad\qquad\qquad\qquad\qquad$ kmax = 0.1 md
22X (6.8 mm) PPL $\qquad\qquad\qquad\qquad\qquad\qquad\qquad\qquad\qquad\qquad$ k90 = 0.1 md

C. YFU# 4804 1492.6 ft. Calcitic dolomite; skeletal-fusulinid packstone-grainstone. A well preserved fusulinid (right) forms a grain-supported framework together with a highly altered codiacean algal fragment (F-G, 2-3) and other unidentified skeletal grains. Most grains have isopachous marine cements. Calcite cement occurs as blocky crystals which partially fill BP/BC pores.

San Andres Fm $\qquad\qquad\qquad\qquad\qquad\qquad\qquad\qquad\qquad\qquad$ Ø = 10.7%
Shoal $\qquad\qquad\qquad\qquad\qquad\qquad\qquad\qquad\qquad\qquad$ kmax = 59.0 md
22X (6.8 mm) PPL $\qquad\qquad\qquad\qquad\qquad\qquad\qquad\qquad\qquad\qquad$ k90 = 23.0 md

D. YFU# 4936 1601.0 ft. Calcitic dolomite; fusulinid-peloid packstone-grainstone. Calcite cement fills irregular BP/BC pores between altered fusulinids (D-F, 4-6 and B-F, 7-11) and other large skeletal grains (lower left). Dolomite crystals which project into voids have clear overgrowths (E-F, 10-11 ad G6-9). An isolated intraskeletal pore (E-F5) is preserved within a fusulinid. Isolated dolomite rhombs appear to "float" in calcite cement at A11.

San Andres Fm $\qquad\qquad\qquad\qquad\qquad\qquad\qquad\qquad\qquad\qquad$ Ø = 10.1%
Shoal $\qquad\qquad\qquad\qquad\qquad\qquad\qquad\qquad\qquad\qquad$ kmax = 35.0 md
22X (6.8) PPL $\qquad\qquad\qquad\qquad\qquad\qquad\qquad\qquad\qquad\qquad$ k90 = 24.0 md

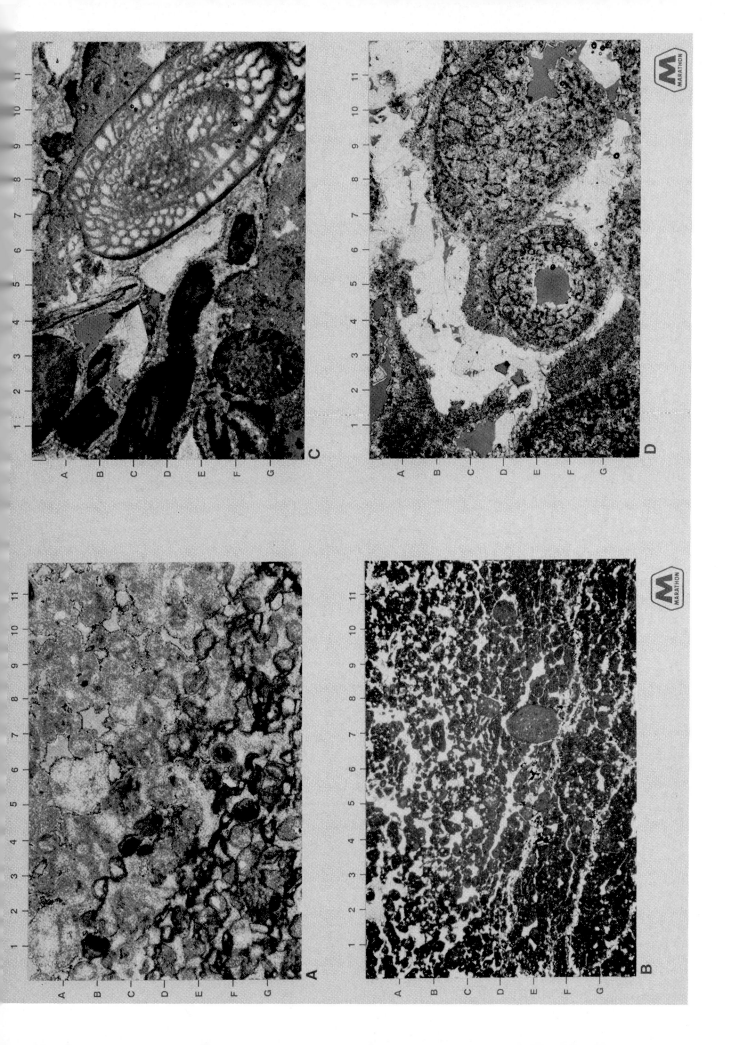

Figure 12.

A. YFU# 3566 1783.4 ft. Dolomite; peloid-fusulinid wackestone. Molds of fusulinids are preserved in patches of dense dolomite within porous, peloid-bearing matrix. Many molds appear isolated from permeability pathways represented by BC and mBC pores surrounding dense patches. Short fractures (A-G, 1-2) may locally contribute to sample permeability. Calcite cement (white) fills vug developed around solution-enlarged fusumolds at F9-10.

San Andres Fm $\emptyset = 12.5\%$
Subtidal $kmax = 359.$ md
 $k90 = 306.$ md

B. YFU# 2908 1361.2 ft. Dolomite; peloid-skeletal wackestone-packstone. Fusulinids locally define a packstone texture, particularly where grains are not leached (near C2). Numerous isolated fusumolds are distributed within a very dense dolomite matrix. A network of irregular, low-amplitude stylolites and microstylolites gives this sample a microbrecciated appearance (A-G, 10-11).

San Andres Fm $\emptyset = 10.1\%$
Shallow Subtidal $kmax = 2.2$ md
 $k90 = 0.8$ md

C. YFU# 1915 1199.2 ft. Silty dolomite; coated grain-intraclast packstone-grainstone, grading in part to stromatolitic boundstone. Highly altered, grain-rich fabric contains numerous coated grains and intraclasts visible as tiny to relatively large, white grains throughout. Iron stain (B-F, 8-9) results from oxidation of pyrite apparently associated with organic-rich laminae and nodules. A fracture is visible at A-B, 1-10.

Grayburg Fm $\emptyset = 10.7\%$
Intertidal $kmax = 31.0$ md
 $k90 = 24.0$ md

D. YFU# 3923 1068.2 ft. Silty dolomite; fusulinid-intraclast packstone. Fusulinids, most of which have been leached to form molds (C-D, 2-3), and partly to completely leached small intraclasts (platy molds at D-E,6-7), constitute the principal macroscopic allochems in a burrowed matrix. Rare patchy dolomite cement (A-B, 1-3 and G2-3) does not significantly impact overall reservoir quality as indicated by core analyses.

Grayburg Fm $\emptyset = 25.4\%$
Shoal Margin $kmax = 384.$ md
 $k90 = 329.$ md

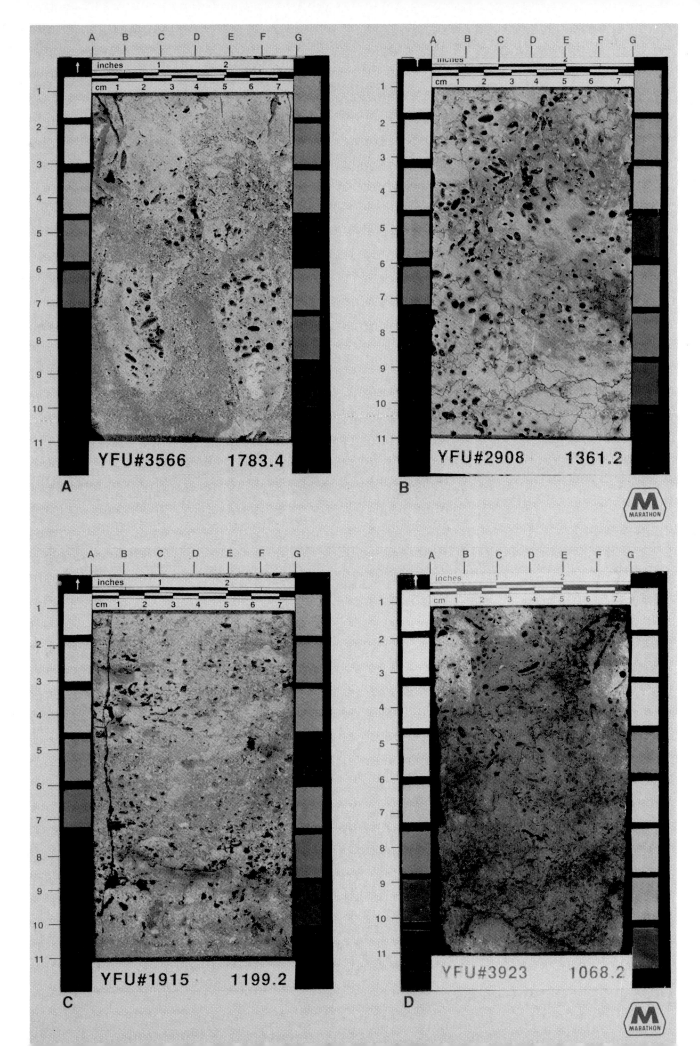

Figure 13.

A. YFU# 5428 1783.5 ft. Silty, dolomitic shale, grading to argillaceous, silty dolomite. Vaguely laminated, clay-rich mudstone is representative of San Andres "shale" marker-bed lithologies. Greenish-gray color in middle and lower portions of the photograph indicate higher clay content. Deformed mud has burrowed appearance at A-C, 10-11. Pyrite cement (near B2 and E2) locally replaces patches of matrix.

San Andres Fm $\emptyset$ = N/A
Restricted Subtidal kmax = N/A
k90 = N/A

B. YFU# 7412 1640.5 ft. Silty, dolomite shale, grading to argillaceous, silty dolomite. Highly deformed mudstone contains clay-rich patches (D3 and F6) as well as irregular laminae (A9-D11). It is likely that most distribution was originally in laminae, having been subsequently deformed by compaction and cementation processes. Greenish-gray color is typical of this San Andres marker-bed lithology. Lenses of pyrite are visible at B-C6 and C5-E6.

San Andres Fm $\emptyset$ = 1.5%
Restricted Subtidal kmax = 0.1 md
k90 = 0.1 md

C. YFU# 3525 1414.5 ft. Argillaceous, silty dolomite; stromatolitic mudstone to peloid grainstone. Finely laminated, possibly stromatolitic mudstone and interbedded peloid grainstone in lower half of core grades upward to clayey and brecciated mudstone. Brecciated fabric may be indicative of exposure and partial dessication.

San Andres Fm $\emptyset$ = 5.5%
Shallow Subtidal-Intertidal kmax = 0.1 md
k90 = N/A

D. YFU# 7412; 1711.0 ft. Silty dolomite and silty, dolomitic shale; mudstone. Dense, laminated mudstone at base is sharply overlain by very clayey dolomite to dolomitic shale at contact along A-G8. Dark brown pyrite crystals are scattered throughout the core sample, and are concentrated in laminae at c-G, 8-9.

San Andres Fm $\emptyset$ = 8.3%
Restricted Subtidal kmax = 0.1 md
k90 = 0.1 md

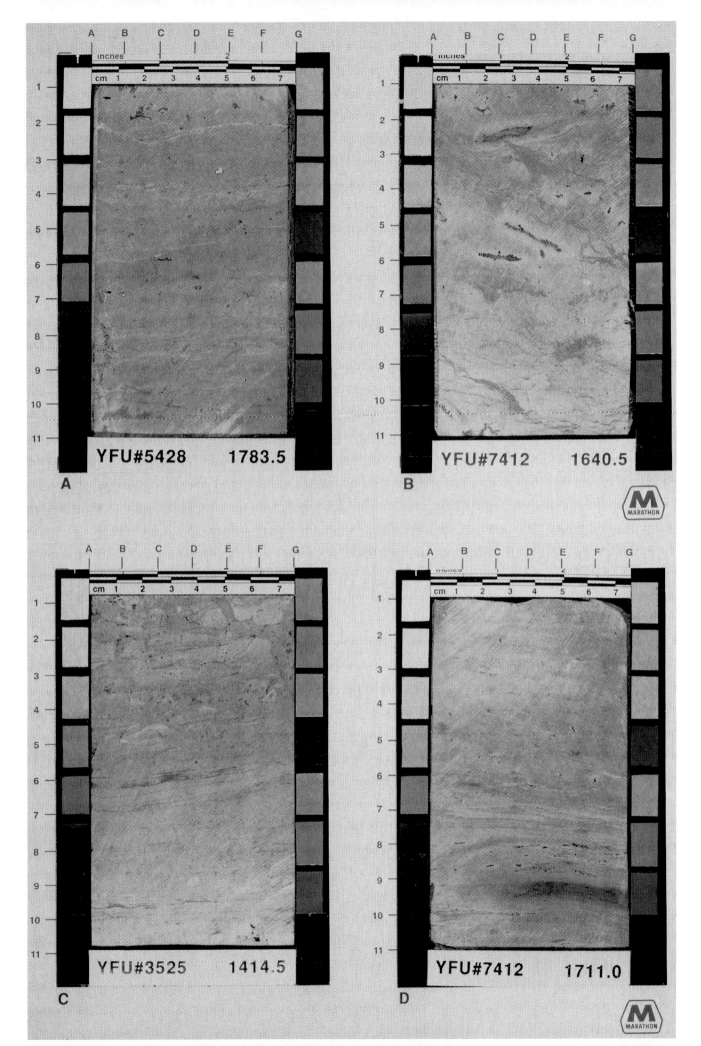

Figure 14.

A. YFU# 9807 1635.5 ft. Silty-sandy dolomite; stromatolitic mudstone to peloid wackestone. Bedded appearance is likely due to poorly preserved stromatolites. Mudstone fabrics such as this are probably better described as stromatolitic boundstones. A single, low amplitude stylolite is visible at A-G, 5-6.

Seven Rivers Fm
Intertidal-shallow Subtidal

$\varnothing = 9.4\%$
kmax = 2.5 md
k90 = 2.4 md

B. YFU# 3733 1190.6 ft. Dolomite; stromatolitic, peloid-coated grain packstone. Fenestral and vuggy porosity is developed within packstone framework. This style of porosity development is typical of grain-rich fabrics which were partly bound by poorly preserved stromatolites (not visible). A large, mudstone intraclast is present at E-G4. A moderate-amplitude stylolite cuts the core a t A-G, 10-11. Permeability analyses indicate that fenestrae and vugs are connected by a framework of BC/mBC pores.

Seven Rivers Fm
Intertidal-Shallow Subtidal

$\varnothing = 13.0\%$
kmax = 498. md
k90 = 328. md

C. YFU# 1915 1191.6 ft. Argillaceous, silty dolomite and argillaceous, dolomitic siltstone; stromatolitic mudstone to boundstone. Irregularly bedded siltstone (A-G, 1-3) overlies dense mudstone and laminated to stromatolitic, silty dolomite (A-F, 8-11). Poorly formed fenestrae and microvugs are developed within the stromatolitic fabric (D-E, 8-9).

Queen Fm
Shallow Subtidal-intertidal

$\varnothing = 5.0\%$
kmax = 0.3 md
k90 = 0.3 md

D. YFU# 5428 1576.0 ft. Silty dolomite; peloid-intraclast wackestone-packstone. Irregular rounded intraclasts are visible as light-colored dolomite grains with otherwise homogeneous, probably burrowed matrix. Wispy microstylolites (B-G, 7-8) render a nodular appearance to the dense fabric, and indicate the presence of insoluble material within the relatively light colored matrix.

Queen Fm
Intertidal-Shallow Subtidal

$\varnothing = 2.1\%$
kmax = 0.1 md
k90 = 0.1 md

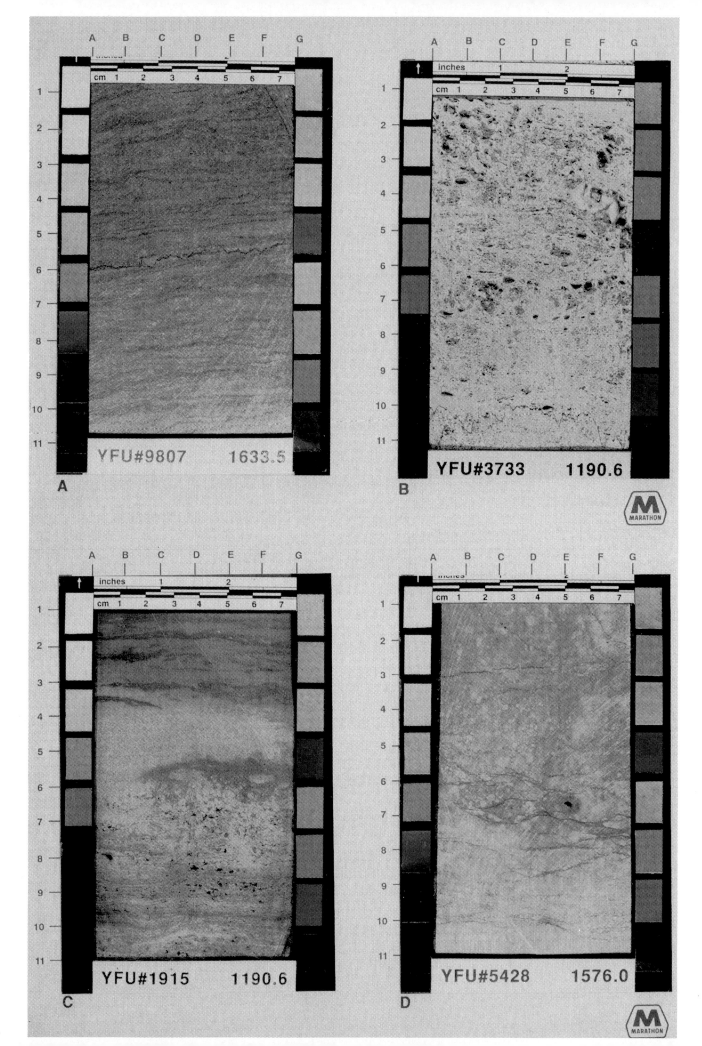

San Andres Formation—

The San Andres Formation is up to 750 ft thick in Yates Field and is the most prolific producing formation. The top of the San Andres is marked by a major erosional unconformity. The San Andres Formation is comprised of three principal lithofacies: (1) west side lagoonal shales, dolomitic, peloid mudstone and wackestone that display high gamma ray and low-porosity log signatures; (2) higher porosity, low gamma ray (except in areas of uranium concentration), dolomitic, peloid-fusulinid packstone and grainstone shoal complexes found dominantly in the deep west side and upper east side of the field; and (3) relatively low gamma ray, low porosity, dolomitic, peloid-skeletal wackestone of the eastern slope and deep ramp. A west-east stratigraphic section illustrates the simplified physiography for the San Andres (Fig. 15) and the position of the three major San Andres lithofacies in the field today. Anhydrite cement in the San Andres is rare, and late calcite cement is restricted to the lower portion of the San Andres.

Of the relatively clean carbonate fraction on the east side of the field, dolomite constitutes more than 96 percent of core rock volume (Fig. 16). Limestone (principally as calcite cement) is the only other significant lithologic/mineralogic component, accounting for about 2 percent of the core rock volume. Other minor components include silt, clay, pyrite, and bitumen. Depositional textures on the east side are dominated by packstone (ps) through grainstone (gs). Peloids are ubiquitous components of all grain-rich textures and probably form the principal constituent of most obliterated, sucrosic dolomite fabrics. Fusulinids (commonly moldic) are highly visible skeletal components of shoal, shoal margin, and shallow subtidal lithofacies. Other common skeletal constituents include molluscs, gastropods, corals, sponges, crinoids, bryozoans, and calcified algae. Pore types are dominated by moldic (mo), fusumoldic (fmo), and vuggy (vg), with a larger proportion of intercrystalline (bc) than micro-intercrystalline (mbc)(Choquette and Pray, 1970). These textures, grain components, and pore types are indicative of open-marine, subtidal deposition on the east side of the field.

On the west side, two lithologies are present: argillaceous dolomites ("shales") shown in the clastics column (core estimate of shale ≥10 percent), and clean dolomites. In the argillaceous dolomites, silt and shale account for approximately 40 percent of the rock volume (Fig. 16), with 40 API gamma ray corresponding to 10 percent argillaceous content. Of the relatively clean carbonate fraction on the west side of the field, dolomite constitutes approximately 95 percent of rock volume. Limestone (principally as calcite cement) is the only other significant lithologic/mineralogic component. Depositional textures on the west side show an increase in mudstone (ms) and wackestone (ws) at the expense of packstone (ps), when compared to the east side. Peloids are ubiquitous grain components. Other common skeletal constituents include molluscs, gastropods, crinoids, and calcified algae. Pore types are dominated by moldic, fusumoldic, and vuggy, with equal amounts of intercrystalline and micro-intercrystalline, and rare fenestral (fe) fabrics. These textures, grain components, and pore types are indicative of shallow marine to peritidal lagoonal deposits.

Because of pervasive dolomitization, porosity within the San Andres is dominated by intercrystalline and micro-intercrystalline pore types, together representing more than half of all pore types observed in core (Fig. 17). The crystalline dolomite packstone and grainstone represent the best reservoir quality rock in Yates Field. Skeletal molds, fusumolds, and vugs

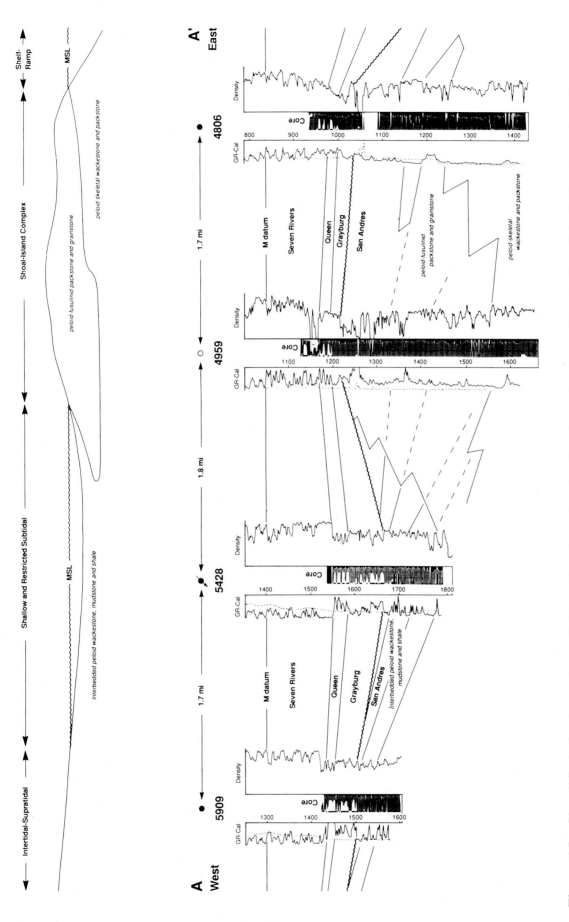

Figure 15. Simplified physiography and depositional setting for the three dominant San Andres lithofacies (upper). West-east stratigraphic section, datumed on the Seven Rivers 'M', illustrating the log responses for the major lithofacies, and the projected time-stratigraphic framework (dashed lines) relative to the three major San Andres lithofacies (lower).

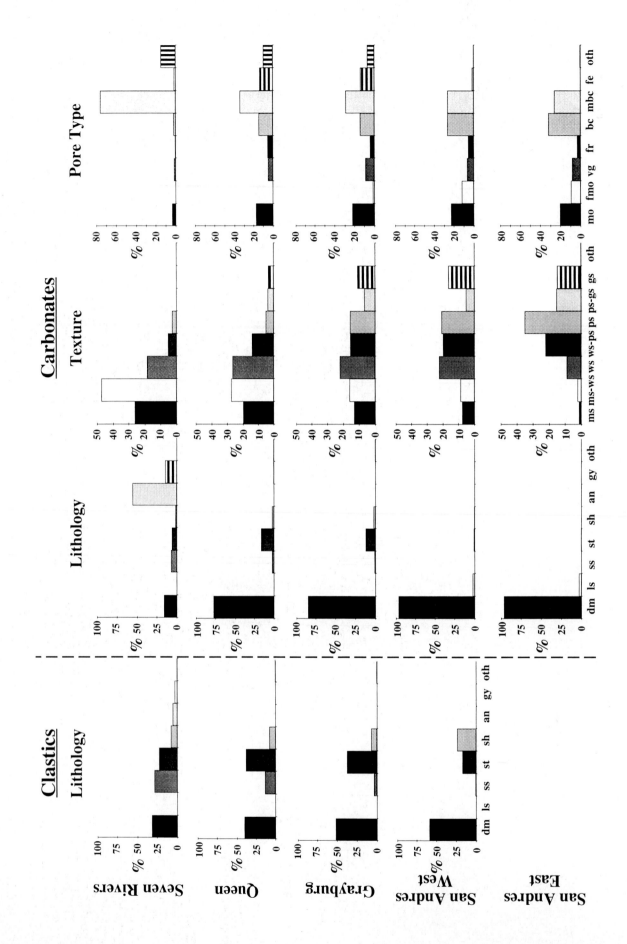

Figure 16. Histograms comparing the lithology, texture, and pore type distributions found in the producing formations in Yates Field.

significantly contribute to pore type heterogeneity in the San Andres. Fractures (fr) represent about 4 percent of the pore types in the San Andres and are variably developed in the San Andres throughout the field (Fig. 18).

Mixed Clastic/Carbonate Environments

Siltstone, fine-to-coarse sandstone, and carbonate mudstone and wackestone of the Grayburg, Queen, and Seven Rivers formations were deposited in a broad range of shallow subtidal to intertidal/supratidal environments. The karsted San Andres surface was onlapped by basal siltstones of the Grayburg during marine transgression. Shallow subtidal conditions were established and sustained during Grayburg deposition across much of Yates Field (Craig, 1988). Grayburg and Queen clastic lithofacies are characterized by dolomitic, laminated-to-massive, fine sandstone and siltstone that are dominantly tabular in geometry (Fig. 19). Residual San Andres paleotopographic highs may have served to locally nucleate intertidal and supratidal (peritidal) environments within this shallow subtidal regime.

Sedimentology and depositional geometry change from the Grayburg-Queen clastic interval to the lower Seven Rivers. The Seven Rivers is composed of variably dolomitic, cross-bedded, and deformed, fine- to coarse-grained sandstone. The wide array of evaporite and sandstone facies in the Seven Rivers Formation reflects sedimentation in a hypersaline to subaerial environment, and consequently represents an end member in the environmental continuum (Fig. 20). The clastic-dominated environments in the Seven Rivers are part of a low-relief, shelf-crest, beach/bar/dune complex that developed along a trend largely coincident with topographic highs over the buried San Andres island complex in the eastern portion of the field (Fig. 21). The clastic system separated shallow subtidal and peritidal sabkha-salina environments to the west from the extensive salina of the Midland Basin to the east.

Grayburg Formation—
The Grayburg Formation has thicknesses that range from less than 10 to 115 ft in Yates Field. The Grayburg is extremely thin over remnant San Andres karst topographic highs, and relatively thick in paleo-karst depressions. On the east side of the field, the top of the Grayburg is locally unconformable.

"Clastic" and "carbonate" lithologies are approximately equally represented in the Grayburg Formation, and the productive intervals in the Grayburg are typically dolomite. Within the "carbonate" portion of the Grayburg, dolomite is the principal component and represents nearly 90 percent of the core data (Fig. 16). The dominant clastic component of Grayburg dolomite is silt. Carbonate depositional textures are equally distributed from mudstone through grainstone, with an increase in mudstone relative to the west side San Andres. Peloids, carbonate mud intraclasts, and coated grains represent the principal allochems within Grayburg dolomite beds. Dominant skeletal constituents include molluscs, calcified algae, and rare fusulinids and crinoids, and stromatolitic mudstone to boundstone is occasionally present. In the Grayburg

Figure 17.

A. YFU#3566 1747.3 ft. Dolomite; fusulinid-peloid packstone. Relatively equal abundances of BC and mBC porosity within highly altered dolomicrospar matrix. The skeletal remnant of a fusulinids visible centered near B4, which retains some intra-skeleta (WP) porosity. Solution enlargement of BC pores into microvugs is apparent in the area of A-C, 1-2. Peloid ghosts are visible throughout (at B6-7).

San Andres Fm	Ø = 15.7%
Shoal	kmax = 653. md
22X (6.8 mm) PPL	k90 = 507. md

B. YFU# 9807 1757.5 ft. Slightly calcitic dolomite; peloid-skeletal packstone. Well developed BC/mBC porosity is distributed throughout a highly altered dolomite matrix. Vague peloids and larger skeletal fragments (probably fusulinids) are discernible despite relatively fine dolospar to dolomicrospar fabric. Larger BC pores (B-C, 3-4 and G5) are well connected by a well developed system of smaller mBC pores.

San Andres Fm	Ø = N/A
Shoal	kmax = N/A
22X (6.8 mm) PPL	k90 = N/A

C. YFU# 3566 1759.2 ft. Dolomite; peloid packstone. Highly altered fine dolospar to dolomicrospar fabric is host to considerable BC and mBC porosity, together with fewer microvugs and small molds. Micritic texture of partially preserved peloids is visible near A3 and F2. Probable pelmolds are present at F7 and G5, which appear in core as "pinpoint" microvugs.

San Andres Fm	Ø = 15.9%
Shoal-Shoal Margin	kmax = 204 md
45X (3.4 mm) PPL	k90 = 133. md

D. YFU# 2433 1482.0 ft. Calcitic dolomite; fusulinid-peloid packstone. High magnification view of dolospar rhombs and associated BC/mBC porosity. Solution enlargement has resulted in a horizontal network of dominantly BC pores at C1-B11 and F3-F9. Calcite cement fills portions of this system and most of the larger vugs whose base appears at A1-A9. Most rhombs have limpid (clear) dolomite overgrowths which are in strong contrast to "dusty" rhomb interiors. Some Rhomb interiors have been partially (B7) to completely leached (C3-4).

San Andres Fm	Ø = 18.8%
Shoal	kmax = 126. md
112X (1.4 mm) PPL	k90 = 62.0 md

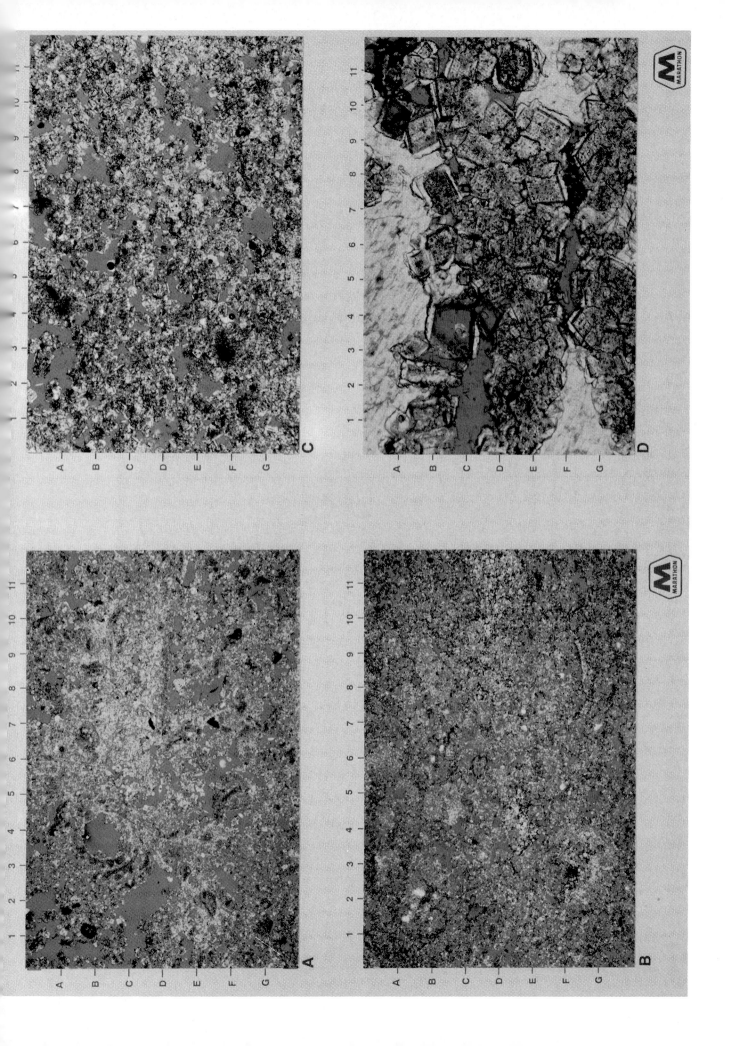

Figure 18.

A. YFU# 1915; 1242.8 ft. Bituminous dolomite; fusulinid wackestone-packstone. Fusulinid molds (FMO pores), some partly lined with dark, bituminous residue, locally define a tightly packed grain framework. The finer grain wackestone matrix (eg. A-D,1-2) contains only rare mBC/BC porosity. Measured permeability in this sample probably results from connected microfractures (eg. near G10-11). Refer also to Plate 37.

San Andres Fm	$\emptyset$ = 14.6%
Shoal Margin-Shallow Subtidal	kmax = 739. md
22X (6.8 mm); PPL	k90 = 316. md

B. YFU# 3923; 1173.0 ft. Dolomite; peloid-skeletal packstone-grainstone. Partially leached skeletal fragments include a thin-shelled mollusc (D2-A7-E11) and probable platy phylloid algae (eg. G1-11). Dissolution of the mollusc shell has produced a partial mold (MO) or shelter (SH) pore directly underneath. Peloids and unidentified skeletal fragments at C-D,5-6 are in various stages of dissolution. Micritic envelopes characterize grains throughout.

San Andres Fm	$\emptyset$ = 27.2%
Shoal	kmax = 1783. md
22X (6.8 mm); PPL	k90 = 649. md

C. YFU# 7412; 1635.2 ft. Dolomite; peloid-skeletal grainstone. Oil stained TOPN fracture within dominantly dense matrix. Comparison of kmax and k90 measurements indicate that core permeability is highly fracture influenced.

San Andres Fm	$\emptyset$ = 10.7%
Shoal	kmax = 53.0 md
	k90 = 0.1 md

D. YFU# 3566; 1769.0 ft. Dolomite; fusulinid wackestone. Oblique-trending (60 degree) POPN fracture (A1 to G-11) with several PONP fractures in dense matrix with FMO porosity. Rare calcite cement fills FMO porosity near C-D10. Most FMO pores remain open, although poorly connected. Core permeability largely results from fracture enhancement.

San Andres Fm	$\emptyset$ = 9.0%
Shoal Margin	kmax = 973. md
	k90 = 730. md

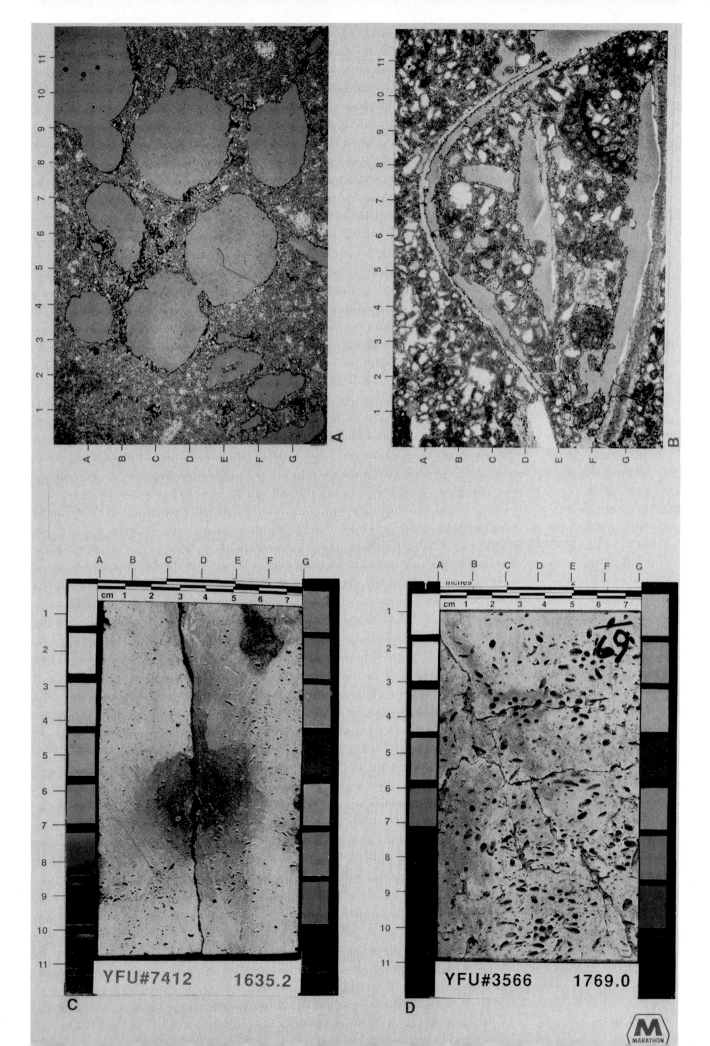

Figure 19.

A. YFU# 2908; 1256.6 ft. Argillaceous, dolomitic siltstone. Coarse siltstone matrix has been variably cemented by dolomite (dominantly in lenses or thin beds; eg. A-G4) and pyrite (as smaller, isolated patches; eg. D2-3 and D8-9). Dolomite cement appears to define crude bedding in otherwise burrowed, relatively homogeneous matrix. Pyrite distribution appears to be more random in nature, and may be related to burrows.

Grayburg Fm $\emptyset = 9.8\%$
Shallow Subtidal $kmax = 0.1$ md

B. YFU# 3923; 1057.0 ft. Silty dolomite; mudstone. Very silty mudstone to dolomitic siltstone grades upward into dense mudstone. Slightly burrowed, laminated fabric is preserved in coarse siltstone in lower portion of core. Lithology grades from siltstone to dolomite over a thin interval represented by A-G,6-8. Pores in overlying mudstone are slightly solution enlarged, and some may represent leached peloids or intraclasts (eg. near B2).

Grayburg Fm $\emptyset = 8.0\%$
Shallow Subtidal-Intertidal $kmax = 0.2$ md
$k90 = 0.2$ md

C. YFU# 3733; 1201.5 ft. Argillaceous, dolomitic siltstone. Faint horizontal lamination is indicated by darker, clayey zones within relatively dense siltstone. Dolomite cement occludes a large part of the mBP pore framework, especially in the lower portion of the core where oil staining is essentially absent. Burrows are visible in thin, dolomite cemented bed at A-G,9-10. Pyrite cement locally cements grain framework (eg. C-D,9-10).

Queen Fm $\emptyset = 7.3\%$
Shallow Subtidal $kmax = 0.1$ md
$k90 = 0.1$ md

D. YFU# 8218; 1614.2 ft. Argillaceous, dolomitic sandstone-siltstone and silty dolomite. Dense fine sandstone to coarse siltstone grades upward into silty, moldic to vuggy, peloid-intraclast mudstone-wackestone. Carbonate intraclasts and peloids are visible within siltstone-sandstone matrix at B-D,8-9. Porosity in this lithology has been mostly occluded by dolomite cement. Some of the vugs developed in relatively dense dolomite above may be solution-enlarged fenestrae.

Queen Fm $\emptyset = 5.8\%$
Shallow Subtidal-Intertidal $kmax = 0.1$ md
$k90 = 0.1$ md

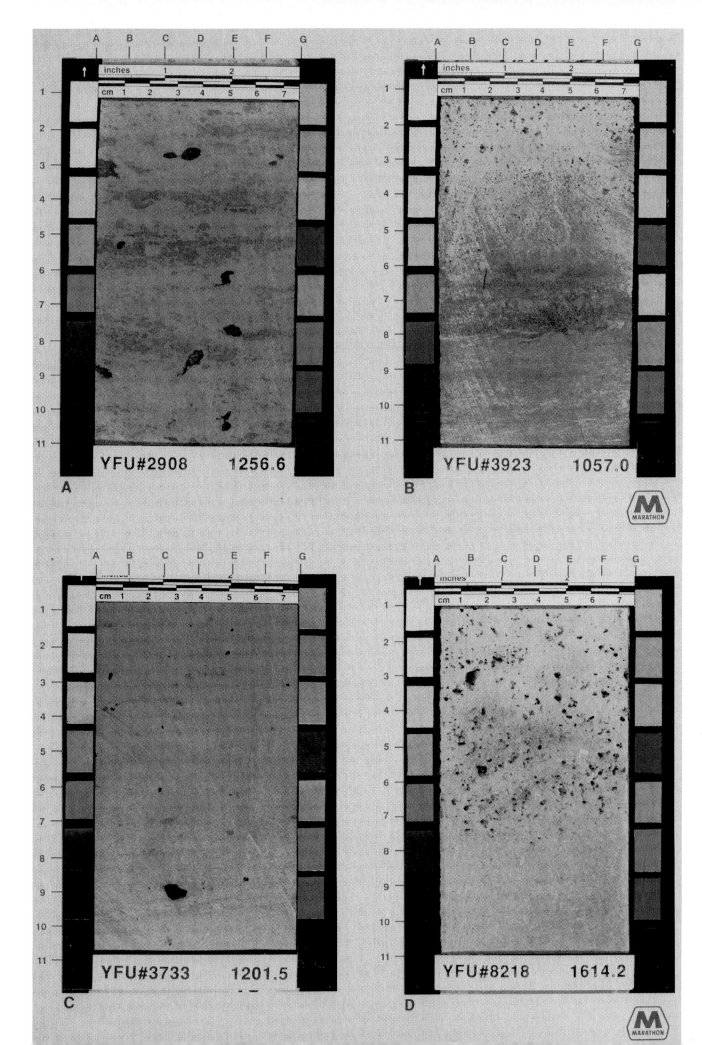

Figure 20.

A. YFU# 2908; 1194.2 ft. Dolomitic, anhydritic gypsum. Massive to nodular gypsum (darker gray crystalline masses) is intimately associated with light to bluish gray, nodular anhydrite. Deformation by displacive growth of crystals masses and nodules is evident in irregular distribution of relatively rare, originally laminated clayey dolomite (eg. F-G,8-10).

Seven Rivers Fm $\emptyset$ = N/A
Sabkha-Salina k_{max} = N/A

B. YFU# 5428; 1546.0 ft. Dolomitic, anhydritic gypsum. Highly deformed dense dolomite mudstone (eg. C-g,4-7) "floats" within a matrix of irregularly bedded gypsum (darker gray) and lighter gray to white nodular anhydrite. Coarse crystal habit of gypsum is particularly visible at A-D,4-5 and D-G8). Fractures developed in mudstone are filled by gypsum.

Seven Rivers Fm $\emptyset$ = N/A
Sabkha-Salina k_{max} = N/A
k_{90} = N/A

C. YFU# 2212; 1141.5 ft. Slightly sandy, anhydritic gypsum. Bladed gypsum crystals (upper portion of photomicrograph) are sharply delineated from ascicular to bladed anhydrite crystals along a contact extending from D1 to G11. The gypsum represents the margin of a near-horizontal vein contained within nodular to massive anhydrite. Such crystalline fabrics rarely possess porosity.

Seven Rivers Fm $\emptyset$ = N/A
Sabkha-Salina k_{max} = N/A
22X (6.8 mm); XN k_{90} = N/A

D. YFU# 2908; 1193.4 ft. Dolomitic, anhydritic gypsum. Bladed gypsum crystals form displacive veins within matrix of dense dolomite mudstone (dark, tabular features). Traces of quartz silt are visible as white specks within mudstone. Small, dark specks in mudstone are pyrite framboids. Vertical microfractures filled by gypsum at F-G,1-2 are apparently formed by displacive crystal growth.

Seven Rivers Fm $\emptyset$ = N/A
Sabkha-Salina k_{max} = N/A
22X (6.8 mm); XN k_{90} = N/A

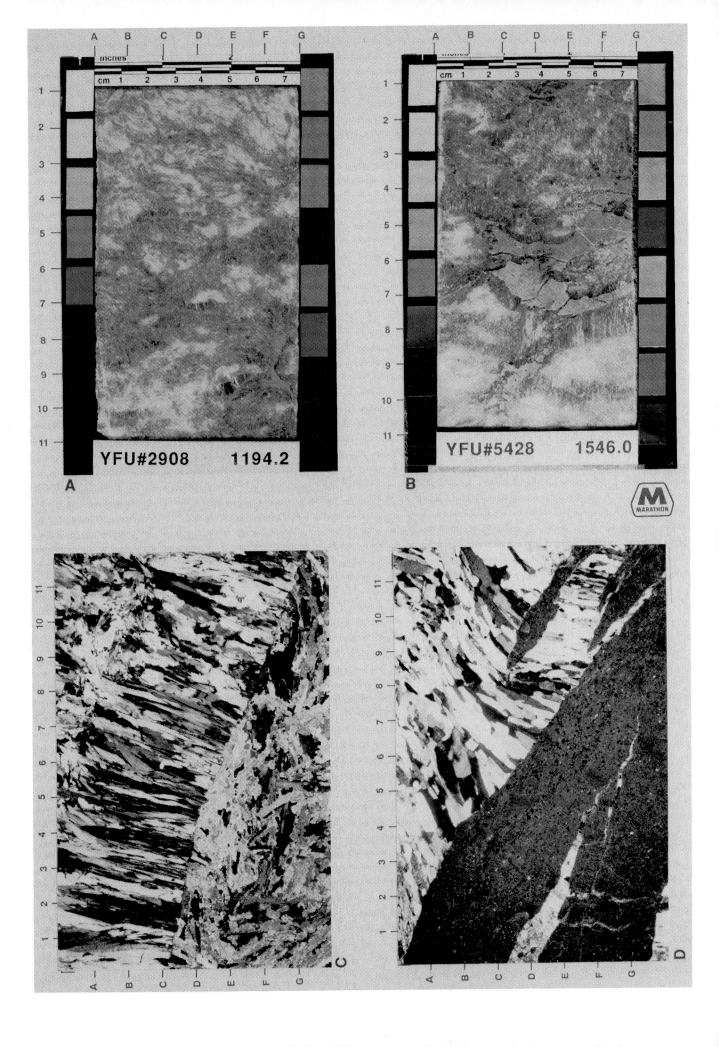

Figure 21.

A. YFU# 8218; 1562.1 ft. Argillaceous, dolomitic sandstone. Irregular bedded fine- to medium-grain sandstone lacks extensive dolomite cementation, with resulting BP/mBP pore system displaying relatively good permeability. Distorted bedding is due in part to burrowing (eg. B-D,2-5), but also is influenced by soft-sediment deformation as indicated by the subtle slip plane along E5-F7. Extracted core sample still retains considerable residual oil saturation indicated by brown stain.

Seven Rivers Fm $\emptyset = 16.7\%$
Shallow Subtidal $k_{max} = 16.8$ md
 $k_{90} = 16.3$ md

B. YFU# 4966; 1238.0 ft. Argillaceous, dolomitic sandstone. Significant soft-sediment deformation has occurred along steep slip faces (eg. D1-F8) in this fine- to medium-grain sandstone. Clayey laminae are preserved partially intact in foundered area represented by E-G,1-5. Dolomite cement locally occludes interparticle pore system (eg. gray area near D4). Spotty dark oil stain reflects non-uniform permeability.

Seven Rivers Fm $\emptyset = 15.0\%$
Shallow Subtidal $k_{max} = 1.8$ md
 $k_{90} = 0.5$ md

C. YFU# 2212; 1117-18 ft. Argillaceous, dolomitic sandstone. Well rounded, very coarse eolian sand grains (B-D,1-11) mark the position of a bedding surface within an otherwise homogeneous, fine sandstone matrix. Darker material within matrix includes detrital clay grains, authigenic clay and dolomite cement. Porosity is dominantly confined to mBP pores not visible at this magnification.

Seven Rivers Fm $\emptyset = 10.9\%$
Sabkha-Salina $k_{max} = 0.3$ md
45X (3.4 mm); PXN $k_{90} = 0.2$ md

D. YFU# 9807; 1618.6 ft. Argillaceous, dolomitic sandstone. Fine grain sandstone consists of a framework of dominantly white quartz and plagioclase feldspar grains, and rather abundant orange-yellow potassium feldspar grains (stained by sodium cobaltinitrate). Apparent dissolution of some feldspar has locally produced "molds" which appear as enlarged BP pores scattered throughout the matrix. The absence of significant dolomite cement has preserved much of the BP/mBP pore framework and associated permeability.

Seven Rivers Fm $\emptyset = 19.1\%$
Shallow Subtidal $k_{max} = 17.4$ md
44X (3.4 mm); PPL $k_{90} = 17.0$ md

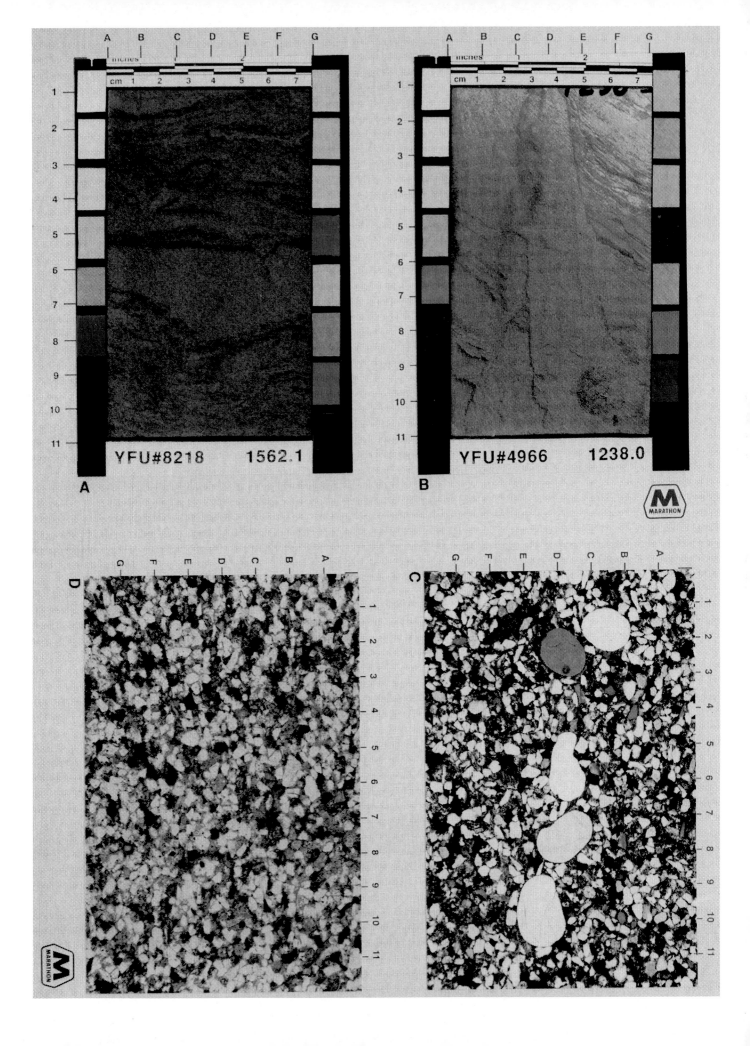

micro-intercrystalline (mbc) porosity is dominant, and fenestral (fe), and vuggy (vg) pore types are also common throughout the field. Fractures represent nearly 4 percent of the pore types in the Grayburg and are largely developed within the carbonate fraction.

Of the clastic portion, siltstone dominates sandstone (Fig. 16). Owing to feldspathic sand and shale content, the minimum gamma-ray cutoff for clastics of the Queen and Grayburg was set at 50 API. Dolomite is an ubiquitous mineralogic component in Grayburg siltstones, largely in the form of cement and less commonly as peloids and intraclasts. The distribution of siltstone in the Grayburg reflects the progressive filling of the topographic depression behind the island complex and onlap of underlying San Andres topographic highs through time. Gross "sand" thickness varies from more than 50 ft to less than 10 ft in eastern portions of the field, coincident with onlap of the buried San Andres island complex.

Queen Formation—

The Queen Formation averages approximately 45 ft in thickness in Yates Field, which is significantly thinner than the roughly 400 ft measured at the type section in the Guadalupe Mountains or on the Central Basin Platform to the north and west of the Yates Field. The dramatic thinning of the Queen in the Yates Field area may reflect distance from the sediment source, depositional patterns along the Central Basin Platform, position on the shelf, local paleotopography, and regional correlation differences.

More than half of the "clastic" portion of the Queen is comprised of sand and silt (Fig. 16), with an approximate 1:3 sand/silt ratio. Reservoir development and production in the Queen of Yates Field are from the sandstones. Dolomite forms a large part of the remaining mineralogic component. The principal Queen clastic lithology is argillaceous, dolomitic siltstone to fine sandstone, with dolomite accounting for up to 40 percent of the mineralogy dominantly in the form of cement and, to a lesser degree, in the form of peloids and intraclasts. The sandstone and siltstone lithofacies within the Queen interval are distributed across the central portion of the field, where gross sand thickness ranges from 25 to 35 ft in thickness. Queen sands tend to thin in the eastern portion of the field over where they onlap the buried San Andres island complex.

Within the "carbonate" portion of the Queen, dolomite is the principal component (Fig. 16). The clastic component of the Queen dolomite beds is dominated by siltstone, with less sandstone and shale (clay). Carbonate depositional textures are predominantly mudstone through wackestone, with significantly less packstone and grainstone than is present in the older Grayburg and San Andres carbonates. Peloids and carbonate mud intraclasts represent the principal allochems within Queen dolomite beds. Skeletal constituents are rare in the Queen. Stromatolitic mudstone to boundstone is recognized in the thin carbonate beds, but is rare. About half of the core data consists of micro-intercrystalline and intercrystalline pore types, and fenestral fabrics are well represented. Fractures constituted nearly 3 percent of the pore types in the Queen and are mostly within the carbonate fraction.

Seven Rivers Formation—

The Seven Rivers anhydrite forms the principal seal for Yates Field; it ranges in thickness from 300 ft to greater than 400 ft. Production from the Seven Rivers in Yates Field is from a clastic "arc" rimming the north, east, and south sides of the field at the base of the Seven Rivers that is informally referred to as the Seven Rivers "P" sand. The Seven Rivers "P" sand ranges in thickness from 0 ft to greater than 100 ft and comprises sandstone, siltstone, dolomite, anhydrite, and gypsum. The Seven Rivers "P" sandstone may be equivalent to the Shattuck Member of the Queen, as defined by Newell and others (1953). Correlations from the type section in the Guadalupe Mountains to the tip of the Central Basin Platform have not been completed, and this correlation remains uncertain. The Seven Rivers "P" sand has a limited areal distribution generally associated with paleotopographic highs. The "P" sand overlies and pinches out laterally into evaporites that were deposited in paleotopographically low areas, and has therefore been considered part of the Seven Rivers Formation (Spencer, 1987). The majority of sandstone and siltstone lithologies are defined by gamma-ray response in the 40 to 100 API range; therefore, a gamma-ray cutoff of 40 API was used for sand.

Nearly 50 percent of the "clastic" portion of the Seven Rivers is comprised of sand and silt, and the remainder is largely dolomite (Fig. 16). The principal lithology reflected by these components is argillaceous, dolomitic sandstone to siltstone, with dolomite cement accounting for as much as 40 percent of the mineralogy in some highly cemented intervals. To a lesser extent, anhydrite can also be a significant cementing agent, particularly in thinly interbedded sequences. The Seven Rivers is characterized by the highest proportion of sandstone to siltstone within the clastic-dominated section. As a result, the Seven Rivers sandstone has better reservoir quality than the Queen and Grayburg clastic intervals.

Within the "carbonate" Seven Rivers interval, anhydrite, gypsum, and dolomite are the principal lithologic components, and sand and silt are also present (Fig. 16). Depositional textures are dominated by mudstone through wackestone. Peloids and intraclasts form the principal allochems in the dolomites. The rare occurrence of skeletal constituents attests to the stressed water conditions that developed within the peritidal to supratidal environment, and the presence of stromatolites supports a peritidal environment. Approximately half of the pore types visible in whole core are micro-intercrystalline and intercrystalline pore types, and the remainder are variably distributed among moldic, fusumoldic, vuggy, and fenestral pore types, supporting a peritidal to supratidal depositional environment.

SECONDARY CALCITE

Dolomite matrix permeability in the San Andres Formation is a function, in part, of the complex relationship between porosity, pore type, and volume percent calcite. Calcite is a secondary pore-filling cement that generally decreases matrix permeability. Fracture permeability is also affected by fracture-filling calcite, but the relationship is more difficult to quantify. Since recovery efficiency and fluid flow are related to matrix and fracture permeability, understanding and mapping the volume percent and spatial distribution of calcite should help to delineate areas that have low recovery efficiencies, abnormal fluid flow, or relatively low gravity drainage rates. These areas can then be evaluated for alternate stimulation techniques or horizontal drainholes to improve recovery efficiency.

Calcite Characteristics

Late calcite in Yates Field is a coarse (up to several centimeters), equant spar that is dominantly pore filling and exists principally as poikilotopic, crystalline masses in dolomicrospar and dolospar fabrics. Late calcite can also occur as "dogtooth" spar crystals lining larger vugs, caves, and fractures. Early meteoric cements and calcite cements replacing dolomite matrix have been described petrographically, but are volumetrically insignificant. Macroscopic descriptions of whole core indicate that calcite is present in all pore types, including fractures, and empirical observations indicate that it preferentially fills larger pores.

Calcite Distribution

Top of Calcite Structure—
The "Top Calcite" was typically defined by the first interval of three continuous feet of calcite where at least one foot was ≥5 percent, or by the first occurrence of a 10 percent calcite foot. In some cases secondary calcite was present to the top of the core, placing the Top Calcite at an elevation somewhere above the top of the core. In other cases no calcite was present in the core, placing the Top Calcite somewhere below the deepest core sample. Structure contours were constrained to honor these maximum and minimum elevation limits. In a few cores a very different, probably early meteoric, calcite spar was observed quite high in the San Andres and was excluded from the mapped calcite data. A field-wide lower limit of the calcite-cemented interval has not been observed in cores.

A relationship exists between the Top Calcite and the Grayburg structure map (Fig. 22). The three distinct geographic clusters are present in the data and are probably the result of some combination of stratigraphic and structural controls on calcite distribution during the time of emplacement. By using linear regression analysis on each separate cluster, three equations were derived that allowed prediction of the Top Calcite from the top Grayburg by geographic region in the field. The Top Calcite structure map incorporates all core data points in the field and uses the Grayburg structure map to predict Top Calcite values away from core control for every well with a Grayburg top.

Percent Calcite Distribution Slice Maps—
Twelve 20-ft elevation slice maps from +1180 through +961 were constructed to illustrate the lateral calcite heterogeneity in the field (Fig. 23). The visual estimates of percent calcite were made on whole core stained by a dual calcite-ferroan carbonate stain. Estimates are to the nearest 5 percent of total volume and were made by the same geologist for all cores. Percent calcite map values for each slice map represent the arithmetic mean volume percent calcite over a 20-ft interval. If the core intersected only a portion of the slice, then maximum and minimum

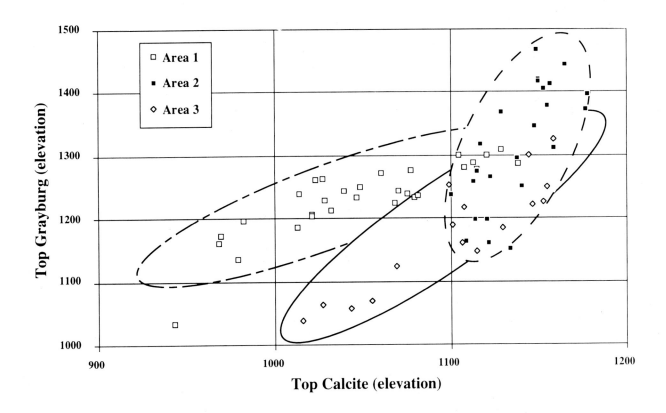

Figure 22. First occurrence of secondary calcite (Top Calcite) vs. Top Grayburg Formation from wells with logs and cores. Three distinct regions exist as a function of geographic position in Yates Field.

percent calcite values were calculated as range values. A minimum calcite value was estimated by assigning a value of zero to every foot not cored, summing all 20 ft, and dividing by 20. A maximum calcite value was calculated using the same method but assigning 30 percent, the maximum percent calcite described in core, to every noncored foot. Contours are constrained to fall within the calculated range.

In order to establish a contour trend for each slice map, the intersection of a given elevation contour from the Top Calcite structure map, with the base elevation from the appropriate 20-ft slice, was set at 1 percent calcite and contoured to follow the Grayburg structure contour trend. Other percent calcite contours follow the trend established by the 1 percent calcite line. This mapping procedure provides a logical method to guide contouring between the limited control points and results in a powerful predictive tool. In general, the amount of secondary calcite increases with depth, but shows pronounced vertical and lateral heterogeneity.

Three-Dimensional Calcite Distribution—
Values for percent of secondary calcite from all cores were integrated with percent calcite estimates from log analysis using GEOLOG's Multimin program loaded in StrataModel. The

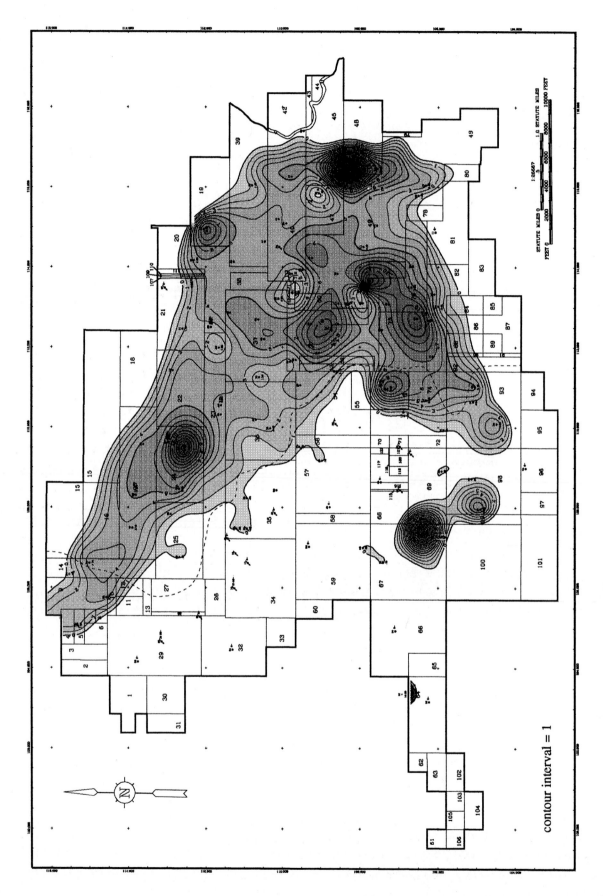

Figure 23. Percent secondary calcite 20 ft elevation slice map. Contour interval is 1% calcite. Dark areas show high secondary calcite.

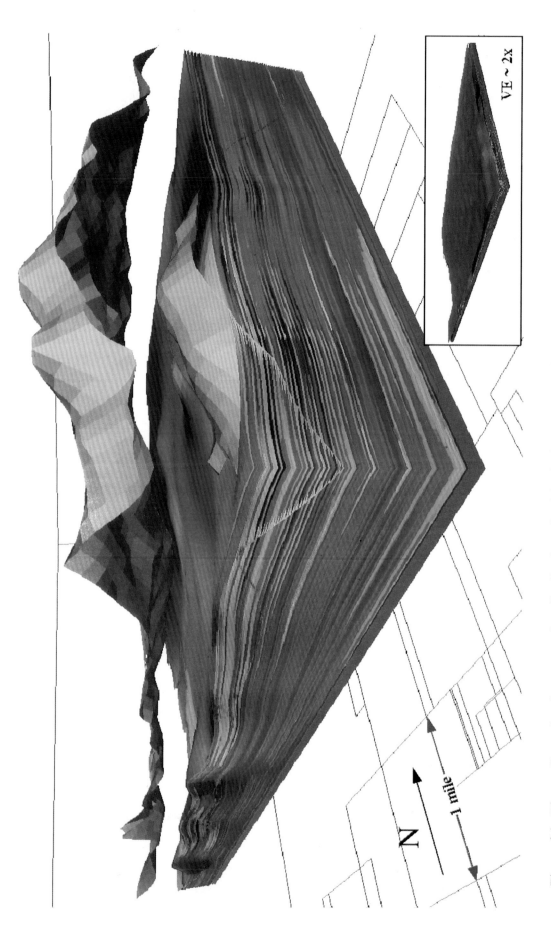

Figure 24. Three-dimensional, present-day distribution of calcite in the NW part of Yates Field. Color range is 0 to 15% calcite (blue to red). The top of the San Andres is the upper gray surface, and the top of High-Frequency Sequence 1 is the lower gray surface. Tract boundaries are shown at the bottom of the figure. Note: 1) calcite crosses depositional stratigraphy; 2) calcite structural dips are less than stratigraphic structural dips; and 3) calcite is distributed in layers of high and low calcite, and varies spatially within a given layer. These observations support the interpretation of episodic-hydrocarbon fill, and multiple-phase calcite precipitation related to oil biodegradation. Inset shows the same view, with a vertical exaggeration of ~2x.

Top Calcite structure grid was used to constrain the upper limit of the data in 3-D. The resulting 3-D model illustrates (1) the distribution of secondary calcite relative to stratigraphy (Fig. 24); (2) the distribution of secondary calcite relative to present-day structure; (3) the vertical and spatial variation in secondary calcite volume; and (4) the relationship between pore-plugging calcite and matrix porosity. In addition, porosity and calcite can be used in conjunction with pore type to predict matrix permeability in 3-D.

Model for Calcite Origin

One model for the origin of calcite is based on a hypothesis of episodic-fill and multiple-phase calcite precipitation related to biodegradation of the oils. This model helps to explain the observed distribution and vertical heterogeneity in volume percent calcite in Yates Field.

Data—
Calcite samples from three cores were examined using cathodoluminescence petrography and carbon and oxygen isotopic analyses. Results from these analyses indicate at least two phases of calcite precipitation (Fig. 25). Gas Chromatography/Mass Chromatography and Gas

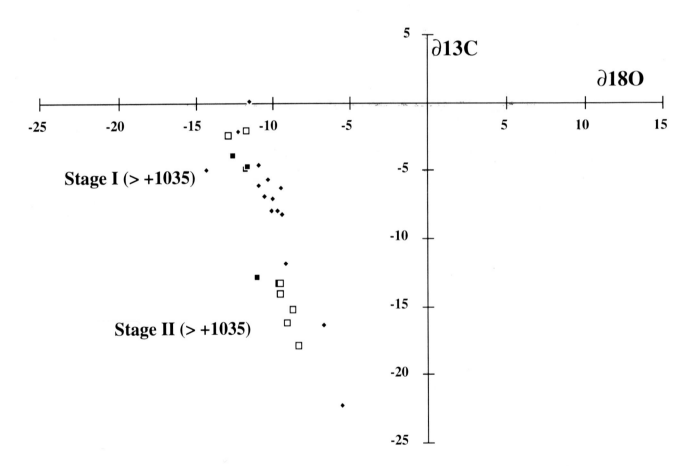

Figure 25. Stable carbon and oxygen isotopes for pore-filling calcite in Yates Field. The data are from three different wells indicated by different symbols.

Chromatography analysis of oils from Yates Field indicate that there is a single oil type in the field that was generated by a marine source rock with moderate clay content. This single oil is variably biodegraded throughout the reservoir, with extensive biodegradation at an elevation of +1035 to +940, the lower limit of the sample data set.

Interpretation—

The characteristic light isotopic values, association with biodegraded oils, and distribution in the reservoir indicate that the formation of much of the calcite is related to the process of oil biodegradation. A histogram of mean percent calcite by biodegradation level (Fig. 26) supports the hypothesis that percent calcite increases with increased biodegradation. The biodegradation process begins when bacteria present in the aquifer beneath the oil column metabolize hydrocarbons through sulfate reduction. A by-product of this process is CO_2. The released CO_2 is dissolved in formation water as HCO^{-3} and later can combine with calcium to form $CaCO_3$ (calcite). As a result of this precipitation process, calcite "layers" were probably deposited in a relatively flat profile, mimicking to some degree the topography of the oil-water contact. In that the first occurrence of calcite was probably deposited as a relatively flat surface, the Top Calcite structure map provides a crude estimate of structural deformation since calcite emplacement.

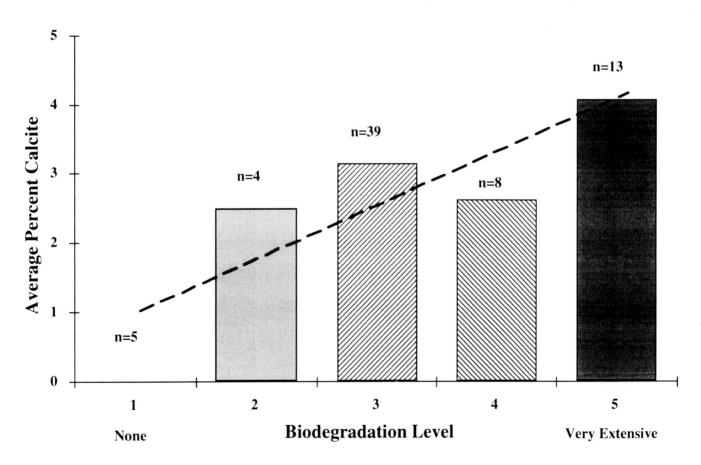

Figure 26. Relationship between average percent secondary calcite and oil biodegradation level.

The tectonic history of the Permian Basin and Central Basin Platform supports the theory of multiple phases of oil expulsion, migration, and filling of the Yates Field reservoir (Craig, 1990). The first stage of oil expulsion and migration probably occurred in the late Guadalupian-Ochoan. Expulsion and migration following the first stage was likely episodic, with phases of active migration and fill represented by low calcite precipitation rates and volumes, and phases of passive migration and fill represented by increased biodegradation and high calcite precipitation rates and volumes. A phase of secondary remigration of the oils occurred following structural tilting of the Central Basin Platform. The later calcite may mark this late-fill episode. This episodic, multiple-fill process is partly responsible for the vertical and lateral heterogeneity of percent calcite seen in the reservoir today. A working model for calcite emplacement helped to improve the understanding and prediction of calcite distribution in Yates Field.

Effect of Calcite on Reservoir Quality

There is an empirical relationship between matrix permeability, porosity, pore type, and percent secondary calcite. To help quantify this relationship, a study was designed that controlled the parameters of pore type, porosity, and percent calcite. Sixty core plugs (3" long by 1" diameter) were cut, extensively cleaned, and measured for porosity and permeability at Marathon's Petroleum Technology Center and by Core Laboratories, using several porosity and permeability measurement techniques. Pore type was estimated macroscopically in the whole core and plug, and by thin-section point count. The volume percent of pore-filling calcite was estimated macroscopically in the whole core and plug, by thin-section point count, and by X-ray Diffraction (XRD).

The present-day volume of calcite relative to present-day porosity is important to know. In other words, 5 percent calcite will not significantly affect permeability in a 30 percent porosity rock, but will dramatically affect permeability in a 10 percent porosity rock. To address the issue of calcite as a percent of porosity, the Calcite Porosity Ratio is used on the horizontal axis of all plots:

$$\text{Calcite Porosity Ratio} = \frac{\%\text{Calcite} + \%\text{Porosity}}{\%\text{Porosity}}$$

If calcite is assumed to be completely pore filling, the numerator of the Calcite Porosity Ratio approximates "pre-calcite" porosity. Pre-calcite porosity is the post-dolomitization matrix porosity that existed prior to later cementation by pore-plugging calcite. Dividing the numerator by porosity normalizes each sample by present-day porosity. The Calcite Porosity Ratio can be used to calculate the percent of pre-calcite porosity currently plugged by calcite. For example, for any porosity value, a ratio of one indicates that calcite is not present and porosity is unchanged. A ratio of two [percent Ca = percent Ø] indicates that 50 percent of the pre-calcite porosity is currently plugged. A ratio of three [percent Ca = 2(percent Ø)] indicates that 67 percent of the pre-calcite porosity is currently plugged, and so forth.

Calcite has an effect on porosity and permeability as a function of pore type. A matrix of 12 plots (Fig. 27) is intended to be compared both by column and by row. Column comparisons demonstrate changing relationships as a function of increasing percent moldic porosity, and row comparisons demonstrate the effect that percent calcite has on porosity and matrix permeability.

The left column (plots "a"-"d"; Fig. 27) has Boyle's Law Porosity on the vertical axis. Within each plot in the left column, porosity decreases as percent calcite increases (larger Calcite Porosity Ratio). When the data are separated into three pore-type classes ("b"-"d"), the scatter in the data is significantly reduced. For a given Calcite Porosity Ratio, each successive plot shows that porosity decreases as percent moldic pores increase. The middle column (plots "e"-"h"; Fig. 27) has Klinkenberg permeability (k) on the vertical axis. Within each plot in the middle column, permeability decreases as percent calcite increases (larger Calcite Porosity Ratio). When the data are separated into three pore-type classes ("f"-"h"), the scatter in the data is significantly reduced. For a given Calcite Porosity Ratio, each successive plot shows that permeability decreases as percent moldic pores increase. The right column (plots "i"-"l"; Fig. 27) also has Klinkenberg permeability (k) on the vertical axis. In addition, three porosity classes, $\emptyset < 10$ percent, 10 percent $< \emptyset < 20$ percent, and $\emptyset > 20$ percent, are represented in each plot by different symbols.

The data indicate that matrix permeability is related to pore type, pore-plugging calcite, and porosity. The best reservoir rocks in Yates Field are the nonmoldic pore-type rocks with a Calcite Porosity Ratio less than 2. If the Calcite Porosity Ratio exceeds 2, only nonmoldic pore types have permeability >1 md, and if the Calcite Porosity Ratio exceeds 3.5, porosity is typically less than 10 percent and permeability is typically less than 1 md, regardless of pore type. To illustrate the negative effect of calcite on permeability, a nonmoldic rock with 30 percent calcite and a porosity of 15 percent (Calcite Porosity Ratio = 3) would have a permeability in the reservoir today of 1 to 30 md. This same rock with no calcite would have a 45 percent "pre-calcite" porosity (15 percent today + 30 percent plugged; Calcite Porosity Ratio <2) and permeability well in excess of 200 md. This relationship exists for three principal reasons: (1) the tortuosity of the pore network increases as the variability in pore size and pore shape increases, causing decreased permeability; (2) moldic and mixed pore systems have lower porosities than nonmoldic rocks, and therefore typically have lower matrix permeabilities; and (3) it takes less pore and pore-throat-plugging calcite per unit pore volume to decrease permeability in a mixed or moldic rock, because of pore throat size inhomogeneity.

The San Andres whole core data show the same general relationships as the empirical core plug data. In rocks with calcite cement present, geometric mean permeability drops approximately one order of magnitude between nonmoldic and moldic rocks for a given Calcite Porosity Ratio, and there is a subtle decrease in permeability between intercrystalline (bc) and micro-intercrystalline (mbc) rocks within a given pore type class. There is greater scatter in the whole core data relative to the experimental core plug data, owing to (1) the variability in cleaning and measurement techniques over the years, (2) the significantly increased amount of data, and (3) the error inherent in percent calcite estimation using only whole core visual estimates, without supporting thin-section point counts and XRD analysis. Understanding calcite distribution and its relationship to permeability can help guide operations decisions such as acid

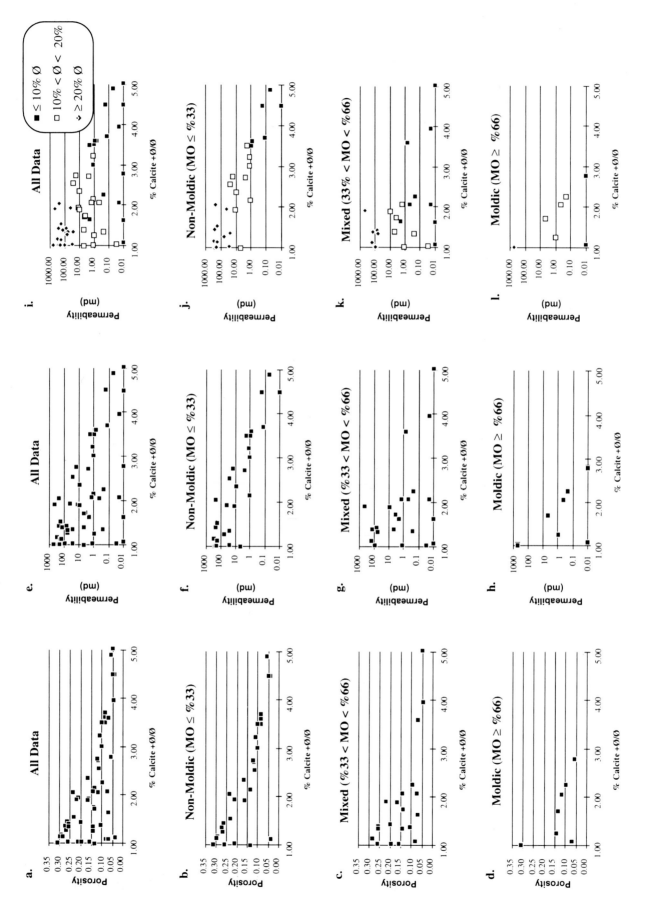

Figure 27. Experimental core plug data of porosity vs. Calcite Porosity Ratio (a-d), permeability vs. Calcite Porosity Ratio (e-h), and permeability vs. Calcite Porosity Ratio coded by porosity (i-l).

stimulations, horizontal drainhole selection and evaluation, and well performance evaluation. In addition, since a value for matrix density is important for accurate porosity calculations from the bulk density log, understanding the distribution of calcite improves porosity calculations.

FRACTURE AND CAVE SYSTEM

A significant reservoir component of the San Andres Formation in Yates Field is an extensive fracture network. Early regional fractures in Yates Field were significant because they acted as conduits for solution enlargement during karstification and resulted in the development of vuggy and cavernous porosity along joints. These near-vertical, open fractures in the field enhance permeability by several orders of magnitude. Of primary importance to Yates Field reservoir management strategy is the influence that these fractures have on fluid flow in the reservoir and their contribution to the dual-porosity system. Fracture intensity and orientation data are the key to field modeling, volumetric analysis, and oil column management. Fracture information is currently being used in the oil column stabilization program to optimize the location of deep-water reinjection and horizontal drainholes. Ongoing studies are using FRACMAN discrete element modeling software to generate stochastic realizations of fracture distribution, which can then be fed into reservoir simulation models (Curran, pers. comm.).

Fracture Genesis

Three major phases of fracture development are recognized in Yates Field, including an early phase associated with regional fracture and fault trends, a later phase associated with the formation of the field anticline, and a final phase associated with differential compaction of shales and muddy carbonates, principally on the west side of the field.

Early Fracture Development and Regional Influence—
Although San Andres time was relatively quiet tectonically in the Permian Basin, two late Paleozoic periods of tectonic activity appear to have influenced early regional fracture development and orientation in Yates Field: (1) Late Mississippian to Early Pennsylvanian, and (2) Late Pennsylvanian to Early Permian (middle Wolfcampian)(Hills, 1970; Font, 1985). During the Late Mississippian-Early Pennsylvanian period of activity, the principal east-west stress direction resulted in a fault system composed of dominant left-lateral faults, striking N50°W to N65°W, and conjugate right-lateral faults, striking N55°E to N80°E. A lesser stress field, oriented N65°E during this period, is evidenced by principal fold axes striking N35°W in the region. During the Late Pennsylvanian-Early Permian period of activity, the principal stress field shifted to north-south, resulting in a strike-slip fault system composed of dominant right-lateral faults, striking N30°W, and conjugate left-lateral faults, striking N30°E (Hills, 1970; Font, 1985).

Fault lineaments for both periods trend in the NW-SE and NE-SW directions. Hills (1970) has suggested that the jagged pattern observed along the NW-SE trending eastern edge of the Central Basin Platform may have resulted from north-northeast shears which caused northwest faulting. Although Late Permian time was quiet in terms of regional tectonics, fractures and normal faults may have been activated over pre-existing planes of weakness (Hills, 1970). Early-formed fractures in Yates Field, apparently established along planes of weakness set up in earlier periods, trend northwest-southeast, with a conjugate northeast-southwest set. Evidence of early fracturing and fracture orientation in Yates Field is discussed below.

Karst Development Related to Early Fractures—A major erosional unconformity separating the San Andres and Grayburg formations in the field indicates that a relative sea-level fall occurred at the end of San Andres time, resulting in an extended period of subaerial exposure (Craig, 1988). The relative sea-level fall resulted in San Andres island emergence and associated cementation, lithification, and subsequent fracture development in the San Andres limestone along pre-existing planes of weakness. The early regional fractures acted as conduits for solution enlargement, resulting in the development of an extensive karst system and associated vuggy and cavernous porosity. The presence of an extensive open cave and fracture network is indicated by bit drops during drilling (the longest having been over 20 ft), significant porosity and caliper log anomalies, and single-well production volumes far greater than a reasonable drainage radius. There is a relationship between cave distribution and sequence stratigraphy in Yates Field. Each of the four high-frequency sequences was capped by a subaerially exposed island complex on which separate cave lenses formed (Tinker and others, in press). The separate cave lenses formed as a result of meteoric processes, and the island complexes and associated cave lenses aggraded and prograded from west to east (Fig. 28).

An important economic issue in carbonate reservoirs is the effect of subaerial exposure on porosity and permeability. Cavernous porosity forms beneath major unconformities in Yates Field. When examined by stratigraphic cycle, the data show that cycles with abundant caves tend to have lower total porosity than the cycles with few caves. However, since Yates Field is relatively shallow and most caves have not undergone mechanical collapse, the extensive network of open vertical caves and solution-enhanced fractures increases total vertical permeability of the system.

Calcite Fill of Early Fractures—Other evidence for early fracturing comes from interpretation of a Formation MicroScanner (FMS) log in a medium radius horizontal well. A total of 671 fractures were picked over the horizontal interval of 1550 ft (Merkel, pers. comm.). One dominant fracture orientation was observed at N40°E, and a secondary conjugate set was observed at N50°W. Seventy-two calcite-filled fractures, trending primarily N40°E, were picked on the FMS. Since the fractures pre-date calcite fill, and calcite is thought to have been precipitated relatively early in the burial history of the field, the filled fractures apparently formed quite early. A probable paragenetic sequence is:

- early NE/SW fractures with a conjugate NW/SE set

- solution enhancement related to subaerial exposure

- burial

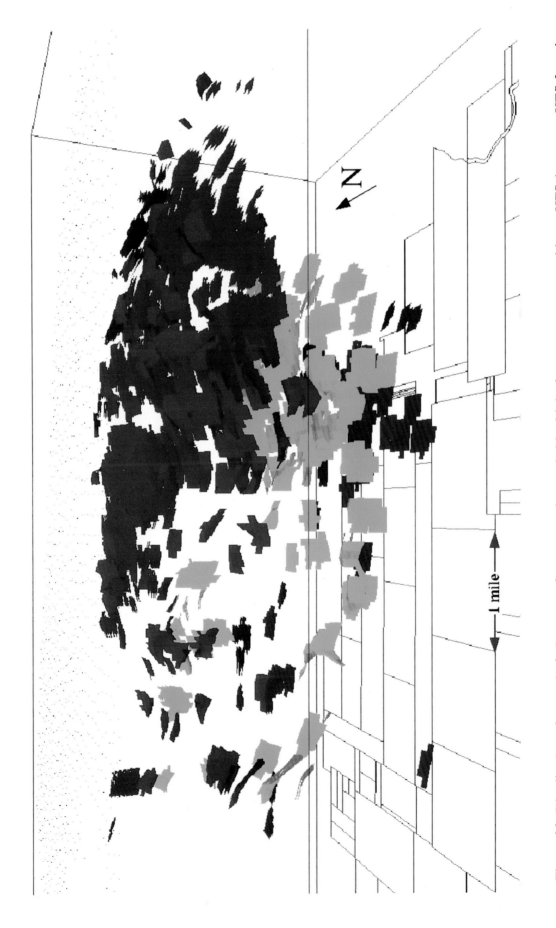

Figure 28. Distribution of caves in 3-D space. Caves are color-coded by high-frequency sequence: blue = HFS 1, green = HFS 2, and red = HFS 3 and 4. Tract boundaries are shown at the bottom of the figure. Dots at the top of the figure represent wellbore control.

- local drape structure and related faulting

- oil migration into the reservoir

- oil biodegradation and calcite precipitation, partially filling early fractures

- later fractures developed along pre-existing regional planes of weakness

Later Fracture Development—
Local fractures were formed in association with formation of the field anticlinal structure. FMS (Formation MicroScanner) and FIL (Fracture Identification Logs) orientation data indicate that these fractures tend to strike parallel or perpendicular to the strike of the anticlinal limb where they occur. A third set of fractures is probably associated with compaction of the argillaceous dolomites and dolomitic shales in the upper San Andres on the west side of the field. These fractures affect the more brittle dolomites between the shales, but typically do not completely penetrate the more ductile shales. A preferred orientation of these fractures is not obvious.

Fracture Characteristics

Core Fractures—
Examination of core indicates that most naturally occurring fractures in Yates Field are high angle. Ninety percent of the data has dips greater than 60° (Fig. 29), and nearly half of the data has dips that exceed 80°. Natural fractures in Yates Field are those fractures that:

- are filled with gouge or vein material

- are unfilled, but lie parallel to filled fractures in the same core piece

- may have preferential oil staining in the fracture, but not in the surrounding matrix

- have slickensides

- have clean, fresh, planar surfaces and are accompanied by parallel incipient fractures,

- are relatively high angle (60°-90°)

Within the classification of natural fractures, a subset of "effective fractures" has been defined for mapping purposes as those fractures that completely penetrate the core and have potential for flow beyond the wellbore.

Drilling-induced fractures in Yates Field are those fractures that:

- show none of the properties of natural fractures

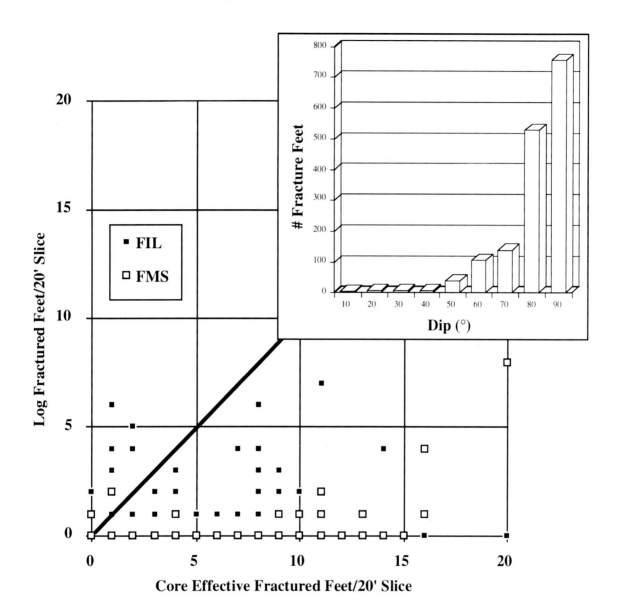

Figure 29. Core count of effective fractured feet averaged over a 20 ft. elevation slice vs. FMS log count averaged over the same interval. Black line represents a 1:1 fit. Inset shows a bar plot of the number of fractured feet in each 10° dip category. Fracture dip is measured from horizontal.

- have dips coincident with bedding planes, if present
- are cuspate or change direction at the core edge
- have faces that fit together perfectly, without evidence of mineralization or oil stain
- are relatively low angle (<60°)

Fracture Aperture—Fracture aperture measured at surface conditions is typically larger than subsurface aperture, because the core is at lower surface pressures and may be altered by the core handling process. When the fracture is partially mineralized, confidence in aperture measurement is greater because the subsurface aperture is preserved at the surface by natural cement. Since aperture typically varies across the core, an "average" aperture was measured using an illuminated hand lens with a built-in micrometer. Fracture apertures measured in core, compared to those recorded by the FMS tool, are nearly an order of magnitude greater (±1 mm vs ±0.1 mm).

Downhole In-Situ Fractures—

The most reliable logging tools for in-situ fracture delineation are borehole imaging devices. Schlumberger's Formation MicroScanner (FMS), Formation MicroImaging (FMI), and Fracture Identification Logs (FIL) have been used with mixed success to detect fractures in Yates Field. The early FMS tool was a 2-pad device, and the current tool is a 4-pad resistivity device that produces an electrical image of the borehole face. In an 8-inch borehole, pad coverage is approximately 40 percent in a single logging pass, and therefore fracture counts relative to core counts are conservative. Where a resistivity contrast exists between a fluid-filled fracture and the rock, it is possible to detect filled or open fractures on the borehole image. Data from the FMS can be processed to provide information on fracture azimuth, true dip, and, if mud filtrate resistivity (R_{mf}) is known, aperture. The FMI is an 8-pad resistivity device, in which, in an 8-inch borehole, pad coverage is approximately 80 percent.

The FIL tool is a 4-pad resistivity device. In an 8-inch borehole, pad coverage is approximately 25 percent, and therefore fracture counts relative to core are conservative. The FIL tool was run in several wells on the west side of the field in 1978. The data from these wells were normalized and analyzed with a program that examined for conductive anomalies, such as fractures, coincident on pads 180° apart. The program provides information on fracture azimuth, but not dip or aperture. In order to eliminate anomalies due to differences in lithology, only counts in the nonshaly intervals, where the gamma ray was less than 40 API, were included.

Core Versus Log Fractures—

A comparison of fracture identification techniques between cores and logs shows that core counts are typically higher and possibly more accurate than logs. For comparison, fractured feet (a cored foot containing at least one effective fracture, or a logged foot containing at least one fracture) were summed over a 20-ft interval. This summation was made to minimize discrepancies in core-to-log depth shifts, which indicate that foot-by-foot, core-to-log comparison of fracture counts are not valid. A plot of Core Effective Fractured Feet/20 ft vs Log Fractured Feet/20 ft shows that counts of core-effective fractured feet are greater than FMS counts more than 90 percent of the time, and greater than FIL counts more than 80 percent of the time (Fig. 29). This is largely a result of incomplete wellbore coverage and smaller aperture fractures beyond the resolution of the logging tools.

Relationship of Fractures to Porosity and Lithology

There is a relationship between fracture count, lithology, and porosity in Yates Field. Average fractures per foot are greatest in dolomite mudstone and wackestone, slightly lower in dolomite packstone and grainstone, and significantly lower in shale and sandstone. In dolomite, average core-effective fractures per foot increase up to 8 to 10 percent porosity, and then decrease predictably with increased porosity (Fig. 30). The sand and shale do not show a similar fracture to porosity relationship. The lithology and porosity relationship can be used to predict the gross fracture intensity in areas away from core and log control. In terms of Yates Field formations, the brittle dolomites of the San Andres fracture more readily than the more ductile mixed dolomites and clastics of the Grayburg, which in turn fracture more readily than the more ductile clastics and anhydrites of the Queen and Seven Rivers. This ductility hierarchy has been demonstrated by Stearns and Friedman (1972).

Fracture and Cave Distribution

Mapping Data—
Core and log data were used to map fracture distribution in Yates Field. Core-effective fractured feet were used for mapping, as opposed to total fracture counts, because the effective fractured-feet count from cores is more comparable to log values. Fractured feet provide a qualitative assessment of fracture distribution within the interval, or how the fractures are spaced; whereas, total fractures provide an assessment of fracture intensity, or how many fractures are present in an interval. Since core-effective fractured-feet counts are greater and possibly more reliable than the FMS and FIL counts, the log-fracture counts are used as minimum values in the mapping process. More than 90 percent of the fractures have dips greater than 60°, so no correction for dip was made in mapping.

Fracture Trends and Maps—
Five elevation slice maps of Fractured Feet/20 ft from present-day elevation +1160 through +1061 and a cumulative map of San Andres Fractured Feet/100 ft were constructed (Fig. 31). Based on examination of depositional trends, San Andres cave lineaments, sinkholes and karst depression lineaments, San Andres flexure lineaments, fluid-production trends, and in-situ strike orientations from the medium-radius horizontal well, a regional conjugate trend of N50°W and N40°E was imposed on the data. When taken alone, any one of these data types could have several lineament orientations, but when considered together, the N50°W and N40°E lineament-trend interpretation results in the most consistent representation of data. In general, fractures are more confined to discrete linear zones on the east side of the field and tend to broaden and coalesce on the west side of the field. This occurs because the enhanced later fracturing on the west side of the field, associated with shale compaction, does not follow early regional trends.

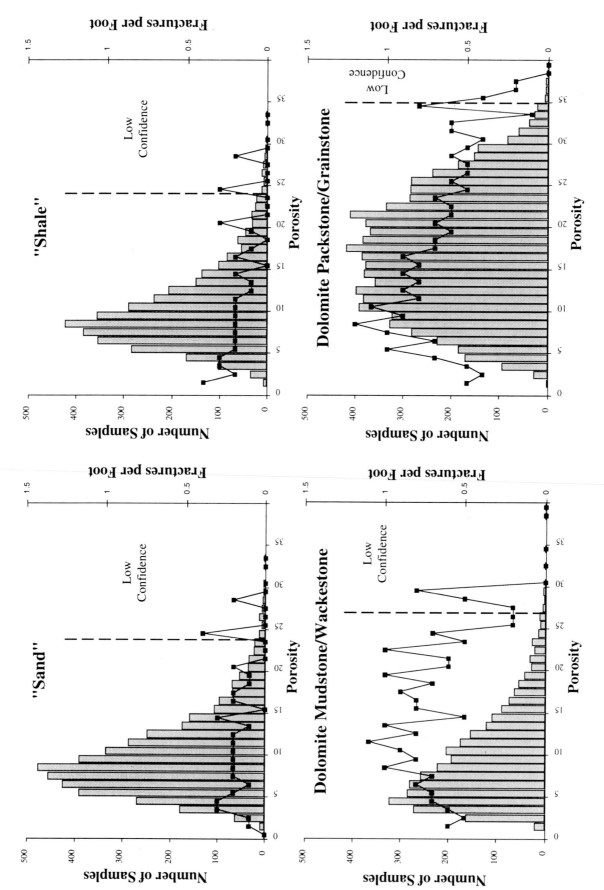

Figure 30. Porosity vs. average fractures per foot (solid line), for four major lithology classes. Bar plot shows the distribution of the number of data points for each porosity unit that went into calculating the average fractures per foot, and is essentially a distribution of porosity for each lithology class.

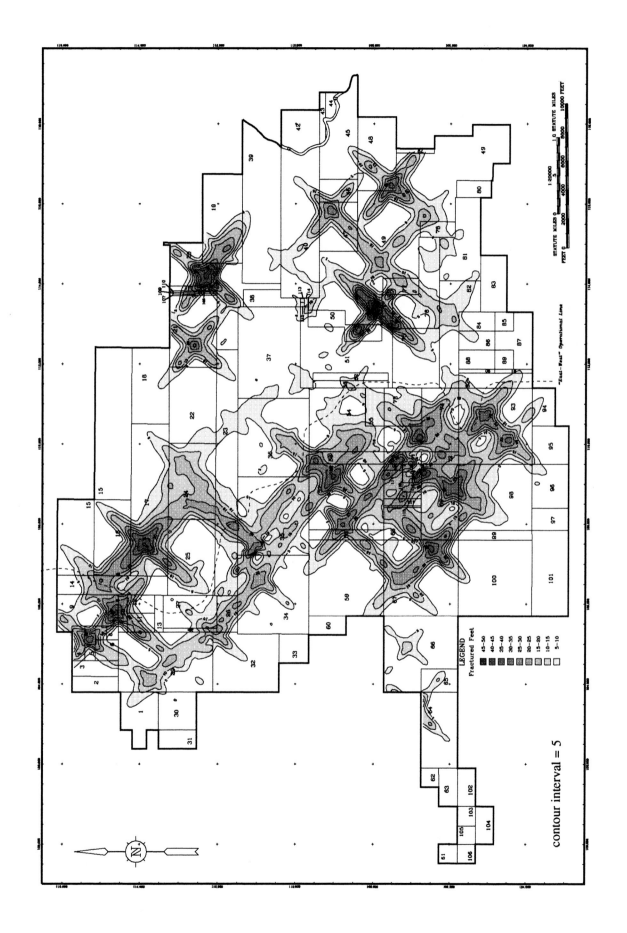

Figure 31. Number of fractured feet in a 20 ft elevation slice map. Contour interval is 5 fractured feet. Dark areas show high fractured feet. NW-SE and orthogonal NE-SW trends were imposed on the data.

Flexure Lineaments—A flexure map on the top of the San Andres shows a composite of early regional and later local structural influence on lineament trends (Fig. 32). Flexure measurement is the rate of change of dip or the second-order derivative. The arrows indicate the significant impact of early regional fracture trends, with an orthogonal pattern clearly demonstrated in the N40°-50°W and N30°-40°E directions.

Fluid Flow Trends—Normalized fluid-production trends indicate fluid-production highs in N50°W and N40°E orthogonal patterns (Fig. 33).

Depositional Trend—Depositional thickness and porosity trends generally follow the N50°W trend in the northern part of the field and the N40°E trend in the southern part of the field. San Andres shoal deposition and the orientation of the shelf-margin break were probably influenced by structural lineaments established prior to San Andres deposition.

Cave Lineaments—Caves were picked by examining logs, cores, and completion and workover reports for all wells in the field. Caves were plotted by present-day elevation slice for seven 20-ft slices from +1280 through +1141. Lineament trends were drawn through the data for each slice, and the N50°W trend is the most dominant.

Karst Features—Post-San Andres karst features were plotted on a map and connected by N70°W and N20°E lineament trends (Craig, 1988). Craig also used San Andres paleotopographic highs and an interpreted fault to support these trends. The same data can be reinterpreted with an alternate N50°W and N40°E lineament interpretation, which illustrates that trends derived from multiple data sets could have several interpretations.

FMS Orientations From Medium-Radius YFU No. 17D5—A rose diagram of the FMS-interpreted fractures from a medium-radius horizontal well shows very pronounced N40°E and orthogonal N50°W trends. The quality of the data in this well and the confidence in the fracture picks are very high.

Regional Trends—The two late Paleozoic periods of tectonic activity resulted in lineament trends of N40°E and orthogonal N50°W.

RESERVOIR DISTRIBUTION

The San Andres Formation dolomite is the principal reservoir in Yates Field and accounts for approximately 90 percent of the storage volume in the reservoir. The siltstone, sandstone, and dolomite of the Grayburg, Queen, and Seven Rivers formations account for the remainder. The relationship of core-to-log porosity is fundamental to the mapping of matrix-quality reservoir

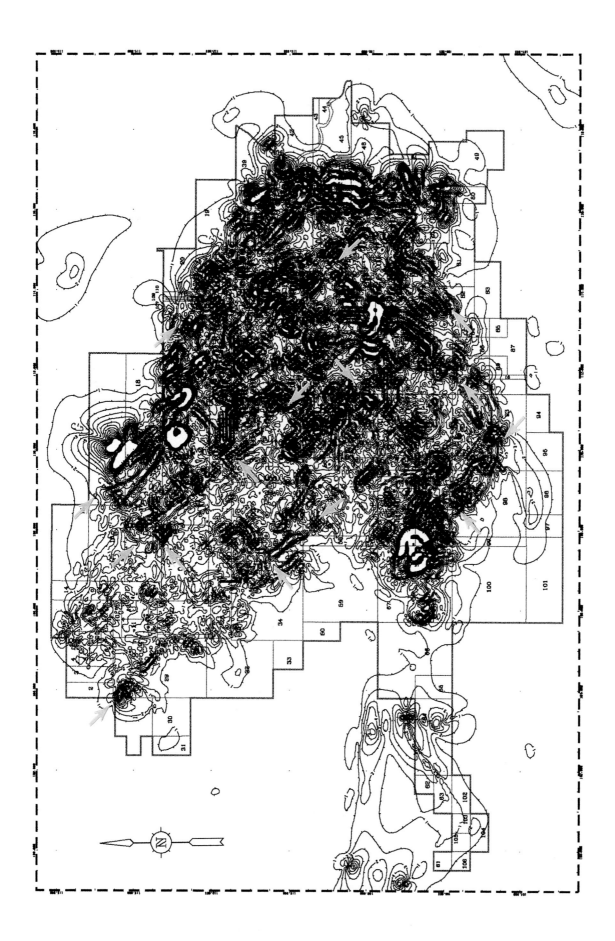

Figure 32. Top of the San Andres structural flexure map (second-order derivative) showing NW-SE and NE-SW flexure lineament trends. Flexure is defined here as the rate of change of dip on the top San Andres surface.

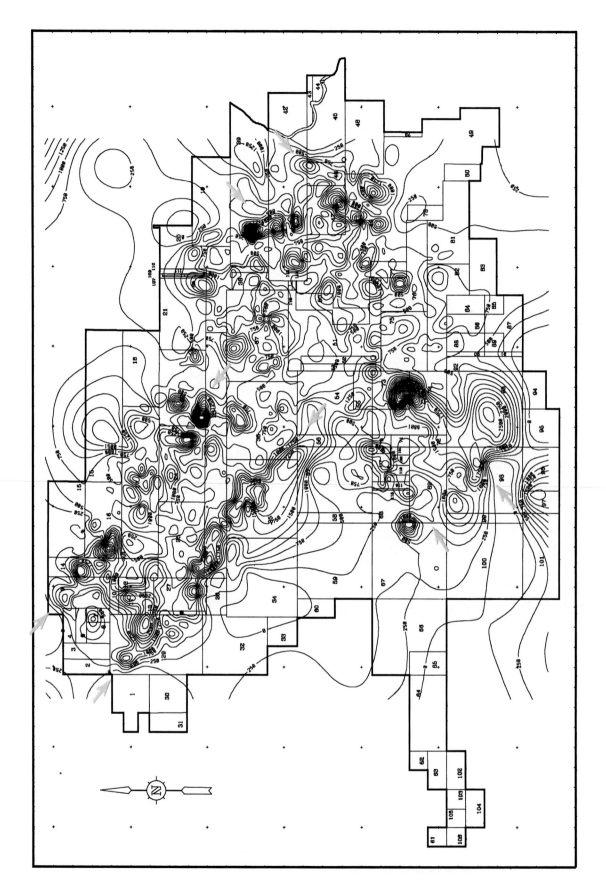

Figure 33. Fluid production map showing linear trends probably associated with fractures in Yates Field. Fluid is defined here as the barrels of oil and water produced in the last 5 years, normalized for oil and water viscosity differences.

intervals and to the calculation of accurate volumetrics. Matrix permeability can be derived as a function of lithology, porosity, pore type, and cementation, and is critical to modeling the reservoir.

Porosity

The matrix density values used to calculate porosity from density logs in the San Andres of Yates Field are shown in Table 2.

The values for core porosity compare reasonably well with the distribution of log-calculated porosity in the San Andres (Fig. 34). Below 20 percent core porosity, the log-calculated values are high relative to core. This is the result of at least three things: (1) bound water in the shale seen as porosity by the density log and not in core; (2) hollow-cored dolomite crystals seen as porosity by the density log and not in core; and (3) oil "baked" into the core during the heat-cleaning process used on some of the cores that resulted in oil "cements" that occlude core porosity relative to log porosity. Above 20 percent core porosity, the log-calculated values are low relative to core. This relationship could be caused by a pervasive uranium anomaly on the east side of the field, which could drive the log-derived bulk density up and the log-calculated porosity down relative to core.

TABLE 2. MEAN MATRIX DENSITY VALUES BASED ON CORE DESCRIPTION

Formation	Gamma Ray	Mean Matrix Density
Seven Rivers "Sand"	$\geq$40 API	2.72
Seven Rivers "Carbonate"	<40 API	2.84
Queen "Sand"	$\geq$50 API	2.73
Queen "Carbonate"	<50 API	2.76
Grayburg "Sand"	$\geq$50 API	2.77
Grayburg "Carbonate"	<50 API	2.80
San Andres "Carbonate"		2.85

Porosity = (Matrix Density - Bulk Density)/(Matrix Density - Fluid Density), where Fluid Density above the GOC = 0.6 and below the GOC = 0.9

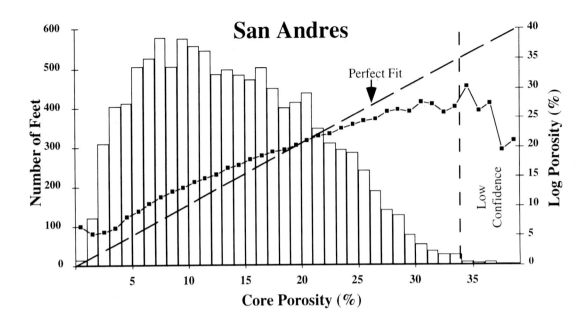

Figure 34. Core porosity vs. log porosity in the San Andres (solid line). The straight dashed line represents a perfect 1:1 fit. Bar plot shows the distribution of the number of data points for each porosity unit that went into calculating the average log porosity, and is essentially a distribution of log porosity.

Permeability

In order to evaluate the massive amount of available data, a "binning" technique was used whereby the arithmetic mean permeability for each 1 percent core porosity increment or "bin" was calculated and plotted on a log scale against core porosity. Permeability from whole core analysis measured 90° to maximum permeability (k90) in the San Andres was used in an attempt to evaluate the matrix permeability, as opposed to the potentially fracture-enhanced permeability (kmax). The data were classed by dominant pore type (Fig. 35). The three classes, in order of decreasing pore connectivity, are:

- Low-Moldic Pore System: ≤33.33 percent "moldic" porosity

- Mixed Pore System: 33.33 percent < "moldic" porosity < 66.66 percent

- Moldic Pore System: ≥66.66 percent "moldic" porosity

As used here, "Moldic" equals the sum of moldic (mo), fusumoldic (fmo), vuggy (vg), and cavernous (ca) pore types. In addition, data in each of the three pore-type classes were separated into two groups: one with no secondary calcite plugging, and one with secondary calcite plugging.

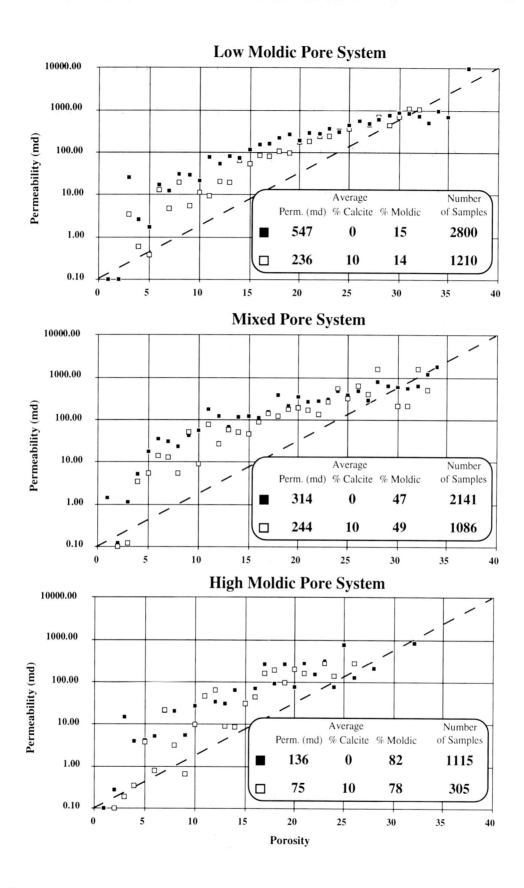

Figure 35. Porosity vs. Log of permeability for three major pore type classes in the San Andres. Low Moldic = moldic porosity ≤ 33%; Mixed = 33% < moldic < 66%; High Moldic = moldic porosity ≥ 66%. A arithmetic average permeability was calculated for each porosity unit ("bin") and then plotted on a log scale. Dashed line is simply a reference line for each plot.

Average porosity and permeability are significantly higher in the low-moldic rocks than in the high-moldic rocks. Within a given pore type class, the slope and the intercept of a line fit through the calcite data and a line fit through the noncalcite data change. In the low-moldic pore-type class, calcite significantly decreases permeability in rocks with <20 percent porosity. In the mixed-moldic pore-type class, calcite significantly decreases permeability in rocks with <15 percent porosity. In the high-moldic pore-type class, calcite significantly decreases permeability in rocks with <10 percent porosity.

The slope and intercept of a line fit through the noncalcite data does not change significantly between the three pore-type classes. The same is true for a line fit through the calcite data in the three pore-type classes. In other words, pore type alone would not alter the equation used to predict permeability from porosity, even though the porosity range is greater in low-moldic rocks than high-moldic rocks. Similar work examining pore-type and pore-geometry relationships to permeability has been reported by Lucia (1983) and Friedman and others (1981).

Reservoir Distribution

The principal reservoir in Yates Field is the dolomitized, subtidal packstone and grainstone of the San Andres Formation on the east side of the field. There is a progressive shift from moderate porosity in the San Andres dolomite on the west side of the field to lower maximum and lower average porosity in the dolomite of the Grayburg, Queen, and Seven Rivers (Fig. 36). This is a function of changing depositional setting from shallow subtidal wackestone and packstone of the San Andres into the peritidal to supratidal microcrystalline, fenestral mudstone and wackestone of the Seven Rivers. The reverse is true in the clastics. The "shales" of the San Andres on the west side of the field are tight, and there is a progressive shift from low porosity in the Grayburg siltstone to higher average and maximum porosity in the Queen and Seven Rivers siltstone and sandstone. This is a function of increasing depositional energy in the Seven Rivers that resulted in better-sorted and coarser-grained sandstones.

San Andres Formation—
The San Andres Formation can be packaged into four high-frequency sequences, each comprised of multiple cycles. The distribution of the four high-frequency San Andres sequences has been examined in 2-D using dolomite reservoir and "shale" maps. In general, the best reservoir rock in a given high-frequency sequence is found in the subtidal packstone and grainstone shoal and shoal margin in the east. Reservoir quality decreases in the west, where carbonates become more muddy and clay-rich. The position of high-quality reservoir moves from deep west to shallow east, as the four major high-frequency sequences aggrade and prograde to the east toward the basin. At the base of each major high-frequency sequence is an argillaceous carbonate that is typically thicker, more continuous, and more clay-rich than other argillaceous carbonates found at the base of cycles within high-frequency sequences. These argillaceous carbonates typically have low core analysis permeability, very few observed fractures in core, and may act as local baffles to vertical fluid flow.

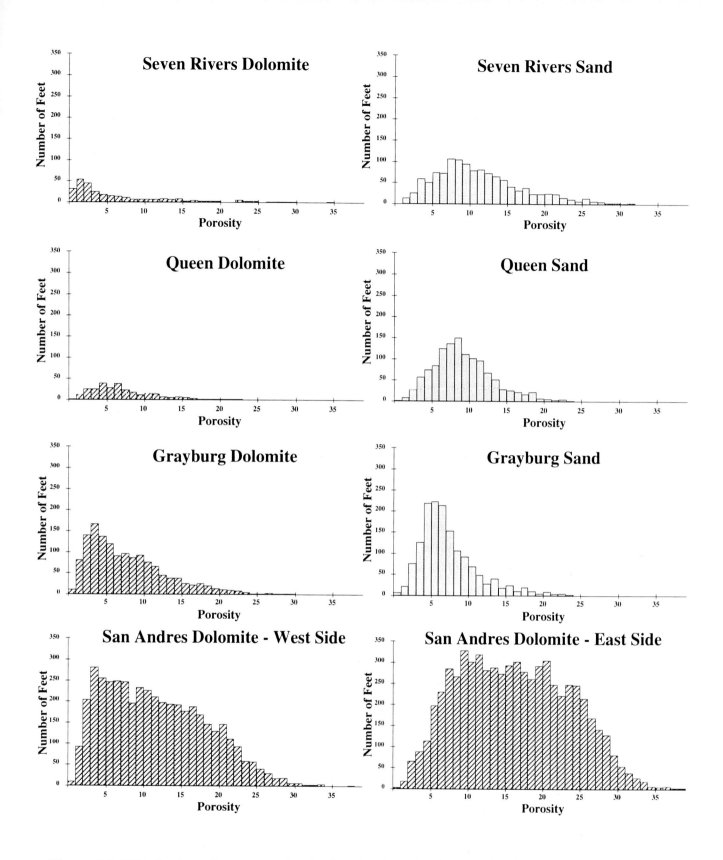

Figure 36. Distribution of core porosity for the clastic and carbonate fraction of each formation. Seven Rivers Dolomite: API < 40; Seven Rivers Sand: API ≥ 40. Queen Dolomite: API < 50; Queen Sand: API ≥ 50. Grayburg Dolomite: API < 50; Grayburg Sand: API ≥ 50.

Grayburg Formation—
The porosity thicks in the Grayburg sand are localized in the central part of the field, and mapped as porosity "pods." Grayburg dolomite porosity is constrained to the eastern and southern part of the field, coincident with underlying San Andres shelf-crest islands. Grayburg dolomite and sand porosity are best developed where the gross sand accumulation is relatively thin. This high-porosity to low gross sand correlation is consistent with the depositional energy and underlying San Andres karst topography. The Grayburg sediments were deposited during a relative sea-level rise onto the San Andres karst surface. The finer-grained sediments were winnowed from the topographic highs and accumulated in the lows of the San Andres karst surface. Alternatively the slightly higher wave energy associated with the San Andres paleo-highs resulted in deposition of cleaner, thinner, grain-rich carbonates, and better-sorted and rounded sands. These higher-energy deposits developed and preserved better porosity than the fine-grained sediments that filled the lows.

Queen Formation—
The porosity thicks in the Queen are located in the central portion of the field. The Queen is composed of four tabular sands with relatively uniform petrophysical properties. Therefore, where the gross sand is thick, porosity is thick. The thick sands were deposited where the accommodation on the platform was greatest. This was behind (west of) the shoal complex of the underlying San Andres, over the San Andres lagoon sediments, where compaction of the carbonate muds and shales was greatest.

Seven Rivers Formation—
Two distinct en-echelon porosity thicks trend NW-SE along the eastern margin of the field and NE-SW along the southern margin of the field. The Seven Rivers porosity thicks coincide with the underlying San Andres island complexes, which remained as subtle topographic highs during Seven Rivers sand deposition. The remnant topographic high is interpreted to have been formed by differential compaction of underlying sediments. The packstone and grainstone carbonates that comprise the San Andres shoal/island complex, and the thin, clastic-rich sediments of the Queen and Grayburg that were deposited over the islands, did not compact as much as the carbonate muds and shales in the lagoons surrounding the islands. The Seven Rivers sands are interpreted to have been deposited as a shoreface, beach, coastal-dune complex on a remnant topographic high feature.

Phi•h Elevation Slice Mapping—
Although stratigraphic porosity analysis is critical to trend analysis and interpretation of depositional history, porosity distribution within a given present-day elevation slice is very useful with reference to present-day fluid contacts. In addition, porosity elevation slice mapping provides a 2-D means to examine San Andres porosity distribution on the east side of the field, where internal San Andres stratigraphic correlations are impossible.

The lateral porosity heterogeneity within an elevation slice is dramatic. The Seven Rivers and Queen sands define an outer perimeter of porosity around the east and southern margins of the field, and the Grayburg defines a relatively nonporous "fairway" between the Queen and the San Andres. The porosity appears somewhat horizontally layered on the west side of the field, sandwiched between shales that act to break up the vertical continuity. Porosity lines up parallel to the relatively tight shale "subcrops," and isolated porosity highs or "bull's-eyes" on the west side of the field are typically windows through the shales into the underlying reservoir. "Bull's-eyes" on the east side of the field often stack and are representative of cavernous or solution-enhanced porosity formed along early vertical joints.

3-D Reservoir Distribution

One of the best ways to address reservoir management and simulation issues is 3-D analysis of the data, because the vertical and spatial resolution of the data can be preserved, and all of the data are integrated in one common environment. The subtleties of reservoir distribution and baffles within the San Andres, and sandstones and dolomites within the Grayburg, Queen, and Seven Rivers, are apparent in 3-D (Fig. 37). The distribution of isolated bodies, based on imposed pay criteria, can be examined and volumes calculated for each discrete body. Complex model operations using existing attributes to calculate new attributes (e.g., permeability from porosity, pore type, and calcite) are straightforward. The model can be "grossed-up" and exported for use in numerical simulation.

APPLICATION

The working knowledge gained, and data accumulated from, the reservoir characterization project have been integrated into the Yates Field dual-porosity reservoir management strategy. The core descriptions and analyses provide quantified data for the evaluation of individual well performance. The cross-section correlations define the stratigraphic framework and, when used in conjunction with the core descriptions and maps, provide predictive tools of area and well performance. The formation maps, San Andres internal shale marker maps, porosity maps, and mean percent calcite elevation slice maps demonstrate the high degree of heterogeneity and complexity that exists in the Yates Field reservoir. This heterogeneity has an impact on matrix storage capacity and gravity drainage efficiency. The fracture elevation slice maps illustrate the spatial and vertical variability in fracture distribution across the reservoir and delineate the high transmissibility conduits from the matrix blocks to the wellbore.

During the transition to the dual-porosity reservoir management plan, the stratigraphic framework and fracture trends were used to evaluate the waterflood on the west side of the field in order to understand fluid movement and water production. Ultimately the waterflood was terminated in favor of the more efficient gravity drainage recovery mechanism. The lithofacies, porosity, and fracture maps provide the tools to model gravity drainage as the gas cap expands from the east side to the west side of the field. The core descriptions, cross sections, maps, and

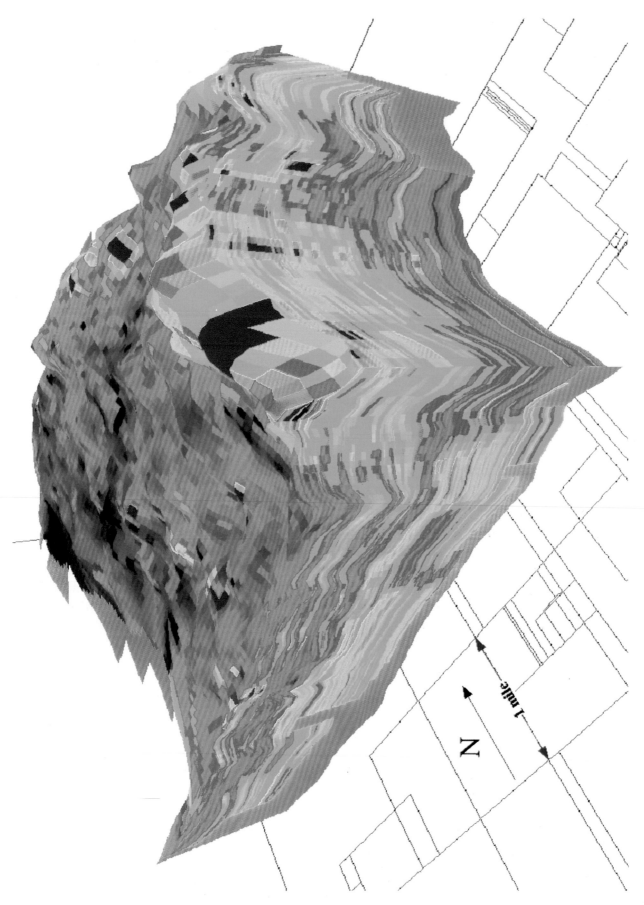

Figure 37. Three-dimensional distribution of matrix porosity within the San Andres Formation. Color range is 0 to 40% porosity (blue to red). Tract boundaries are shown at the bottom of the figure. Low porosity areas (blue) are layered, lagoonal shales on the shallow west side of the field, and ramp deposits on the deep east side of the field. Carbonate packstone and grainstone appear as high porosity (green and yellow) areas in the deep west and shallow east. Caves are shown as extremely high porosity (red) areas.

3-D model are also used to analyze oil-water contact management. These data aid in the placement of water column reinjection wells, evaluation of deeper recompletions in producing wells, and selection of short-radius horizontal drainholes locations.

An understanding of the matrix heterogeneity and fracture distribution allow for better management of the field as a dual-porosity reservoir. The synthesis of the geologic components of reservoir heterogeneity with engineering data and interpretation of field performance has optimized reservoir management of Yates Field and stabilized production.

SUMMARY

The primary goal of the Yates Field reservoir characterization project was to quantify rock matrix and fracture reservoir parameters. The reservoir characterization project provided the data that support the fundamental changes in reservoir management which have resulted in an essentially flat production decline.

Yates Field is composed of dolomite reservoirs of the San Andres Formation and clastic-dominated reservoirs of the Seven Rivers, Queen, and Grayburg formations. San Andres dolomite, which is the principal reservoir lithofacies in the field, is dominated by variably moldic to sucrosic packstone and grainstone deposited in shallow subtidal and shoal-island environments. Clastic reservoir lithofacies include tabular, shallow-subtidal siltstone and sandstone beds in the Grayburg and Queen formations, and fine to coarse sandstone deposited in an areally restricted beach-dune-bar complex within the lower Seven Rivers Formation.

Calcite is a secondary pore-filling cement in Yates Field that can significantly decrease matrix and fracture permeability. In general, the presence of calcite tends to decrease permeability more in highly moldic rocks. Calcite is interpreted as having formed in association with biodegradation at paleo-oil/water contacts. Understanding calcite distribution and its relationship to permeability can help guide operations decisions such as acid stimulations, horizontal drainhole selection and evaluation, and well performance evaluation.

Three major phases of fracture development are recognized in Yates Field: an early phase mimicking pre-San Andres NW-SE and NE-SW regional fracture and fault trends; a later phase associated with the formation of the field anticline; and a final phase associated with shale compaction on the west side of the field. The early regional fractures were significant because they acted as conduits for solution enlargement, which resulted in the development of multiple karst horizons and associated vuggy and cavernous porosity. Fracture relationships to porosity, lithology, and texture are as follows: fractures increase as matrix porosity decreases; carbonates fracture more readily than sandstones, evaporites, or shales; and carbonate mudstone fractures more readily than carbonate grainstone.

One of the best ways to address reservoir management and simulation issues is 3-D analysis of the data, because the vertical and spatial resolution of the data can be preserved, and all of the data are integrated in one common environment. All of the core and log data in Yates Field were integrated into a 3-D geologic model using SGM's StrataModel software. The result was a

6.8 million cell StrataModel of the entire San Andres through Seven Rivers stratigraphic interval in Yates Field that can be used to assist in reservoir management and as input into reservoir simulation.

ACKNOWLEDGMENTS

A project of this magnitude requires the dedicated and integrated efforts of many people. We thank Marathon Oil Company, especially the Mid Continent Region and Petroleum Technology Center, for permission to publish this paper. Thanks to Lisë Brinton and Neil Hurley for critical reviews of the manuscript. Jim Ehrets was integral to the core description effort and stratigraphic correlation, was involved in much of the initial phase of mapping, and wrote the first version of the stratigraphic and depositional sections. Matt Parsley was involved with the bulk of the project, worked on stratigraphic correlations, and wrote much of the first version of the fracture section. Mike Brondos did all of the ZYCOR mapping and was involved with the original StrataModel, along with Mike Uland. Roger Straub and Dick Merkel did the majority of the petrophysics in GEOLOG, and Roger and Loretta Hubert maintain the digital log database. Suzi Thompson maintains the complete engineering and production SAS database in the Mid Continent Region. Ron Manley worked on correlations and mapping in the southern part of the field. Don Caldwell helped to set up the SAS databases at the Petroleum Technology Center. Bob Just described many fractures in core, and Neil Hurley and Dick Merkel worked on many of the FMS interpretations. Thanks to the "core crew" for endless slabbing, staining, photography, and thin-section preparation. Discussions with Gene Wadleigh, Brian Rothkopf, Marv Svaldi, and Brendan Curran helped us to keep a "holistic" perspective of the field. John Barnes, Jim Bowzer, Randy Bruner, Dave Clark, Mike Cooper, Paul Gardner, Ken Thoma, Kevin Williams, and Marvin Woody all supported the project "through thick and thin." To all of you, and to those we missed, we owe our gratitude.

Finally, thanks to Dexter Craig and other earlier workers in Yates Field; this work could not have been done without the excellent foundation laid by you.

REFERENCES

CHOQUETTE, P. W., AND PRAY, L. C., 1970, Geologic nomenclature and classification of porosity in sedimentary carbonates: American Association of Petroleum Geologists Bulletin, v. 54, p. 207-250.

CRAIG, D. H., 1988, Caves and other features of Permian karst in San Andres dolomite, Yates Field reservoir, west Texas, in James, N. P., and Choquette, P. W., eds., Paleokarst: Springer, New York, p. 342-363.

CRAIG, D. H., 1990, Yates and other Guadalupian (Kazanian) oil fields, U.S. Permian Basin, in Brooks, J., ed., Classic Petroleum Provinces: Geologic Society of London Special Publication No. 50, p. 259-263.

DUNHAM, R. J., 1962, A classification of carbonate rock types according to depositional texture, *in* Ham, W. E., ed., Classification of Carbonate Rocks: American Association of Petroleum Geologists Memoir 1, p. 108-122.

FONT, R. G., 1985, Late Paleozoic structural patterns of the Delaware Basin, west Texas and southeastern New Mexico, *in* West Texas Geologic Society Structure and Tectonics of Trans-Pecos Texas Field Conference Guide Book: West Texas Geologic Society Publication No. 85-81, p. 77-80, Midland, Texas.

FRIEDMAN, G. M., RUZYLA, K., AND REEKMANN, A., 1981, Effects of porosity type, pore geometry, and diagenetic history on tertiary recovery of petroleum from carbonate reservoirs: Department of Energy report #11580-5, 217 p.

GARRETT, C. M., JR., AND KERANS, C., 1990, Correlation of the San Andres and Grayburg (Guadalupian) reservoirs, University Lands, Central Basin Platform, *in* Bebout, D. G., and Harris, P. M., eds., Geologic and Engineering Approaches in Evaluation of San Andres/Grayburg Hydrocarbon Reservoirs - Permian Basin: Bureau of Economic Geology, Austin, Texas, p. 49-52.

HILLS, J. M., 1970, Late Paleozoic structural directions in southern Permian Basin, west Texas and southeastern New Mexico: American Association of Petroleum Geologists Bulletin, v. 54, p. 1809-1827.

KERANS, C., LUCIA, F. J., AND SENGER, R. K., 1994, Integrated characterization of carbonate ramp reservoirs using Permian San Andres Formation outcrop analogs: American Association of Petroleum Geologists Bulletin, v. 78, p. 181-216.

LUCIA, F. J., 1983, Petrophysical parameters estimated from visual descriptions of carbonate rocks: a field classification of carbonate pore space: Journal of Petroleum Technology, v. 35, p. 629-637.

NEWELL, N. D., RIGBY, J. K., FISCHER, A. G., WHITEMAN, A. J., HICKOX, J. E., AND BRADLEY, J. S., 1953, The Permian Reef Complex of the Guadalupe Mountains region, Texas and New Mexico: Freeman and Company, San Francisco, 236 p.

SARG, J. F., AND LEHMANN, P. J., 1986, Lower-middle Guadalupian facies and stratigraphy, San Andres-Grayburg formations, Permian Basin, Guadalupe Mountains, New Mexico, *in* Moore, G. E., and Wilde, G. L., eds., Lower and Middle Guadalupian Facies Stratigraphy and Reservoir Geometries, San Andres-Grayburg Formations, Guadalupe Mountains, New Mexico and Texas: Society of Economic Paleontologists and Mineralogists, Permian Basin Section, Publication No. 86-25, p. 1-36.

SPENCER, A. W., 1987, Evaporite facies related to reservoir geology, Seven Rivers Formation (Permian), Yates Field, Texas: Masters Thesis, University of Texas, Austin, Texas, 124 p.

SONNENFELD, M. D., AND CROSS, T. A., 1993, Volumetric partitioning and facies differentiation within the Permian Upper San Andres Formation of Last Chance Canyon, Guadalupe Mountains, New Mexico, *in* Loucks, R. G., and Sarg, J. F., eds., Carbonate Sequence Stratigraphy: Recent Developments and Applications: American Association of Petroleum Geologists Memoir 57, p. 435-474.

STEARNS, D. W., AND FRIEDMAN, M., 1972, Reservoirs in fractured rock, *in* King, R. E., ed., Stratigraphic Oil and Gas Fields - Classification, Exploration Methods and Case Histories: American Association of Petroleum Geologists Memoir 16, p. 82-106.

TINKER, S. W., EHRETS, J. R., AND BRONDOS, M. D., in press, Multiple karst events related to stratigraphic cyclicity: San Andres Formation, Yates Field, west Texas, *in* Budd, D. A., Saller, A. H., and Harris, P. M., eds., Unconformities in Carbonate Strata - Their Recognition and the Significance of Associated Porosity: American Association of Petroleum Geologists Memoir.

WARD, R. F., KENDALL, C. G. ST. C., AND HARRIS, P. M., 1986, Upper Permian (Guadalupian) facies and their association with hydrocarbons - Permian Basin, west Texas and New Mexico: American Association of Petroleum Geologists Bulletin, v. 70, No. 3, p. 239-262.

WESSEL, G. R., 1988, Shallow stratigraphy, structure, and salt-related features, Yates Oil Field Area, Pecos and Crockett counties, Texas: Ph.D. Thesis, Colorado School of Mines, Golden, Colorado, 144 p.

FLUID-FLOW CHARACTERIZATION OF DOLOMITIZED CARBONATE-RAMP RESERVOIRS: SAN ANDRES FORMATION (PERMIAN) OF SEMINOLE FIELD AND ALGERITA ESCARPMENT, PERMIAN BASIN, TEXAS AND NEW MEXICO

F. JERRY LUCIA, CHARLES KERANS, AND FRED P. WANG

Bureau of Economic Geology
The University of Texas at Austin, Austin, Texas 78713-8924

ABSTRACT

In carbonate-ramp reservoirs, stacking of rock-fabric facies within a high-frequency, sequence stratigraphic framework provides the most accurate framework for displaying the distribution of petrophysical rock properties of porosity, permeability, relative permeability, and capillarity. Rock-fabric facies are defined on the basis of grain and crystal size and sorting, interparticle porosity, separate-vug porosity, and the presence or absence of touching vugs. Outcrop geostatistical studies of the Algerita Escarpment suggest little spatial correlation of permeability within rock-fabric facies, and petrophysical properties can be averaged at rock-fabric-facies scale. An outcrop reservoir model has been constructed by mapping rock-fabric facies and using average petrophysical values for each rock-fabric facies. Experimental waterflood simulations show that performance depends upon the stacking of the rock-fabric facies, the dense layers, and the location of production and injection wells.

In the subsurface, high-frequency cycles can be observed in cores and calibrated with wireline log response. Grain and crystal size and sorting and separate-vug porosity can be determined from gamma-ray, porosity, acoustic, and resistivity logs. Permeability profiles can be calculated using rock-fabric-specific transforms between interparticle porosity and permeability. A reservoir model of part of the Seminole San Andres Field was constructed using these methods. Three-dimensional waterflood simulations using this model result in a more realistic display of remaining oil saturation than the traditional layered model and show the importance of thin, dense mud layers in controlling vertical migration.

INTRODUCTION

Major aspects of characterizing fluid flow in carbonate reservoirs include averaging wellbore data and extrapolating these data to the interwell region. Outcrops of producing formations provide an opportunity to develop methods for both scale averaging of petrophysical data and extrapolation of vertical data horizontally. Applying the results of outcrop investigations to subsurface reservoirs leads to the development of new methods and techniques for constructing a 3-D image of permeability, porosity, and saturation.

Fundamental to describing carbonate reservoirs is defining fabrics and textures that can be quantified in petrophysical terms. Rock-fabric/petrophysical relationships can be defined on the basis of grain and crystal size and sorting, interparticle porosity, separate-vug porosity, and the presence or absence of touching vugs (Lucia, 1983)(Fig. 1). Grain and crystal size and sorting terms are from Dunham's (1962) classification, except for a modification of packstone. To clearly describe the petrophysical characteristics of packstone, a distinction must be made between intergranular areas that are at present completely and incompletely filled with mud. Incomplete filling of intergranular areas (grain-dominated packstone) results in petrophysical characteristics similar to those of grainstone; whereas, complete filling (mud-dominated packstone) results in characteristics of wackestone and mudstone.

The stratigraphic framework is fundamental to extrapolating well data to the interwell environment. The stratigraphic framework should be defined in terms of rock-fabric facies in order to quantify the framework in petrophysical terms. Sequence stratigraphic concepts are used to guide description of the rock-fabric facies within a chronostratigraphic framework and are fundamental to ranking key correlation surfaces and, ultimately, in constructing the 3-D framework.

LAWYER CANYON OUTCROP STUDIES

The outcrop study area is located on the Algeria Escarpment in the Guadalupe Mountains, Texas and New Mexico. Excellent exposures of the San Andres Formation are found along the entire 14-mile extent of the Algerita Escarpment (Fig. 2). Facies present include distal outer-ramp to basinal cherty mudstone, outer-ramp fusulinid wackestones and packstones, and ramp-crest peloid-ooid wackestones, packstones, and grainstones (Kerans and others, 1994).

The Lawyer Canyon reservoir window, 2700 ft long and 150 ft high, is located in the ramp-crest facies tract. The window is composed of nine upward-shallowing, high-frequency cycles (cycles)(Fig. 3)(Kerans and others, 1994). Cycles 1-3 are prograding dolograinstone-capped and occasionally tidal-flat-capped cycles. Cycles 4-6 are backstepping dolowackestone- and dolopackstone-capped cycles. Cycles 7-9 are prograding dolograinstone-capped cycles. Dense dolomudstones of varying degrees of continuity are present at the base of the cycles.

All cycles have a diagenetic overprint - principally cementation and compaction, dolomitization, and selective dissolution of grains. Cementation and compaction have reduced porosity and permeability to near zero in the basal mudstones of each cycle, while leaving the

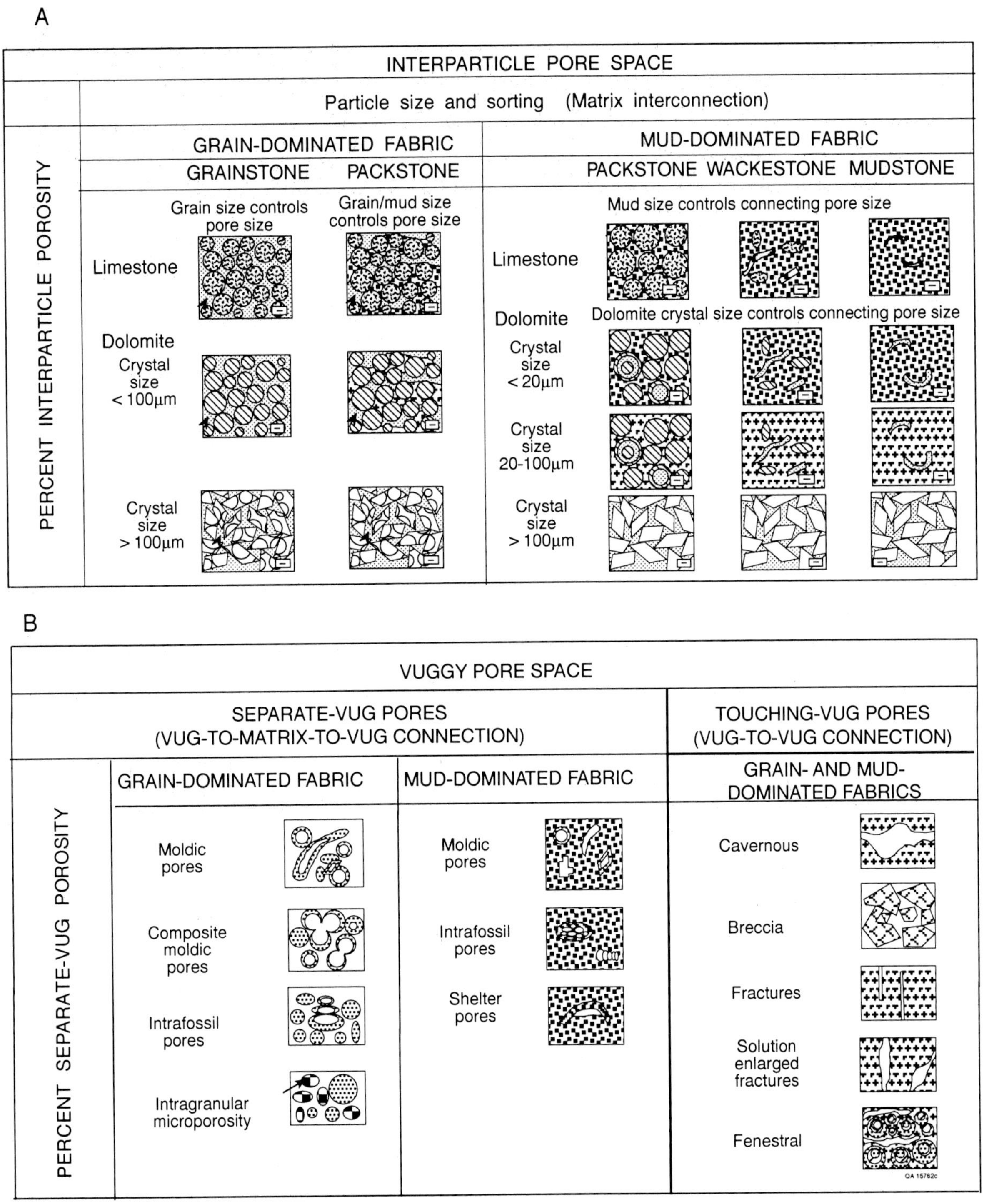

Figure 1. Rock fabric/petrophysical classification of carbonate pore space. (A) Classification of intergranular and intercrystalline pore space based on size and sorting (Senger and others, 1993). (B) Classification of vuggy pore space based on vug interconnection (Wang and Lucia, 1993).

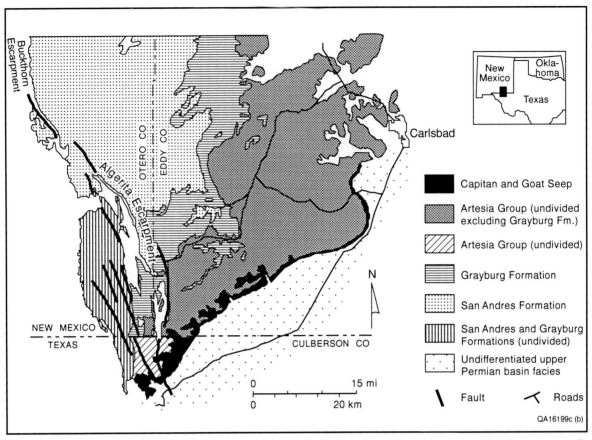

Figure 2. Geologic map of Guadalupe Mountains area showing the location of the Algerita Escarpment San Andes outcrops (Kerans and others, 1994).

other fabrics porous and permeable. All the cycles are dolomitized, and the dolomite replaces the precursor limestone with crystals that closely mimic the depositional fabric. Selective dissolution of grains is concentrated in cycle 7 (Hovorka and others, 1993).

Thin-section study defines six rock-fabric groups based on porosity/permeability relationships (Lucia and others, 1992). Three groups (grainstone, grain-dominated packstone, and mud-dominated packstone-wackestone) are characterized by intergranular and intercrystalline pore space. Two groups - moldic and highly moldic grainstones - have large volumes of separate-vug (moldic in this example) porosity in addition to intergranular porosity. One group is composed of mudstones that have been compacted and cemented to form a dense and tight fabric.

The fabrics described have specific porosity/permeability transforms (Lucia and others, 1992). The average porosity values for grainstone, grain-dominated packstone, and mud-dominated fabrics are similar (12, 14, and 10 percent), but the average permeability values are an order of magnitude different (17 md, 2.5 md, and 0.04 md)(Fig. 4). The moldic grainstones have higher porosity values and lower permeability values than would be expected if all the porosity were intergranular. In general, the permeability varies with the amount of moldic porosity (Fig. 4).

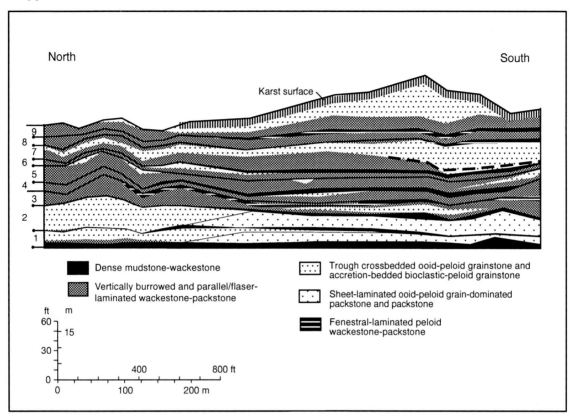
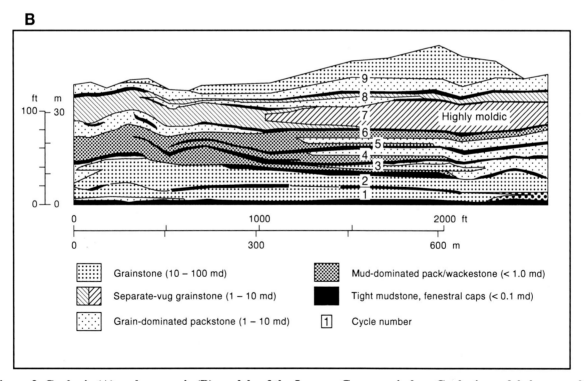

Figure 3. Geologic (A) and reservoir (B) models of the Lawyer Canyon window. Geologic model shows cycles (HFC's) and key facies. Reservoir model is based on rock-fabric flow units and shows general average permeability data. For more specific data, see Lucia and others (1992) and Senger and others (1993).

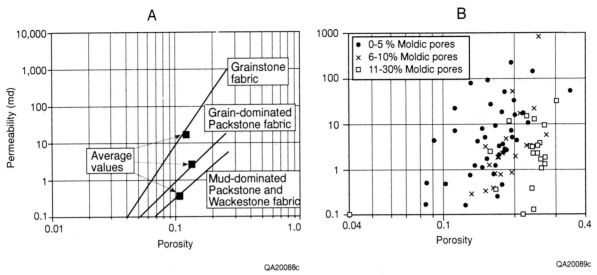

Figure 4. Porosity/permeability relationship for (A) nonvuggy rock-fabric types and (B) separate-vug (moldic) pore type. The nonvuggy pore types have rock-fabric-specific transforms and the moldic grainstones have transforms related to the volume of moldic pores.

Detailed permeability measurements have been made in the grainstone unit of cycle 1, in the moldic grainstone unit of cycle 7, and in cycle 9 grainstone (Senger and others, 1993; Grant and others, 1994). Detailed permeability studies have also been done in the middle San Andres at Lawyer Canyon (Eisenberg and others, 1994). All of these studies show horizontal and vertical variograms with short correlation ranges and large nugget effects, indicating small-scale, near-random variability of permeability. Senger and others (1993) demonstrated that, with such near-random permeability distributions, average petrophysical properties can be used to characterize each rock-fabric unit. However, at a larger scale, a long-range correlation structure is indicated and should be maintained.

The flow model for the Lawyer Canyon window (Fig. 3) was constructed by overlaying the rock-fabric units on the stratigraphic framework and assigning the units average porosity and permeability values. The result is a geologically constrained description of the spatial distribution of petrophysical properties.

Several 2-D waterflood experiments were conducted using this model (see Senger and others, 1993; Kerans and others, 1994). These experiments show the large impact of the distribution of rock-fabric facies, particularly low-permeability mudstone layers, on performance predictions. One experiment demonstrates the importance of the correct spatial permeability distribution by comparing simulation results, using the outcrop model, with results using a simulated subsurface model constructed by linear interpolation of cycles and petrophysical data between the ends of the model. The difference in the predicted recovery is about 13 percent of the original oil in place, or 48 percent recovery from the interpolated model and 35 percent from the outcrop model (Figs. 5 and 6).

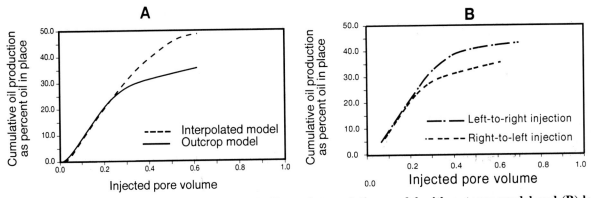

Figure 5. Sweep efficiency plots comparing (A) linear interpolation model with outcrop model and (B) left-to-right to right-to-left injection experiment.

The second experiment (Figs. 5 and 6) demonstrates how the position of wells relative to the spatial distribution of permeability affects recovery by comparing recovery by flooding from left-to-right and flooding from right-to-left. In both cases the highly permeable, discontinuous, cycle 9 grainstone body channels injected fluid until its termination, where it crossflows down across the basal mudstone of cycle 9 into the underlying flow units, trapping unflooded oil in underlying units upstream. More oil is trapped when flooded from right-to-left than when flooded from left-to-right (Fig. 6c), because the downstream termination of cycle 9 grainstone is farther from the producer in Figure 6d than in Figure 6c.

SEMINOLE SAN ANDRES UNIT

The Seminole San Andres Unit lies on the northern Central Basin Platform immediately south of the San Simon Channel (Fig. 7). It covers approximately 23 square miles and contains more than 600 wells. Discovered in 1936, the field is a solution-gas reservoir with a small initial gas cap. Original oil in place is estimated to be 1100 MSTB (Galloway and others, 1983). Waterflooding was initiated in 1970 using alternating rows of 160-acre inverted 9-spot patterns. Infill drilling occurred in 1976, converting the pattern to a mixed 80- and 160-acre inverted 9-spot. A second infill program took place in 1984 to 1985, converting the pattern to an 80-acre inverted 9-spot. CO_2 flooding began in 1985.

Seismic data suggest that Seminole is one of several isolated platforms built during the lower San Andres composite sequence that became linked with the rest of the San Andres platform during progradation of the upper San Andres composite sequence. Core data reveal that the lower 750 ft of the San Andres contains skeletal grainstone and packstone and an open-marine fauna comparable to that of the latest Leonardian retrogradational sequence set of the lower San Andres composite sequence found on the Algerita Escarpment. The highstand systems tract is represented by (1) 300 ft of fusulinid wackestones and packstones, and (2) 150 ft of upward-shallowing, peloidal, shallow subtidal to peritidal cycles. The upper 350 ft of the San Andres at Seminole is largely anhydritic peritidal deposits.

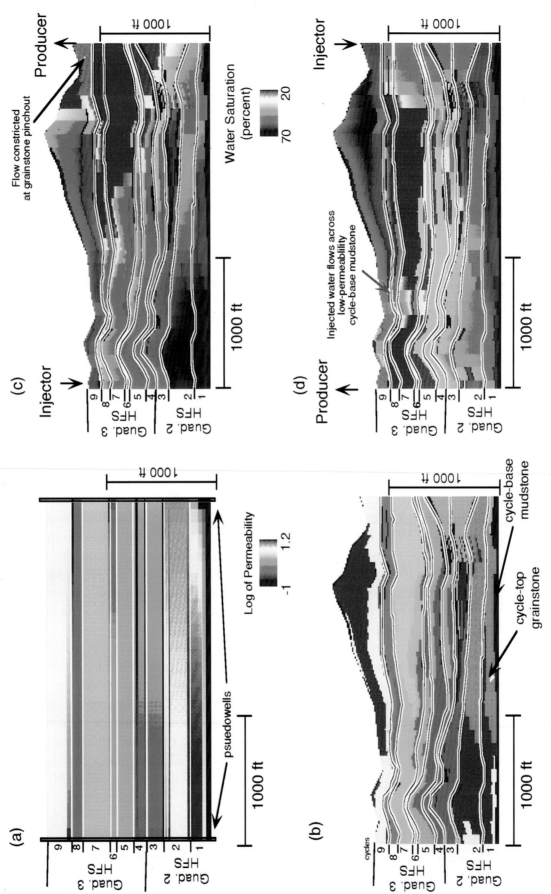

Figure 6. Lawyer Canyon flow models. (a) A linear interpolation of permeability data taken from two pseudowells on each end of the Lawyer Canyon window. (b) The rock-fabric permeability model based on continuous outcrop data. (c) Left-to-right injection experiment showing water saturation after 40 years of injection and crossflow point at downflow termination of high permeability in cycle 9. (d) Right-to-left injection experiment showing water saturation after 40 years of injection and crossflow point at downflow termination of high permeability in cycle 9.

(Lucia et al., 1995)

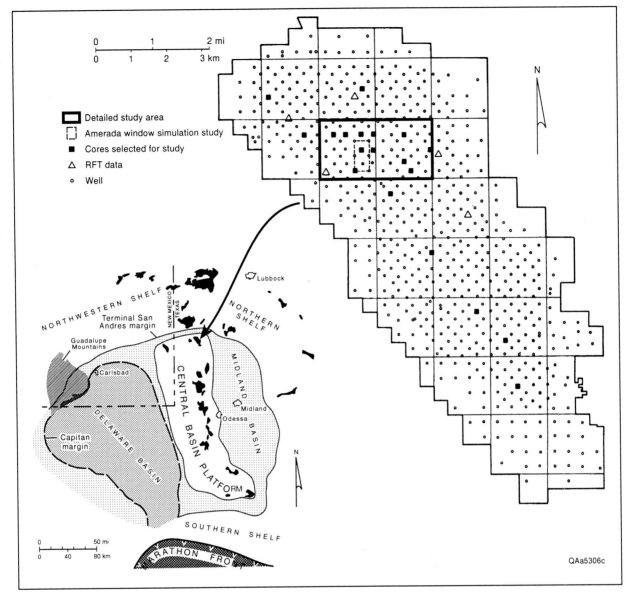

Figure 7. Location map of the Seminole field in the Permian Basin, West Texas, and the location of the two-section study area (bold outline).

A detailed reservoir characterization study was done on two sections of the Seminole San Andres Field. A total of 11 cores covering the reservoir were described in detail. Through an iterative process of description and correlation, 11 high-frequency cycles (HFCs) were confidently identified in cores and uncored wells. The lower three HFCs are 20- to 30-ft-thick units containing low-permeability, fusulinid-peloid, mud-dominated dolostones coarsening upward into permeable, crinoid-fusulinid-peloid, grain-dominated dolopackstones. The dolomite crystal size is between 20 and 100 microns (medium), and mud-dominated, medium crystalline dolostone and grain-dominated dolopackstone have a common porosity/permeability transform (Fig. 8a). However, the mud-dominated fabric has a lower average porosity than does the grain-dominated

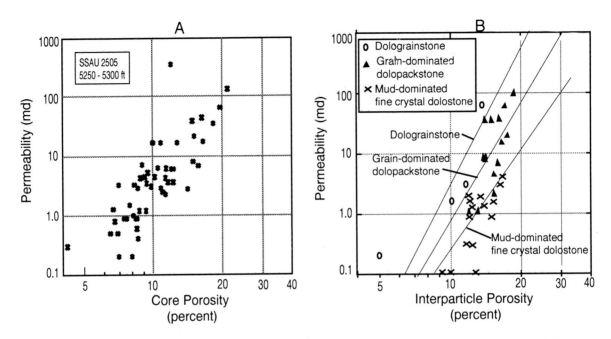

Figure 8. Porosity/permeability cross plots for (A) cycle 11 grain-dominated dolopackstone and mud-dominated medium crystalline dolostone and (B) cycles 1–9 dolograinstone, grain-dominated dolopackstone, and mud-dominated fine crystalline dolostone.

packstone, resulting in lower permeability values for the mud-dominated fabric. Where the dolomite crystal size is less than 20 microns (fine), the mud-dominated dolostones have a different porosity/permeability transform from grain-dominated dolopackstones (Fg. 8b).

The core description, flow units, gamma-ray log, and a comparison between core-analysis and log-calculated porosity and permeability values for HFC 11 are shown in Figure 9, and the description of the complete core, HFCs, and flow units from the Amerada Hess SSU 2505 well are shown in Figure 10. Flow units in cycles 10, 11, and 12 are defined by rock fabric first and porosity second (Fig. 9). Two rock-fabric flow units are defined in cycle 11: a lower low-permeability unit (11b) composed of mud-dominated, medium crystalline dolostone; and an upper high-permeability unit (11a) composed of grain-dominated dolopackstone. A laterally persistent dense zone is located within the lower unit and defines an important barrier to vertical flow. Therefore, in descending order, cycle 11 is divided into four flow units: a grain-dominated dolopackstone (11a); a mud-dominated, medium crystalline dolostone (11ba); a dense, mud-dominated dolomudstone (11bb); and a mud-dominated, medium crystalline dolostone (11bc). Core photographs of HFC 11 in the SSAU 2505 well are shown in Figure 11.

The upper 9 HFCs (cycles 1-9) record progradation of the ramp-crest facies tract over the outer ramp during lower San Andres composite sequence progradation. These HFCs are typical upward-shallowing cycles, with basal mudstones and wackestones grading upward into grain-dominated packstones and grainstones. True crossbedded ooid grainstones are rare, but grain-dominated packstones are common. Cycles 1-5 shoal at least locally into fenestral peritidal caps (Fig. 10).

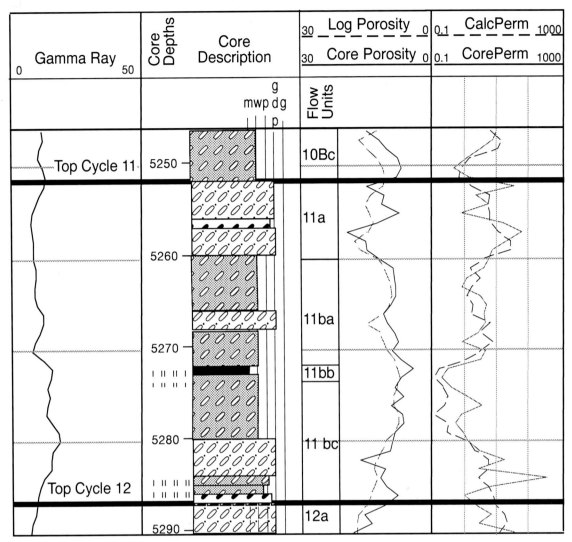

Figure 9. Cycle 11 core description, rock-fabric flow units, core and log-calculated porosity and permeability, and gamma ray log, Amerada Hess SSAU 2505 well. See figure 10 for legend.

In these upper 9 cycles, dolomite crystal size mimics the precursor limestone, resulting in three petrophysical classes: dolograinstone, grain-dominated dolopackstone, and mud-dominated fine crystalline dolostone (Fig. 12). Porosity/permeability crossplots using routine core-analysis data show considerable scatter. However, porosity/permeability transforms for dolograinstones, grain-dominated dolopackstones, and mud-dominated dolostones can be defined using data obtained by careful analysis (Fig. 8).

The core description, flow units, gamma-ray log, and a comparison between core-analysis and log-calculated porosity and permeability values for HFCs 1-4 in the SSAU 2505 well are shown in Figure 13. Flow units are defined by the vertical stacking of rock-fabric facies. Cycle 2 has two flow units defined by a lower mud-dominated, fine crystalline dolostone and an upper grain-dominated dolopackstone; the cycle has a thin fenestral cap. Cycle 3 also has two units

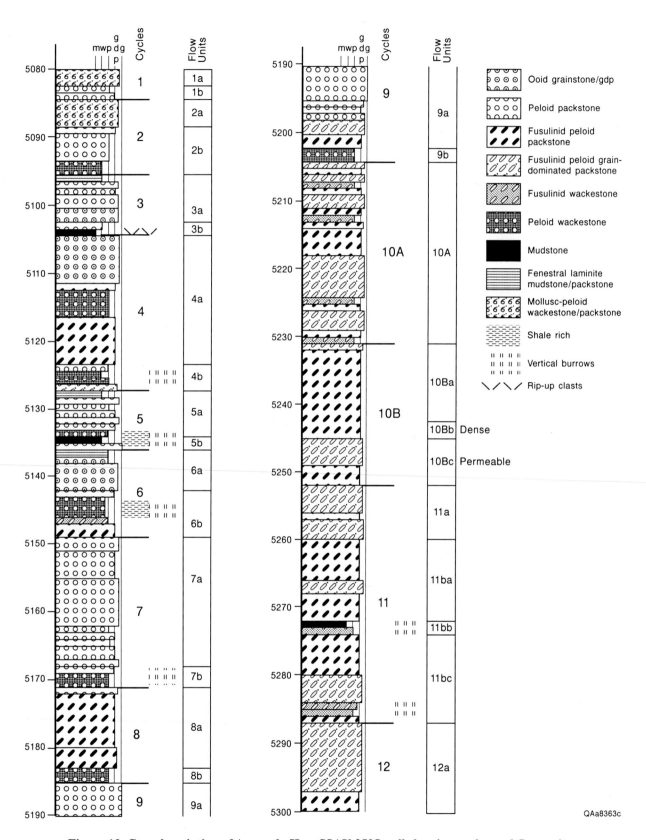

Figure 10. Core description of Amerada Hess SSAU 2505 well showing cycles and flow units.

Figure 11. Photographs of core slabs of Amerada Hess SSAU 2505 well, depths 5,248–5,290 ft, including all of cycle 11.

Figure 11. Continued.

Figure 11. Continued.

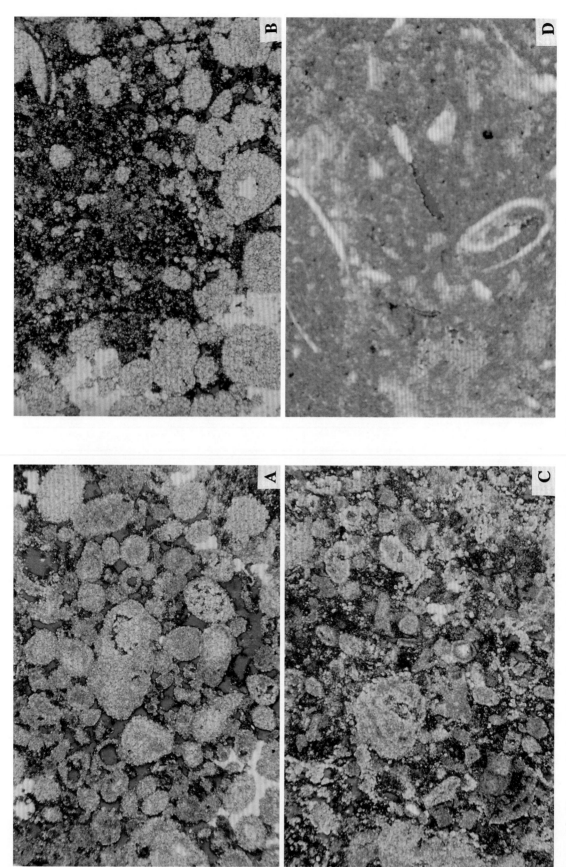

Figure 12. Plane polarized photomicrographs impregnated with blue dye of (A) dolograinstone with intergranular pore space, (B) grain-dominated dolopackstone with intergranular pore space and intergranular dolomitized micrite, (C) grain-dominated dolopackstone with intergranular pore space and intergranular dolomitized micrite, and (D) dolowackestone. Width of photomicrographs is 3.4 mm.

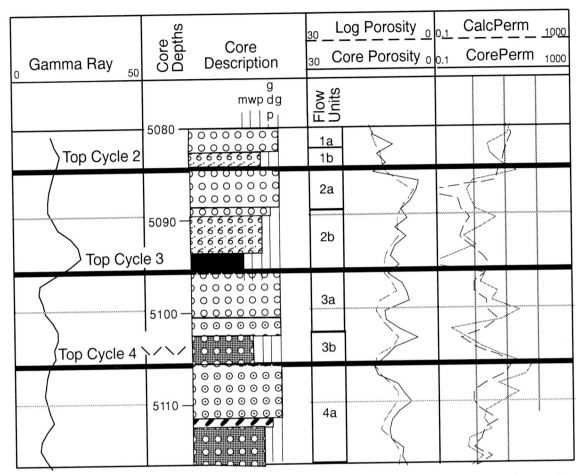

Figure 13. Cycles 2–4 core description, rock-fabric flow units, core and log-calculated porosity and permeability, and gamma ray log, Amerada Hess SSAU 2505 well. See figure 10 for legend.

defined by mud-dominated, fine crystalline dolostone overlain by a grain-dominated dolopackstone. Flow units within HFCs 9, 8, 7, 6, 4, and 1 are also defined by rock-fabric facies. Core photographs of HFCs 1-4 in the SSAU 2505 well are shown in Figure 14.

To construct the reservoir model, the core data were calibrated with log data, using neutron, density, acoustic, and resistivity logs. Total porosity was calculated using the neutron, density, and acoustic logs. Separate-vug porosity was calculated using a calibration of separate-vug porosity to total porosity and acoustic transit time (Lucia and Conti, 1987)(Fig. 15a). A Z-plot of total porosity, water saturation, and rock fabric was used to define rock-fabric fields (Fig. 15b). Core permeability was correlated to interparticle porosity (total porosity minus separate-vug porosity for three petrophysical/rock-fabric classes)(Fig. 15c).

A reservoir model of the two-section study area was constructed by defining 39 rock-fabric flow-unit layers within each well and correlating flow units between wells constrained by the HFC stratigraphic framework. Porosity, permeability, and water-saturation values obtained from log calculations were averaged within each rock-fabric unit at each well location, and the average values interpolated between wells using StrataModel, a 3-D geocellular modeling program

Figure 14. Photographs of core slabs of Amerada Hess SSAU 2505 well, depths 5,083–5,114 ft, including cycles 1–4.

Figure 14. Continued.

Figure 14. Continued.

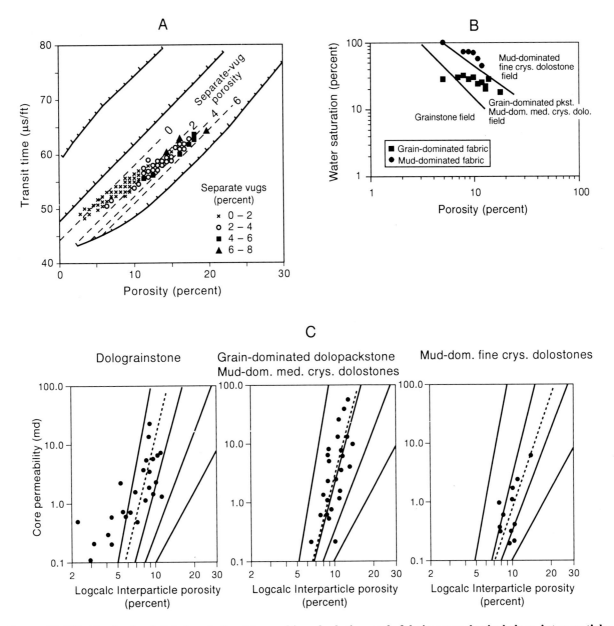

Figure 15. Wireline log/rock fabric relationship used in calculating rock-fabric petrophysical class, interparticle porosity, and permeability. (A) Relationship between transit time and separate-vug porosity from thin-section point counts. (B) Relationship between rock fabric, water saturation, and total porosity. (C) Rock-fabric-specific porosity/permeability transforms compared with generic particle size fields of Lucia (1983).

(SGM). Results are shown in Figure 16. The rock-fabric model is a reasonably accurate representation of the spatial distribution of petrophysical properties. Simulation of waterflooding in one 80-acre pattern shows early water breakthrough in highly permeable layers and bypassing and trapping of high oil saturation in less permeable layers (Fig. 16). More traditional models that average petrophysical data over a few layers show more piston-like displacement and very little, if any, bypassing or crossflow trapping. The original gas-oil contact was at 1700 ft subsea,

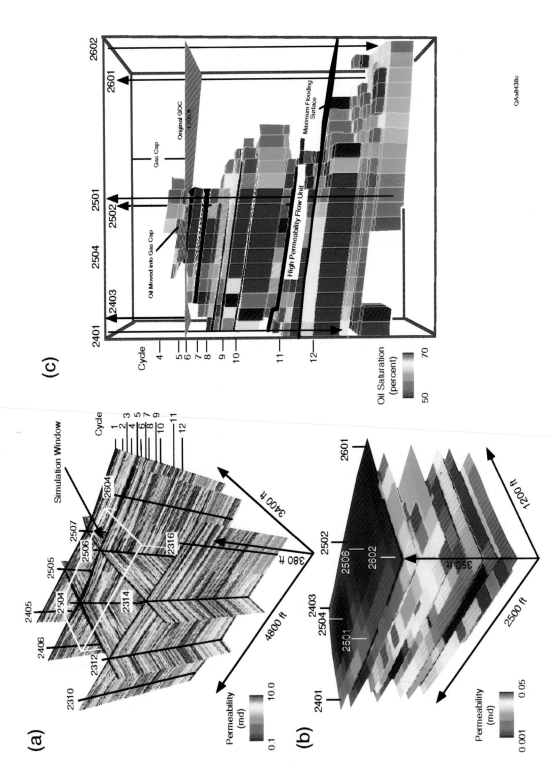

Figure 16. Seminole flow model results. (a) Permeability distribution using a 39-layer rock-fabric model. (b) Model of mud layers in 39-layer model added to control vertical gas flow. (c) Model of remaining oil saturation after 6 years of waterflooding using 39-layer rock-fabric model. The dense mud layers appear as less-than-50-percent oil saturation layers. The position of the original gas cap is shown, and injected water preferentially flows into this region. Note the preferential flooding of high-permeability flow unit 11a at the top of cycle 11 bypassing oil saturation in underlying cycle 12 and being confined by the overlying dense flow unit 10 bc in cycle 10.

and the low oil saturation above cycle 6 (Fig. 16) is due to high gas saturation and the preferential flow of injected water through the gas cap. Oil also can be seen moving into the gas cap.

A history match of primary gas production, using the rock-fabric model flow simulations, required using a ratio of vertical and horizontal permeability of 0.04 rather than between 1 and 0.1, as measured in cores (Wang and others, 1994). The error in the model is the omission of thin, dense mudstone layers. Wireline logs average more than 2 to 3 ft and average out the thin, tight mudstone layers. The mudstone layers were modeled separately (Fig. 16) and input into the rock-fabric model. Although mud layers are discontinuous and are 1- to 2-ft thick in the cores, they had to be enlarged for entry into the computer model. With the inclusion of the dense mud layers, gas production could be matched by using a kv/kh of 0.3 (Wang and others, 1994).

CONCLUSIONS

In carbonate-ramp reservoirs, stacking of rock-fabric facies within a high-frequency, sequence stratigraphic framework provides the most accurate framework for displaying the distribution of petrophysical rock properties of porosity, permeability, relative permeability, and capillarity.

Experimental waterflood simulations using data from the Lawyer Canyon outcrop show that performance is dependent upon (1) stacking of rock-fabric facies (spatial permeability distribution), and (2) location of production and injection wells relative to rock-fabric stacking patterns.

Waterflood simulations using data from the Seminole Field show that (1) a more realistic image of remaining oil saturation is generated using the rock-fabric model constrained by HFC stratigraphy than by using traditional layered models, and (2) realistic kv/kh values can be used in the simulation model if discontinuous barriers to vertical flow (thin, dense mud layers) are included in the reservoir model.

ACKNOWLEDGMENTS

This research was done at the Reservoir Characterization Research Laboratory of the Bureau of Economic Geology, The University of Texas at Austin, and was funded by industrial sponsors Agip, Amoco, Arco, British Petroleum, Chevron, Conoco, Exxon, Fina, JNOC, Marathon, Mobil, Phillips, Shell, Texaco, Total, and Unocal, and by U.S. Department of Energy contract no. AC22-89BC1440. Published with permission of the Director, Bureau of Economic Geology, The University of Texas at Austin.

REFERENCES

DUNHAM, R. J., 1962, Classification of carbonate rocks according to depositional texture, *in* Ham, W. E., ed., Classifications of Carbonate Rocks - A Symposium: American Association of Petroleum Geologists Memoir 1, p. 108-121.

EISENBERG, R. A., HARRIS, P. M., GRANT, C. W., GOGGIN, D. J., AND CONNER, F. J., 1994, Modeling reservoir heterogeneity within outer-ramp carbonate facies using an outcrop analog, San Andres Formation of the Permian Basin: American Association of Petroleum Geologists Bulletin, v. 78, no. 9, p. 1337-1359.

GALLOWAY, W. E., EWING, T. E., GARRETT, C. M., JR., TYLER, N., AND BEBOUT, D. G., 1983, Atlas of major Texas oil reservoirs: The University of Texas at Austin, Bureau of Economic Geology Special Publication, 139 p.

GRANT, C. W., GOGGIN, D. J., AND HARRIS, P. M., 1994, Outcrop analog for cyclic-shelf reservoirs, San Andres Formation of Permian Basin: stratigraphic framework, permeability distribution, geostatistics, and fluid-flow modeling: American Association of Petroleum Geologists Bulletin, v. 78, no. 1, p. 23-54.

HOVORKA, S. C., NANCE, H. S., AND KERANS, C., 1993, Parasequence geometry as a control on porosity evolution: examples from the San Andres and Grayburg formations in the Guadalupe Mountains, New Mexico, *in* Loucks, R. G., and Sarg, J. F., eds., Carbonate Sequence Stratigraphy: Recent Developments and Applications: American Association of Petroleum Geologists Memoir 57, p. 493-514.

KERANS, C., LUCIA, F. J., AND SENGER, R. K., 1994, Integrated characterization of carbonate ramp reservoirs using Permian San Andres Formation outcrop analogs: American Association of Petroleum Geologists Bulletin, v. 78, p. 181-216.

LUCIA, F. J., 1983, Petrophysical parameters estimated from visual descriptions of carbonate rocks: a field classification of carbonate pore space: Journal of Petroleum Technology, March 1983, p. 626-637.

LUCIA, F. J., AND CONTI, R. D., 1987, Rock fabric, permeability, and log relationships in an upward-shoaling vuggy carbonate sequence: The University of Texas at Austin, Bureau of Economic Geology Geological Circular 87-5, 22 p.

LUCIA, F. J., KERANS, C., AND SENGER, R. K., 1992, Defining flow units in dolomitized carbonate-ramp reservoirs: Society of Petroleum Engineers Annual Meeting, Washington, D.C., SPE 24702, p. 399-406.

SENGER, R. K., LUCIA, F. J., KERANS, C., FOGG, G. E., AND FERRIS, M. A., 1993, Dominant control on reservoir-flow behavior in carbonate reservoirs as determined from outcrop studies, *in* Linville, W., ed., Reservoir Characterization III: Proceedings, Third International Reservoir Characterization Technical Conference, Tulsa, Oklahoma, p. 107-150.

WANG, F. P., AND LUCIA, F. J., 1993, Comparison of empirical models for calculating vuggy porosity and cementation exponent of carbonates from wireline log responses: The University of Texas at Austin, Bureau of Economic Geology Geological Circular 93-4, 27 p.

WANG, F. P., LUCIA, F. J., AND KERANS, C., 1994, Critical scales, upscaling, and modeling of shallow-water carbonate reservoirs: Society of Petroleum Engineering Permian Basin Oil and Gas Recovery Conference, Midland, Texas, SPE 27715.

RESERVOIR CHARACTERIZATION AND THE APPLICATION OF GEOSTATISTICS TO THREE-DIMENSIONAL MODELING OF A SHALLOW RAMP CARBONATE, MABEE SAN ANDRES FIELD, ANDREWS AND MARTIN COUNTIES, TEXAS

DENNIS W. DULL
Texaco Exploration and Production, Inc.,
500 N. Loraine St., Midland, TX 79701

ABSTRACT

The Grayburg/San Andres formations of the Permian Basin of west Texas are known to contain almost 50 percent of the original oil in place, with remaining mobile oil exceeding 15 billion barrels. This large volume of residual, mobile oil is the impetus for enhanced oil recovery projects. The Mabee San Andres Field is currently under a CO_2 miscible flood targeting high residual oil saturation, which, after waterflooding, is about 75 percent of the original oil in place.

The San Andres at Mabee Field is an upward-shoaling, regressive-progradational carbonate sequence that draped porous and permeable dolomites over paleostructure of late Mississippian to early Pennsylvanian age. The cyclical nature of the reservoir is illustrated by varied facies interpreted as multiple glacio-eustatic sea-level fluctuations typical of the San Andres in the Permian Basin of west Texas. The detailed description of cores from 45 wells and over 1200 thin sections revealed vertical and lateral heterogeneity that was critical to the exploitation of the reservoir. Six major facies have been identified: supratidal, oncolite, subtidal, ooid, sandstone, and open marine. The facies are representative of a sabkha-type environment of deposition similar to that found in the present-day Persian Gulf. Three distinct cycles were recognized based on the facies distribution and reservoir performance. The zonation is based on the interpretation of three recognizable cycles that constitute the upward-shoaling sequence that comprises the San Andres Reservoir at the Mabee Field.

A major problem with carbonates such as the San Andres Formation is the generally poor correlation between porosity and permeability. This is the result of facies variation that controlled subsequent diagenesis and dolomitization. In addition, a limited amount of permeability data is available from cores, and extrapolation of this information to the entire reservoir is problematic. At Mabee Field extrapolation of permeability data was accomplished with Gridstats (a Texaco in-house, three-dimensional personal computer-based geostatistical program). Variograms of the permeability were computed to quantify the vertical and horizontal spatial variability. The normalized porosity logs were converted to permeability using a linear transform from the regression of the crossplot of core porosity versus permeability. Within the Gridstats program, the "hard" data (core permeability), "soft" data (normalized porosity logs transformed to permeability), and the correlation coefficient (variance observed in the core data) were kriged to produce a three-dimensional distribution of permeability. A three-dimensional distribution that includes both the spatial variability and data variance is a powerful reservoir evaluation tool. The ability of Gridstats to quickly and easily generate cross sections through the three-dimensional model has been used to evaluate completions, pattern conformity, daily fluid production versus permeability-height, drill depths, and horizontal laterals from existing vertical injection wells.

The accuracy of the three-dimensional model was tested by comparing (1) data extracted from the model to the input core data, and (2) the model to past and present reservoir performance.

INTRODUCTION

The Mabee Field, discovered in October 1943, covers an area of 12,800 acres. The field is located east of the Central Basin platform in the central portion of the Midland Basin (Fig. 1). The Mabee Field produces from the San Andres Formation of Permian-Guadalupian age. The San Andres/Grayburg formations are the most prolific oil-productive formations in the Permian Basin. The total primary production exceeds 10 billion barrels, with ultimate primary, secondary, and tertiary adding 3.8 billion barrels (Tyler and Banta, 1989).

The Mabee Field has produced over 100 million barrels of oil and is currently producing about 7000 BOPD (barrels of oil per day), of which 3000 BOPD is the incremental response to the CO_2 miscible flood. The producing interval at the Mabee Field occurs in the upper part of the San Andres Formation at an average depth of 1430 m (4700 ft). The reservoir is 25 to 40 m (80-132 ft) thick, with an average porosity of 10.5 percent and a geometric mean permeability of 1 md. The field contains 272 injection wells and over 400 producing wells. Approximately one-half of the field is now under a CO_2 miscible flood.

In 1988 Texaco assembled a team of geologists and engineers to begin the detailed description of the reservoir at the Mabee Field for assessing the feasibility of initiating a miscible CO_2 flood. In 1989 a CO_2 Department was created for the express purpose of characterizing the San Andres and other reservoirs for the injection of this expensive solvent, CO_2.

The emphasis of this paper is to (1) describe the reservoir characterization used in developing an integrated three-dimensional model of the San Andres in the Mabee Field; (2) examine model validity, qualitatively and quantitatively; and (3) demonstrate how the reservoir characterization has improved the operation of the enhanced oil recovery project.

RESERVOIR DESCRIPTION

The detailed geologic description of the cores was critical to the determination of lateral and vertical continuity of the reservoir. Nearly 1525 m (5000 ft) of core and 1200 thin sections were examined and described from 45 wells. Figure 2 is a map of the Mabee Field showing the location of these cored wells and the other 30 wells in which only the core analysis is available.

From the core descriptions, facies correlations were made on a series of four north-south and four east-west structural and stratigraphic cross sections (Fig. 2). The cross sections were used to develop the stratigraphic framework for the three-dimensional model. The facies were critical in the recognition of porosity and permeability relationships that would lead to the establishment of three major zones for the producing interval.

STRUCTURE

San Andres production in the Mabee Field is due to drapage and compaction of porous and permeable San Andres dolomites over a deep Ordovician-Ellenburger structure. This structure

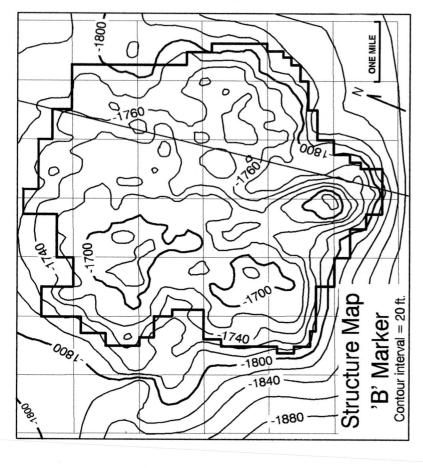

Figure 3. Structure map on top of the "B" marker shale.

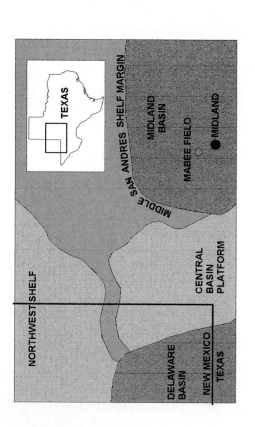

Figure 1. Index map showing the location of the Mabee field.

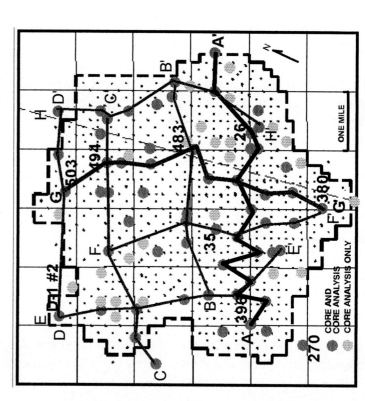

Figure 2. Map of the Mabee field showing the Unit outline, location of cross sections, and core locations.

trends NW-SE in the southwest portion of the field. The structure is bound by upthrown faults on the north and south, with the crest occurring on the western margin of the field. Structural faulting occurred in the late Mississippian-early Pennsylvanian time with the subsequent deposition of younger sediments across it. Structural closure observed on the east side of the field is apparently due to drapage and compaction over a localized carbonate buildup in the underlying Clear Fork Formation. The San Andres Reservoir has 10 to 20 m (30-60 ft) of structural closure with a maximum pay thickness of 40 m (132 ft). Figure 3 is a structure map on top of the "B" marker shale.

STRATIGRAPHY

The San Andres at the Mabee Field is 300 m (900 ft) thick. The lower 150 m (500 ft) of the formation is composed of limestone. This is overlain by 130 m (400 ft) of dolomite with some interbedded sandstone. Deposited on top of the San Andres is 70 m (220 ft) of nonproductive dolomite, anhydrite, and sandstone of the Grayburg Formation. Below the San Andres is 400 m (1200 ft) of sandstone, equivalent to the Brushy Canyon sandstone of the Delaware Basin (Todd, 1976).

FACIES

Based on the examination of 13 cores, Friedman and others (1990) divided the San Andres Reservoir at the Mabee Field into three facies: ooid grainstone (intertidal) facies, mudstone/wackestone (subtidal back barrier) facies, and an upper-intertidal/supratidal facies. However, upon examination of all the available core, geologists at Texaco recognized six distinct facies important to the understanding of reservoir performance and the geologic history. The six facies are supratidal (anhydrite-rich, permeability barrier responsible for trapping the oil), oncolite/pisolite, subtidal, ooid grainstone, sandstone, and open marine (Dull and others, 1991; Horvath and Miller, 1994). The San Andres at Mabee Field is interpreted to be a sabkha depositional environment, as found along the Trucial coast in the present-day Persian Gulf near Abu Dhabi (Todd, 1976). The facies at Mabee are similar to those described by Todd (1976) in his description of the upper San Andres in the Midland Basin.

The reservoir at the Mabee Field is divided into three zones (Fig. 4). Zone 1 is capped by a very thin, clay-rich, stratigraphic marker known as the "B," which is easily identified on the logs by its characteristic high radioactive gamma-ray response. The "B" marker is a bentonitic shale that varies from black to gray or tan and is only one to two inches thick. Below the "B" marker the supratidal facies is composed of dolomite, nodular anhydrite, and stromatolitic laminae. Beneath the supratidal facies is a mixture of subtidal mudstone to wackestone to peloidal packstones and subtidal ooid packstone to grainstones. Zone 2 is composed primarily of a sandstone and ooid facies. The sandstone facies, except on rare occasions when porosity reaches 15 percent, is impermeable, nonreservoir rock. The sandstone facies is identified on the logs by its associated high gamma-ray response when compared to the clean, low gamma ray of the ooid

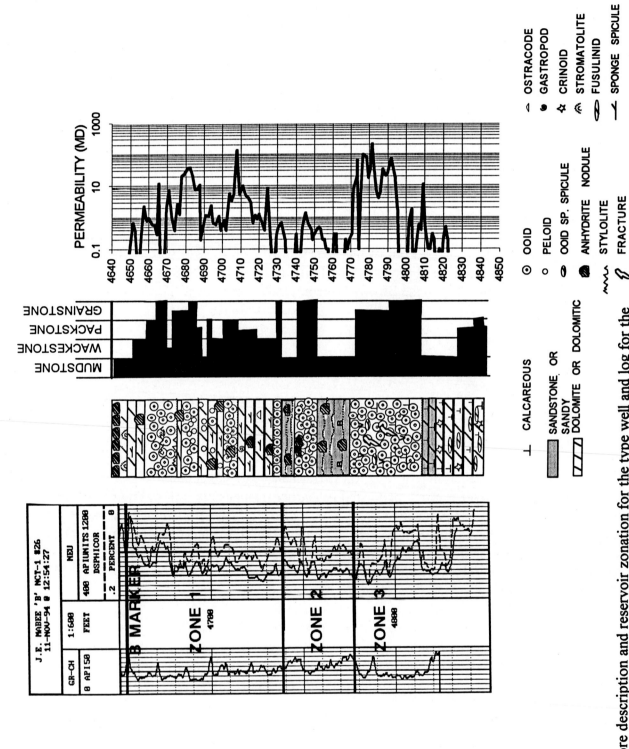

Figure 4. Core description and reservoir zonation for the type well and log for the Mabee field, J.E. Mabee 'B' NCT-1 #26, sec. 17, Twp. 2 N, Blk. 39. (Depths are in feet. GR-CH; gamma ray cased hole, NEU; neutron counts in API units, DSPHICOR; core porosity)

facies. Zone 3 is dominated by the ooid facies, vuggy porosity, solutioning, fractures, and high porosity and permeability in the upper portion. Open marine fusulinid wackestones comprise the lower portion.

Zone 3 typically produces high volumes of water with significant hydrogen sulfide, but has also produced considerable amounts of oil. However, Zone 3 is not being flooded because contrasting high permeability makes it a potential CO_2 thief zone. The intervals currently being flooded are zones 1 and 2 (Fig. 4).

Supratidal Facies

The supratidal facies is a cream-colored, finely crystalline dolomudstone with abundant anhydrite. The facies is characterized with abundant stromatolitic laminations, wavy shale lamina, desiccation cracks, and fenestral "bird's-eye" vugs that are commonly filled with anhydrite (Fig. 5A and 5B). Thin rip-up mudclasts and quartz sand of eolian origin are common in the mudstones. Rare allochems such as pellets, shell fragments, sponge spicules, and pisolites are found in the supratidal facies. The occurrence of these allochems in the supratidal is probably the result of storm deposits or seasonal tides. Anhydrite occurs as individual nodules, coalesced nodules, and microcrystalline pore-filling (Fig. 6A). "Chicken wire" anhydrite is found as thin beds or lenses, from one to two feet thick, with minor amounts of dolomite (Fig. 6B). The supratidal facies has rare fenestral porosity, serves as a barrier to flow, and is the seal for the reservoir. This facies occurs in all three zones but is most abundant at the top of zone 1.

Oncolite/Pisolite (Intertidal) Facies

This facies is characterized as a packstone to grainstone where the major allochems are pisolites and oncolites (Fig. 7A and 7B). The oncolite grains range in size from 2 to 6 mm. The nucleus of the algal grains ranges from shell fragments to dark, organic-rich clays. Occasionally clasts are encrusted with algal laminations. The algal grains are characterized by crenulated layers, while others are continuous concentric bands. The layers often contain quartz silt found dispersed through the layers. Anhydrite is common and locally occurs as thin beds. The oncolite/pisolite facies is confined to zones 1 and 2. Because of its very low porosity and permeability, this facies is nonreservoir and serves as a flow barrier.

Subtidal Facies

The primary reservoir at the Mabee Field is the subtidal facies. It consists of brown, finely crystalline dolomudstones, dolowackestones, and dolopackstones. The major allochems are pellets, with sponge spicules being the major skeletal component (Fig. 8A and 8B). Gastropods, pelecypods, ostracodes, and bivalves comprise the remainder of the allochems of this facies (Fig. 9A). The abundance of wispy shale laminations is indicative of quiet-water deposition (Fig. 9B). These shale laminations commonly occur at the base of zone 1 and are characterized by a high gamma-ray response on the logs, as is shown on the type log in Figure 4. The porosity

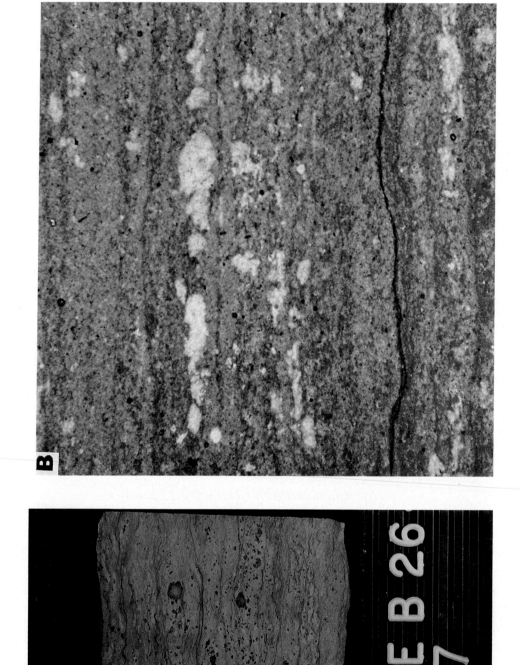

Figure 5. Supratidal facies at the top of Zone 1 that forms the reservoir seal, B-1 # 26, 4647 ft. A.) Slab photo and B.) thin section photomicrograph of dolomudstone showing stromatolitic laminations and fenestrae filled with anhydrite.

Figure 6. Supratidal facies the top of Zone 1. Slab photographs of A.) "chicken wire" anhydrite with some dolomite from well #503, 4688 ft. B.) Stromatolitic dolomudstone with anhydrite nodules that have disturbed the algal layers from well #349, 4670 ft. Dark material is indigenous organic matter. Scale is in centimeters.

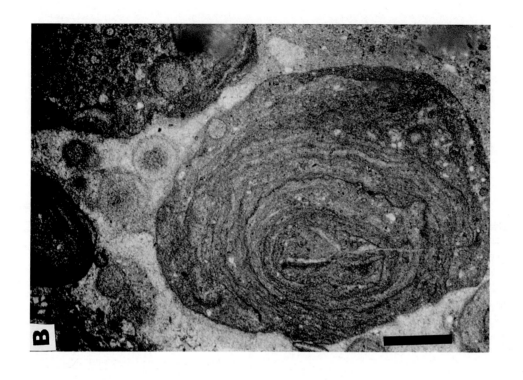

Figure 7. A.) Slab photograph of pisolites cemented in anhydrite, D-1 #2, 4746 ft. and B.) thin section photomicrograph of oncolite with crenulated layers that contain some quartz silt. Note also pellodial grapestone and ooids from #503, 4785 ft. The bar scale is 1 mm.

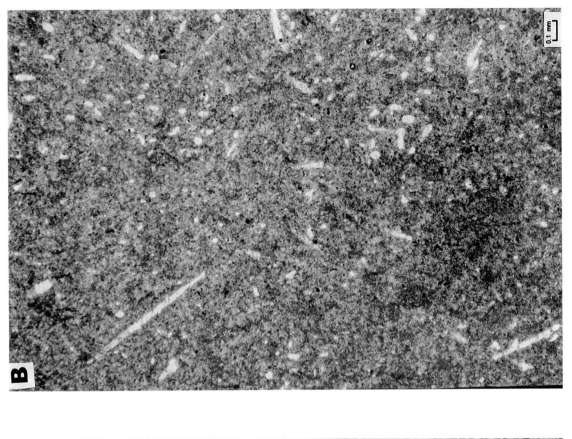

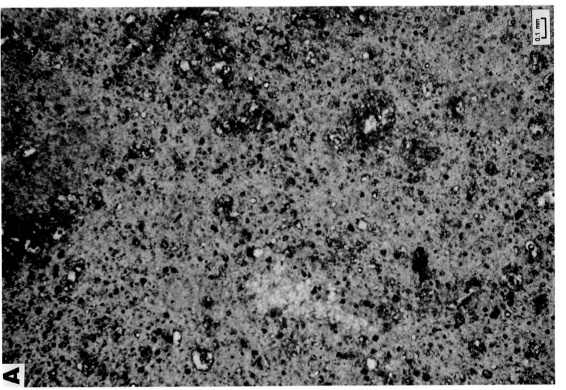

Figure 8. Thin section photographs of subtidal facies, B-1 #26. A.) peloidal packstone, 4646 ft and B.) spicular-peloidal wackestone with anhydrite filling the sponge spicule molds, 4731 ft.

Figure 9. A.) Slab photograph of skeletal wackestone (pelecypods, sponge spicules, and bivalves) from #483, 4730 ft. Note abundant moldic porosity and large skeletal voids are filled with anhydrite. Scale is in cm. B.) peloidal packstone with black shale layer from B-1 #26, 4692-3 ft.

is mainly intercrystalline and moldic, with some vuggy porosity (Fig. 10A and 10B). The bulk of the moldic porosity is spicular and pelletal molds (Fig. 8B). The large fossil molds, such as those of thin-shelled pelecypods and gastropods, were preferentially filled with microcrystalline anhydrite (Fig. 9A). Porosity ranges from 2 to 23 percent, with permeability from 0.1 to 39 md. The average porosity is 10 percent and the average permeability is 0.5 md, with the highest porosity and permeability occurring in zone 3. The occurrence of higher porosity and permeability in zone 3 is primarily due to less pore-filling anhydrite. Figure 11 contains tables that compare porosity and permeability by facies and zone. Anhydrite also occurs as nodules, microcrystalline cements, and as replacement of calcite and aragonite grains (Fig. 12A and 12B). The amount of anhydrite decreases with depth and to the southern portions of the field as the distance from the supratidal environments and the paleoshoreline increases.

The occurrence of dark brown to brownish orange indigenous organic matter is common in many of the pore spaces (Fig. 10B). According to geochemical reports, the organic matter is not mature enough to have been the source of the San Andres oil. The source of the oil has been identified as the underlying Leonardian/Wolfcampian sediments (Ramondetta, 1982).

Ooid Facies

The other major reservoir facies in the Mabee Field are primarily composed of ooid packstones and grainstones (Fig. 13A). The existence of ooid shoals and barriers indicates a depositional environment of high energy. Lower energy oolitic wackestones (Fig. 13B) and transitional sandy ooid packstones comprise the remainder of the ooid facies (Fig. 14A and 14B). The ooids exhibit some low-angle crossbedding and are commonly well sorted. The ooids range in size from 0.3 to 0.7 mm, with rare preservation of the concentric banding due to the intense dolomitization (Fig. 15A). Where the ooids have been well preserved, the nucleus of the ooids is composed of sponge spicules, skeletal fragments, and occasionally mudclasts (Fig. 15B). In zone 2 the nucleus is often made of quartz silt and sand (Fig. 14B).

The porosity of the ooid facies is predominately interparticle, with some moldic porosity due to the selective leaching of the ooids (Fig. 16A). The porosity of the ooid facies averages the same as the subtidal facies, 10 percent, but generally has a higher permeability. The permeability of the ooid facies averages 3 md, with some as high as 484 md. Within the reservoir, the ooids are typically more permeable with increasing depth. The ooids in the upper portion of the reservoir, zone 1, are often tightly cemented with poikilotopic anhydrite, whereas the surrounding subtidal facies is not (Fig. 16B). The ooids in zones 2 and 3 have correspondingly much higher average permeability, 3.29 and 4.48 md respectively, than those of zone 1, 1.11 md (Fig. 11). Higher permeability of the ooid facies of zones 2 and 3 is responsible for the water cycling problem observed during later stages of the waterflood.

Sandstone Facies

The composition of this facies is dominated by a gray, very fine-grained dolomitic sandstone to coarse-grained siltstone cemented with anhydrite. The major allochems are rare ooids. The

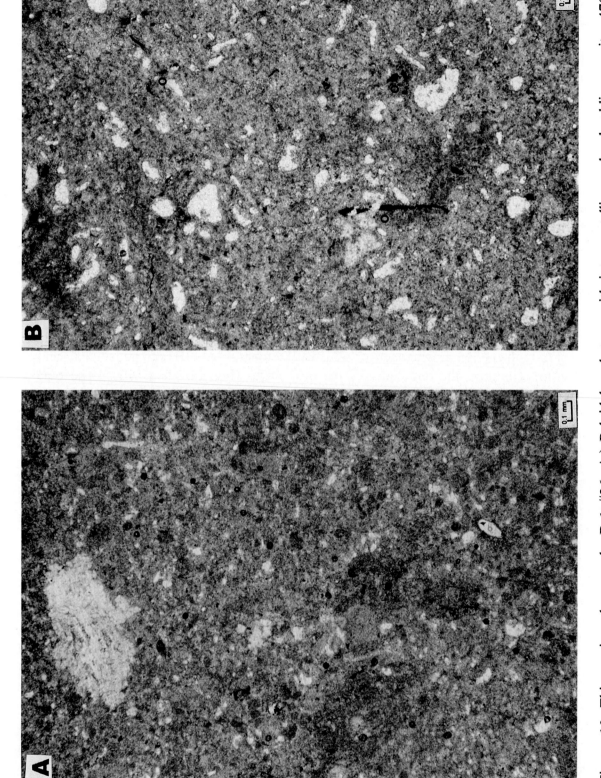

Figure 10. Thin section photographs B-1 #26. A.) Peloidal packstone with intercrystalline and pelmoldic porosity, 4705 ft. Core porosity is 12.2 percent and permeability is 3.4 md. B.) Peloidal wackestone with primarily moldic porosity, 4719 ft. Core porosity is 14.2 percent and permeability is .9 md. Dark areas are indigenous organic matter (o).

POROSITY

FACIES	ZONE	HIGH	LOW*	AVERAGE
SUBTIDAL	1	19.9%	1.3%	9.5%
	2	16.9%	2.4%	8.9%
	3	22.5%	1.8%	11.8%
OOID	1	18.5%	2.7%	9.2%
	2	22.3%	1.2%	10.4%
	3	20.9%	2.8%	11.3%
SANDSTONE	2	15.1%	2.7%	10.1%
	3	19.7%	6.7%	12.8%
OPEN MARINE	3	26.1%	4.9%	11.9%

PERMEABILITY (MD)

FACIES	ZONE	HIGH	LOW*	AVERAGE
SUBTIDAL	1	37.4	0.1	0.71
	2	37	0.1	0.35
	3	39	0.1	0.84
OOID	1	135	0.1	1.11
	2	187	0.1	3.29
	3	484	0.1	4.48
SANDSTONE**	2	26	0.1	0.36
	3	5	0.1	0.42
OPEN MARINE	3	31	0.1	1.09

* Permeability cutoff.
** Majority of the samples fall below the .1 md cutoff. Zone 2 sandstones with high porosity and permeability are productive.

Figure 11. Tables of average core porosity and permeability by facies and zone.

Figure 12. Slab photographs from B-1 #26 of A.) peloidal wackestone with burrows filled with anhydrite, 4671 ft and B.) peloidal packstone with anhydrite nodules and some open fractures stained with oil, 4708-9 ft.

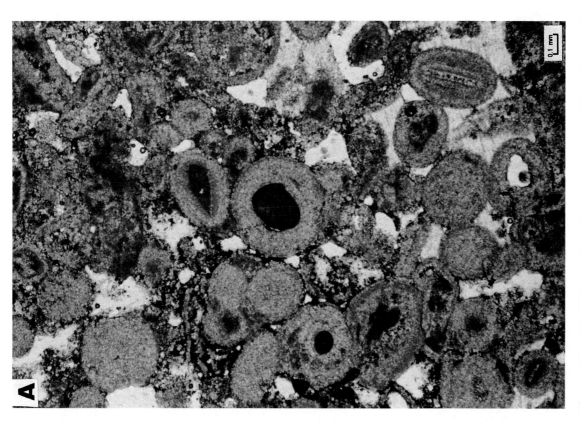

Figure 13. A.) Thin section photograph of ooid packstone from B-1 #26, 4683 ft. B.) Slab photograph of ooid wackestone with abundant ooid molds from B-1 #26, 4694-5 ft.

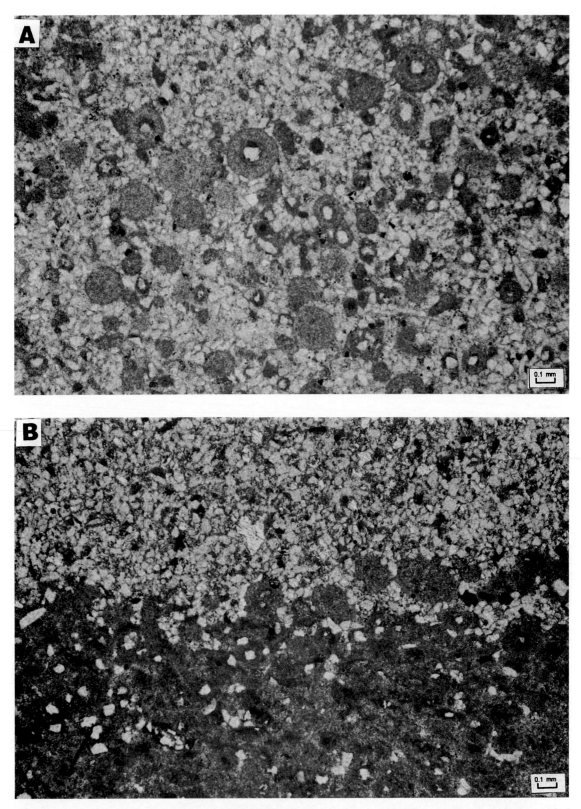

Figure 14. A.) Thin section photograph of sandy ooid packstone. Note that many of the ooid nuclei contain quartz grains. B-1 #26, 4750 ft. B.) Transitional contact between ooid packstone and quartz sandstone, B-1 #26, 4753 ft.

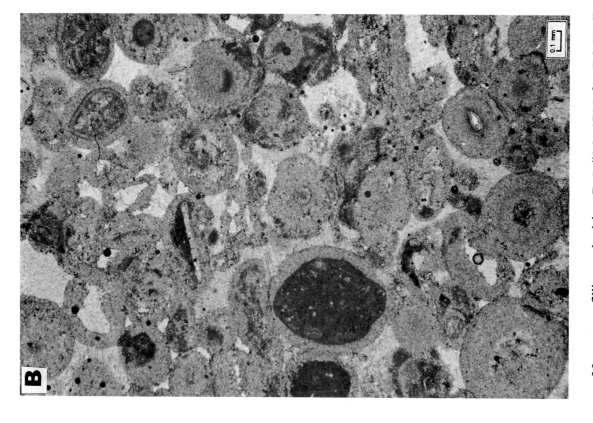

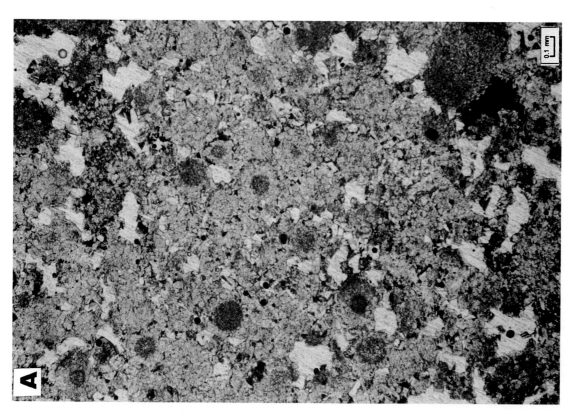

Figure 15. A.) Thin section of highly recrystallized ooid grainstone. Note pore-filling anhydrite. B-1 #26, 4783 ft. B.) Well preserved ooid grainstone with nuclei composed of mudclasts, skeletal fragments, and sponge spicule. B-1 #26, 4775 ft.

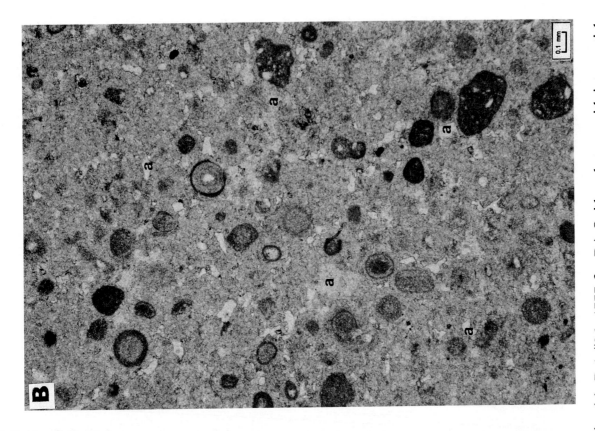

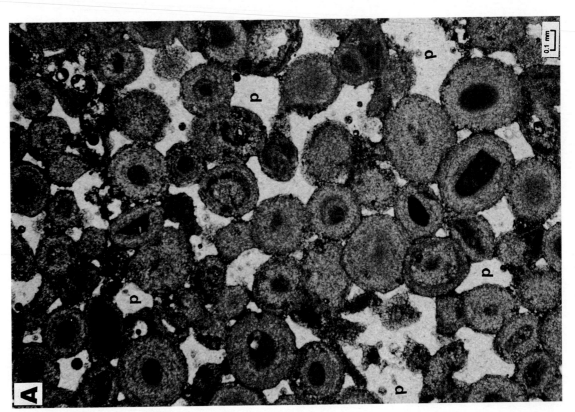

Figure 16. A.) Ooid grainstone with good interparticle porosity (p), B-1 #26, 4777 ft. B.) Ooid packstone with interparticle and intercrystalline porosity occluded with microcrystalline anhydrite (a), B-1 #26, 4757 ft.

sands attain a thickness over 30 ft in zone 2. Bedding varies from planar-bedded to cross-bedded (Fig. 17A and 17B) and is commonly bioturbated (Fig. 18A). This sandstone facies, confined to zones 2 and 3, is typically nonreservoir and acts a barrier to fluid migration. The porosity averages 10 to 12 percent, but, due to the fine-grained texture and an abundance of anhydrite and dolomite cement, has very low permeability, often less than 0.1 md (Fig. 18B). The sandstones are oil-productive when they are thick and have porosity of more than 15 percent. These productive sandstones are primarily confined to zone 2 and are mainly located in the north-central and northeastern portions of the field.

The origin of the sand is believed to be eolian in nature and deposited and reworked in a subtidal environment. During sea-level lowstands, the eolian sands could migrate as dunes across the dry tidal flats, where they were subsequently reworked, during the next marine transgression, into the subtidal as sheet sands and tidal channel sands between the ooid shoals.

Open Marine Facies

The open marine facies, which is composed of a tan to brown dolomitic wackestone (Fig. 19A), is confined to zone 3 and is within the high water saturation portion of the transition zone. The major allochems are fusulinids, bryozoans, and crinoids (Fig. 19B). The open marine facies is located basinward of the ooid shoals in normal marine waters. The position basinward of the source of the dolomitizing fluids in the supratidal explains the partial dolomitization and transitioning to 100 percent limestone in the lower portion of the San Andres. The porosity is primarily intercrystalline and moldic with some solution-enlarged vugs (Fig. 20A). Good porosity is developed in the dolomitic matrix, averages 11 to 12 percent, but locally is as high as 26 percent. The molds are from the selective leaching of fusulinid and skeletal fragments. The permeability varies from 0.1 to 31 md with an average of about 1 md. Anhydrite is not abundant, but has filled some of the fossil molds and vertical fractures (Fig. 20B).

STRATIGRAPHIC FRAMEWORK

The reservoir at the Mabee Field is an overall upward-shoaling, regressive/progradational sequence. The zonation of the reservoir reflects relationships between porosity and permeability, facies, and reservoir performance. The zonation is representative of three upward-shoaling cycles, with small-scale minor cycles reflecting minor transgressive/regressive cycles. The facies and the observed cyclical nature of the deposition are similar to those described by Silver and Todd (1969) and Todd (1976) in their work on the San Andres in the Midland Basin. Figure 21 is a stratigraphic dip cross section, G-G', showing facies relationship, sequence development, and reservoir zonation. Figure 22 is a stratigraphic cross section, A-A', along depositional strike. See Figure 2 for the location of the cross sections.

Zone 3 is the last apparent major deepening event of this regressive sequence. This reflects a shoaling upward from an open marine fusulinid wackestone at the base to dolowackestones and packstones of the subtidal facies capped with thick ooid grainstones to packstones. These are frequently interbedded with thin silty-dolomites and dolomitic quartzose sandstone and with

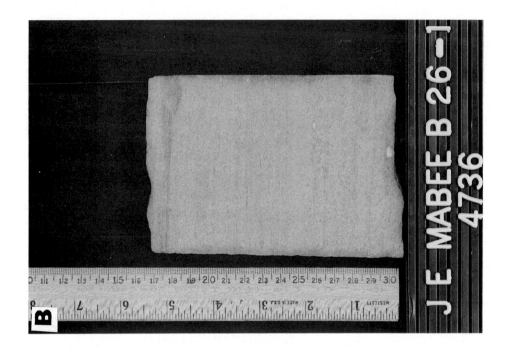

Figure 17. Slab photographs of sandstone facies, B-1 #26, showing: A.) low angle cross-bedding, 4757 ft, and B.) planar-bedding, 4736 ft.

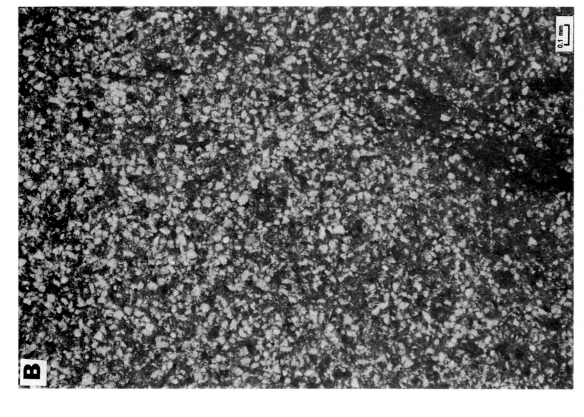
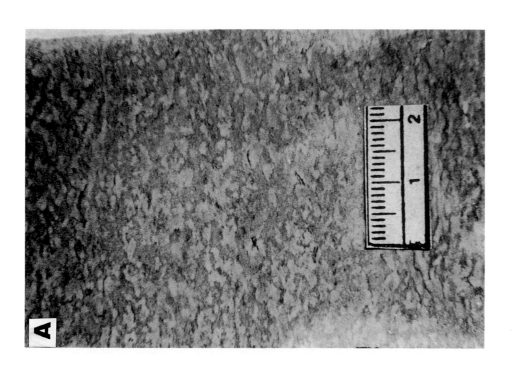

Figure 18. A.) Slab photograph of sandstone facies showing evidence of a highly bioturbated interval, #494, 4752 ft. Scale is in centimeters. B.) Thin section photograph of sandstone facies that is cemented with anhydrite and dolomite from B-1 #26, 4768 ft. Porosity = 11 percent, permeability = 0.1 md.

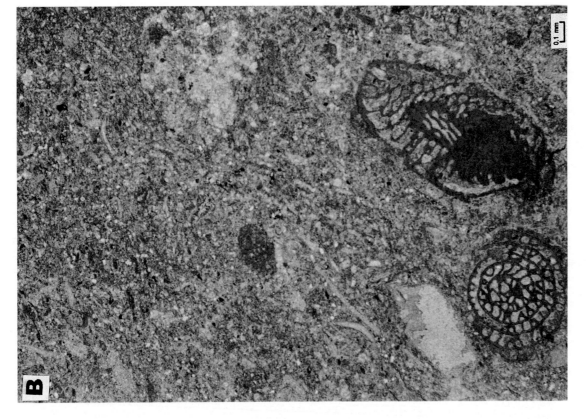

Figure 19. Open marine facies. A.) Slab photograph of fusulinid wackestone, B-1 #26, 4831 ft. B.) Thin section photograph of fusulinid-skeletal wacketone, B-1 #26, 4829 ft.

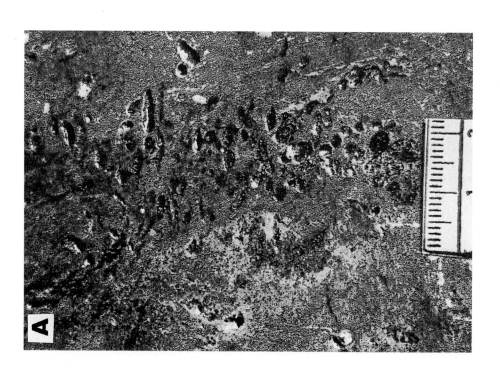

Figure 20. Open marine facies. A.) Abundant fusulinid molds, #35, 4800 ft. B.) Fusulinid molds filled with anhydrite, #270, 4870 ft. Scales are in centimeters.

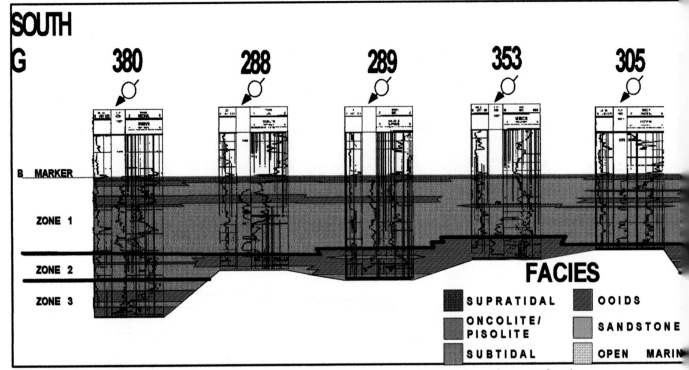

Figure 21. Stratigraphic dip cross section, G-G', using the cores and logs. See Figure 2 for the location of the cross section.

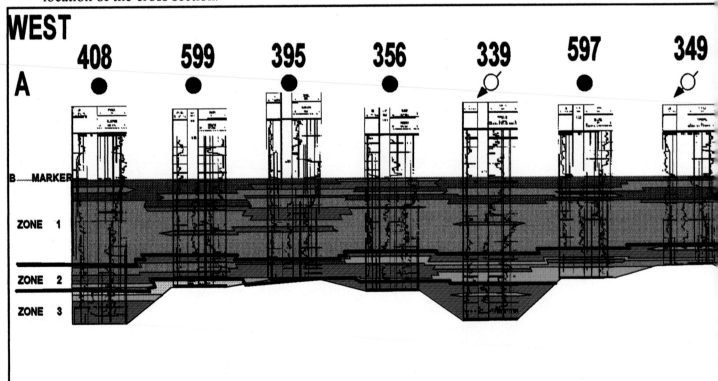

Figure 22. Stratigraphic dip cross section, A-A', using the cores and logs. See Figure 2 for the location of the cross section.

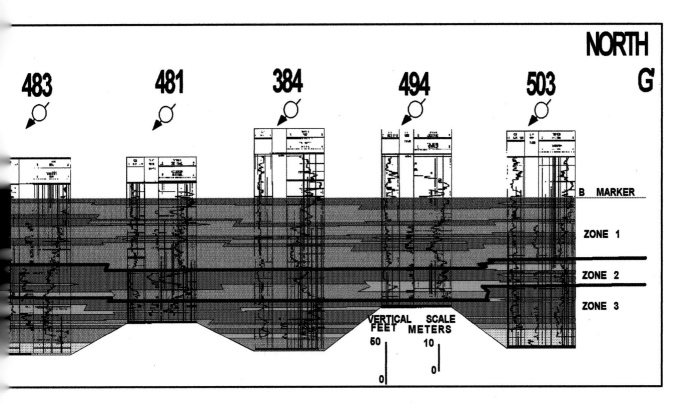
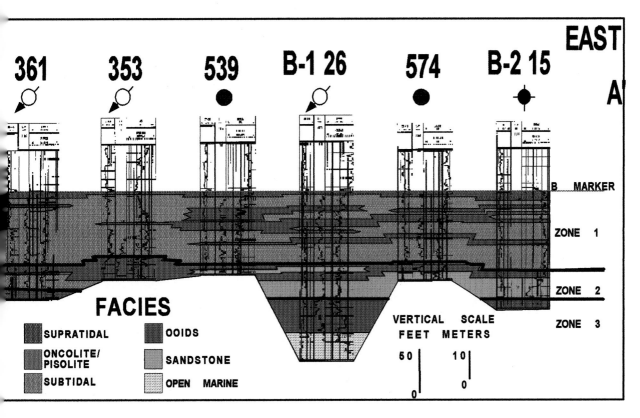

subtidal peloidal wackestones. The exception to this is in the southern edge of the field where sandstones are less prevalent and there is a greater predominance of subtidal wackestones. This cycle is distinguished from the other zones by having solution-enlarged fractures, vuggy porosity, moldic porosity, and less anhydrite (Fig. 23A and 23B). While there is no direct evidence of karst, the observed diagenetic features suggest subaerial exposure and fresh-water influx into the subsurface. The upper portion of zone 3 and the remainder of the reservoir are highly dolomitized. The thickness of the ooid grainstones in zone 3 indicates that there was considerable accommodation space for the aggradation and progradation of the ooid shoals. The sandstones are coarse-grained silt to fine-grained quartz sand that are inferred to have been transported across the tidal flats by prevailing winds into evaporitic pans during regressions and subsequently bioturbated and reworked by marine transgressions. The fact that the sands are thin suggests that the sands were in short supply. They were either being trapped landward or the source of the sand was of a considerable distance as to preclude significant wind transport into the marine environment.

In contrast, zone 2 (Figs. 21 and 22) is a transgressive sequence that reflects an overall deepening of the basin that allowed for the aggradation and progradation of thick ooid packstones and grainstones. These were interrupted by regressive cycles of thick sands and silts of eolian origin that were reworked by subsequent transgressions. There is the obvious lack of subtidal or restricted peloidal wackestones, suggesting that the ooid shoals were not acting as barriers and were being deposited very near the shoreline. Subtidal spicular and peloidal wackestones are much thicker in the southern portions of the field, in the lower portions of zone 2, reflecting the transgressive backstepping of the restricted lagoonal facies behind ooid barrier shoals. The occurrence of oncolite/pisolite facies, locally interbedded with the ooids and sands in the northern portions of the field, suggests close proximity to the paleoshoreline. The cyclical nature of this sequence, reservoir and nonreservoir rock, with highly contrasting permeability, is critical in understanding the past performance of the reservoir and current performance of a low viscosity fluid, CO_2.

Zone 1 (Figs. 21 and 22) is dominated by thick, subtidal, fine-grained peloidal wackestones and packstones that are highly dolomitized. The sequence is described as a regressive sequence reflecting a shoaling upward into the supratidal cap of the reservoir. The obvious lack of sands in this sequence seems to suggest a change in wind direction or that the sands were being trapped landward. The ooid grainstones are generally thin and filled with anhydrite, suggesting little accommodation space during transgressive cycles for the aggradation and progradation of thick ooid sequences. The thick subtidal sequence suggests that thick ooids could occur to the south of the Mabee Field, but, as can be seen on the north-south cross section, G-G' (Fig. 21), the supratidal cap is thinning and the reservoir is potentially lacking a seal.

PROBLEM - POROSITY AND PERMEABILITY

A three-dimensional reservoir model is a powerful and necessary tool for reservoir characterization because of the ability to examine the reservoir architecture spatially. A major consideration in constructing any reservoir model is the accurate assessment of porosity and permeability. Even within a particular facies, diagenetic alteration can create wide variations in

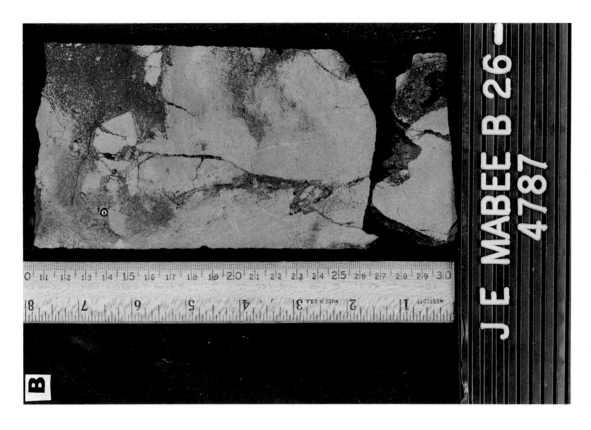

Figure 23. A.) Sandy ooid grainstone with large vugs and oomoldic porosity, B-1 #26, 4785 ft. B.) Sandy ooid grainstone, brecciated with open fractures, B-1 #26, 4787 ft.

pore-throat size and associated permeability. Ghosh and Friedman (1989), in their work on the Mabee Field, showed that primary interparticle and intraparticle porosity of the ooid facies had been lowered due to various amounts of anhydrite growth and dolomite cement. In addition, they also noted that, in some of the ooid grainstones, nearly all interparticle porosity is filled with anhydrite. This leaves only the intercrystalline porosity and associated small pore throats and low permeability. The scatter observed in the crossplots of permeability and porosity by facies, Figure 24, reflects this variation and is one of the main reasons for the observed reservoir heterogeneity. Friedman and others (1990), in their work on the Mabee Field, demonstrated an excellent correlation between porosity and critical pore-throat size obtained from porosimetry and facies. They also showed a good correlation in the upward-shoaling sequence and a crossplot of increasing bulk density and decreasing porosity. A problem arises in the extrapolation of these correlations to the remainder of the log data. The majority of the logs at the Mabee Field are old gamma ray-neutron and gamma ray-compensated neutron porosity logs. The correlation of facies, particularly in zone 1, is difficult because the ooid facies and the subtidal facies are both clean dolomites and indistinguishable with gamma-ray logs (Fig. 4). Another problem that arises in developing a three-dimensional reservoir model is the method of distributing interwell porosity and permeability and preserving the observed heterogeneity of the data.

THREE-DIMENSIONAL MODEL - GEOSTATISTICAL APPLICATION

The problems mentioned above concerning the poor correlation of permeability to porosity and the interwell distributions of these attributes are major hurdles to the development of an accurate reservoir model. The indication of an accurate reservoir model is one that honors all the data, contains all available data, and matches the observed past and present reservoir performance.

An in-house computer program, known as Gridstats, was used to build an integrated three-dimensional geostatistical model of the San Andres Reservoir (Dull and others, 1994). A three-dimensional model of the permeability was constructed with Gridstats that integrates the following data: (1) the structure of all three zones, (2) spatial variability (fractal variogram), (3) the "hard data" or core permeability, (4) the "soft data" - the normalized porosity logs transformed to permeability, and (5) the randomness of the porosity-permeability relationship. The San Andres Reservoir at the Mabee Field, using the same data, was then kriged independently for zones 1, 2, and 3, and the 3 three-dimensional models merged to construct a single model of permeability. The same was done for the porosity model with the exception of not having any "soft data." Figure 25 is a three-dimensional structural model of permeability for the Mabee Field sliced on the 35th y grid row.

Variograms of the core permeability and porosity were done to "quantify" the vertical and horizontal spatial variability. A variogram is a tool that attempts to statistically quantify the spatial variability that is plainly exhibited in the data or assumed from the data based on the geologic interpretation (Fogg and Lucia, 1990). Figure 26 shows the calculation of an "experimental" variogram in one dimension. The "experimental" variogram is calculated from the actual data. The "theoretical" variogram is the curve that best fits or statistically describes the observed spatial variability. The "theoretical" variogram that best describes the porosity and permeability at the Mabee is described as a fractal power-law variogram of 1.1. A fractal is a

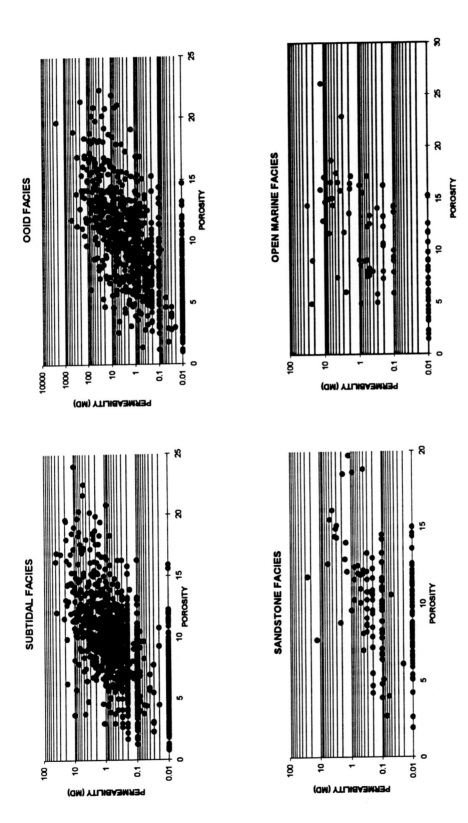

Figure 24. Cross plots of the $\log_{10}$ of the core permeability and core porosity by reservoir facies. Note the obvious higher permeabilities in the ooid facies.

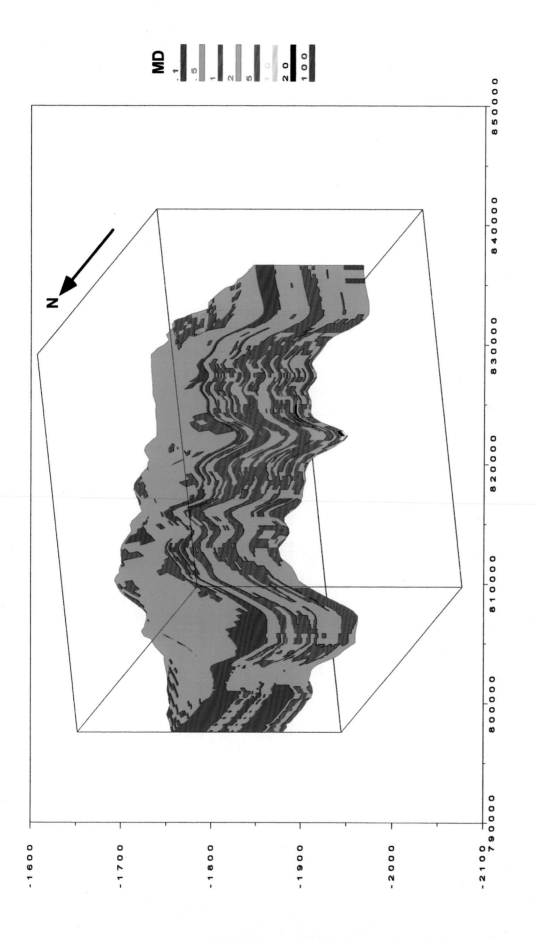

Figure 25. Three-dimensional model of permeability sliced along the y-axis at the 35th gridblock. The view is from the southwest portion of the field. The model is composed of 597,000 grid blocks, is 10 km by 10 km and contains 806 wells. Vertical exageration = 80x. Scales are in feet.

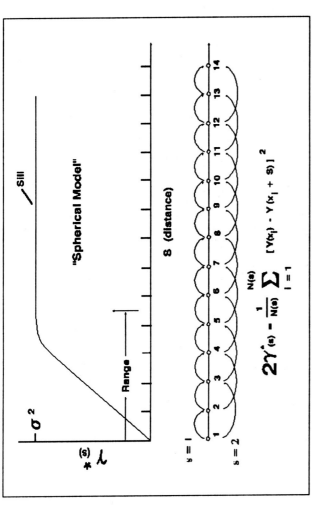

Figure 26. Example of an experimental variogram in one-dimension (after Fogg and Lucia, 1990).

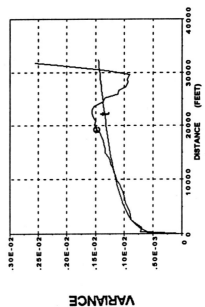

Figure 28. Horizontal variogram of core porosity, (e) experimental variogram, (t) theoretical variogram. Correlation length = 82354 ft. Correlation ratio of vertical to horizontal = 110.0.

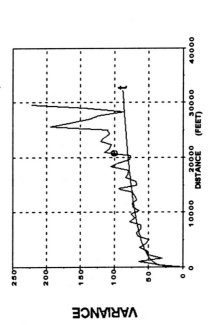

Figure 27. Horizontal variogram of core permeability (e) experimental variogram, (t) theoretical variogram. Correlation length = 62043 ft. Correlation ratio of vertical to horizontal = 110.3.

special type of geostatistics that is scale independent. This means that the spatial variability observed at the wellbore scale is similar to the field scale. Figures 27 and 28 are the "experimental" and "theoretical" variograms of porosity and permeability for the core data. Note that there is a good correlation between the observed spatial variability, or "experimental," and the "theoretical" variogram.

The crossplots of permeability versus porosity show a trend of increasing permeability with increasing porosity, but also contain a great deal of scatter (Fig. 24). To incorporate this data into the model, the porosity logs were first normalized and then transformed to permeability using the linear trend observed in the data. This data is used in the model as "soft data." The "hard data" in the permeability model is the core permeability. The core measurements were all from whole core analyses. In the construction of the porosity model, the core porosity and normalized log porosity were used, but no "soft data." The distinction between the "hard" and "soft" data comes in the three-dimensional kriging algorithm. The "hard data," or core permeability, is preferentially used when in disagreement with the "soft data"; in this case, the porosity logs transformed to permeability data.

To incorporate the scatter that is observed in the permeability and porosity crossplots, the linear correlation coefficient of a line of best fit is calculated and used in the kriging of the permeability model. To illustrate the effects of using the "soft data" and varying the correlation coefficient, 3 three-dimensional models of permeability were constructed. In the first model, only the core permeability was used in the construction. In the other two versions, "soft data" was used and linear correlation coefficients were varied. In addition, the geostatistical three-dimensional distribution of permeability was not confined to zones 1, 2, and 3 to further illustrate the benefits of using the "soft data." In all other respects the models were identical. To illustrate the differences of the three models, a cross section was made in the same location for each model. Figure 29 is a map showing the location of cross section Y-Y'. Figure 30 is cross section Y-Y' through the three-dimensional model that uses only the core permeability and no "soft data." The cross section shows a high degree of lateral continuity of the high permeability intervals and a lack of resolution. In addition, without the control of the "soft data," the permeability is correlated across the zonal boundaries or sequences. Figure 31 is the same cross section; however, in making the three-dimensional model of permeability, the "soft data" and a linear correlation coefficient of 0.9 were used. This cross section shows the high permeability intervals to be much less continuous because the high linear correlation coefficient is indicative of an excellent correlation between porosity and permeability. The cross section also shows much less tendency to extrapolate permeability correlations across zonal boundaries. Figure 32 is the same cross section again, but with a linear correlation coefficient of 0.568. Note that there is more continuity of the higher permeability than in the previous cross section in Figure 31. Also note that the use of the "soft data" enables the Gridstats program to correlate the permeability within zones or sequences.

As with all new tools, verification that results are reasonable and useable with a certain degree of confidence is extremely important. The ultimate test of the geologic model is in using a reservoir simulator to obtain a history match of production, saturation, and pressure. Before the three-dimensional model is used in a reservoir simulator, it can be optimized to reduce the iterations. In creating a three-dimensional interpolation and distribution of petrophysical

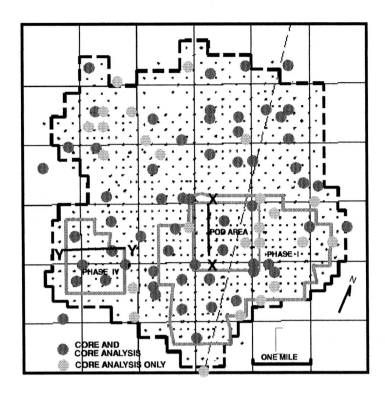

Figure 29. Base map of the Mabee Unit showing the location of core data, Phase IV, POD area, and permeability cross sections X-X' and Y-Y'.

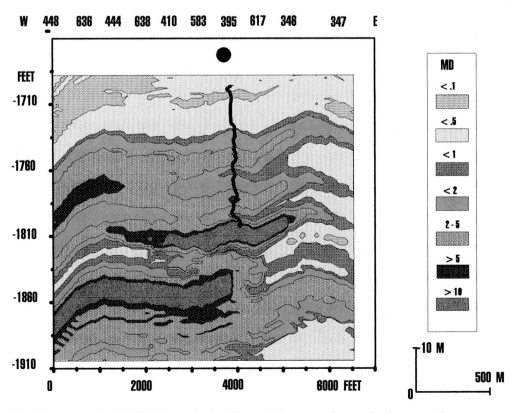

Figure 30. Cross section Y-Y' through the Phase IV area using only the core data

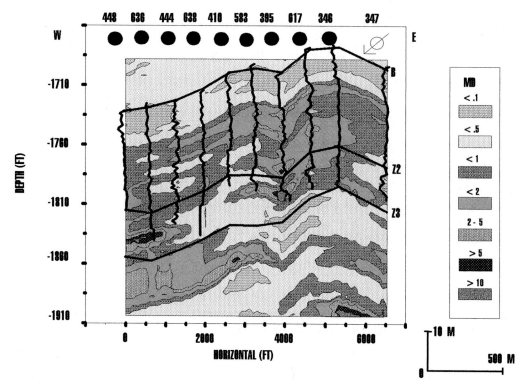

Figure 31. Cross section Y - Y' through the Phase IV area using all the data with a correlation coefficient of 0.9.

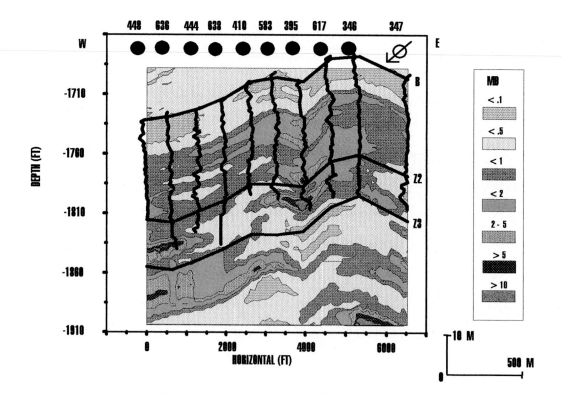

Figure 32. Cross section Y - Y' through the Phase IV area using all the data with a correlation coefficient of 0.568.

properties, such as porosity and permeability, many choices of gridding algorithms and grid sizes exist. One way to be certain that a model is accurately honoring the data is to compare the data input into the model to the data extracted from the completed model. Gridstats allows the user to extract columnar data from the model at the grid nodes. This enables comparison of actual core and log data with the data in the model. Figure 33 shows this comparison for well #380; the KGEOST curve is permeability from the model and the DSMAXPER is permeability from the core. Although the model permeability is not exactly the same as the core, because the actual location of well #380 is not precisely at a grid node, it is in excellent agreement. A comparison of the calculated net reservoir height between the core data and the data extracted from the model was less than one foot. The net reservoir height is the porosity-feet above a permeability cutoff, in this case 0.1 md.

A second method to validate the model is to see if the model is in agreement with observed reservoir performance. In two areas of the field, engineers observed preferential flow between injection and producing wells. The POD area (an area of the field where quick response to CO_2 injection was expected) showed a north-south trend and the Phase IV an east-west trend. The phase numbers refer to the phased implementation schedule of the CO_2 flood. Figure 29 is a map showing the location of these two areas of the field. Horizontal variograms of the permeability and porosity were constructed for each of these areas in four different directions and correlation lengths compared. Correlation lengths are the distances calculated from the geostatistics of the continuity of an element (i.e., permeability and porosity) that exhibits spatial variability. The correlation lengths for the POD area indicate a preferential north-south trend of permeability and porosity (see Figs. 34 and 35). The Phase IV area exhibited a preferential northeast and northwest directional permeability and porosity. This agrees with the observations made by the engineers.

APPLICATION

The development of a three-dimensional reservoir model has provided a very useful tool for evaluating current reservoir performance and predicting the future potential. In the analysis of the reservoir model, more than 300 wells were not completed or only partially completed in zone 2. In addition, 100 wells were not entirely completed through zone 1. Figure 36, a cross section through the three-dimensional model of permeability, also shows that a number of injection wells are not completed in the same zones as the offset producers. To evaluate the potential for deepening the wells, the *kh* or geometric mean permeability-height from the geostatistical reservoir model was used to measure the potential for deepening those wells (*kh* is a measure of the potential for fluids to pass through the reservoir). Interpreted *kh* was crossplotted with incremental daily fluid increase from 19 recent well deepenings. Incremental oil and water are the increases in production. Figure 37 shows a linear trend of increased production in barrels of fluid per day (BFPD) with increased amount of *kh* added. Utilizing the log-log plot of BFPD versus *kh*, 42 producing wells were recommended for deepening.

To further evaluate the performance of the wells at the Mabee Field, a map of total *kh* for zones 1 and 2 was compared to the total fluid production in BFPD. Figure 38 is a bubble map with total *kh* shown as contours. For illustration purposes, only a few of the wells are shown.

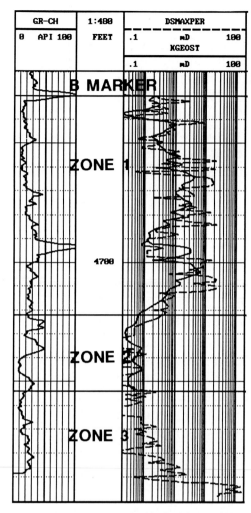

Figure 33. Log of well #380, showing the core permeability (DSMAXK) versus the data from the three-dimensional model (KGEOST).

PERMEABILITY CORRELATION LENGTHS

DIRECTION	PHASE IV	POD AREA
H	33626	19703
E	24300	23290
NE	**79850**	18190
NW	**83390**	22390
N	41080	**30050**
CORRELATION RATIO	111.5	125.7

Figure 34. Relative, directional-permeability correlation lengths from the variograms of the Phase IV area as compared to the POD area.

POROSITY CORRELATION LENGTHS

DIRECTION	PHASE IV	POD AREA
H	83951	17988
E	108200	26570
NE	**128500**	23340
NW	108200	23910
N	83290	**42640**
CORRELATION RATIO	247.8	94.95

Figure 35. Relative, directional-porosity correlation lengths from the variograms of the Phase IV area as compared to the POD area.

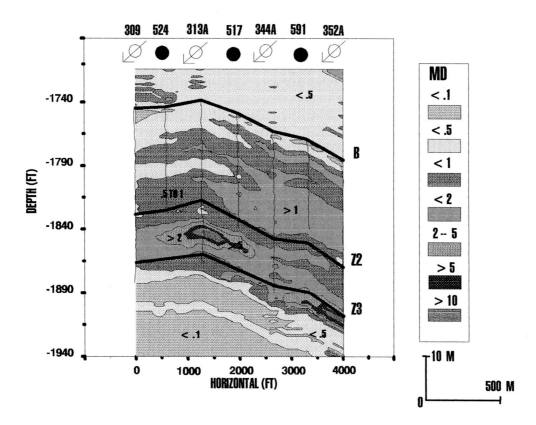

Figure 36. Cross section X - X' through the POD area using all the data illustrating non-conformance of the completions of some of the injection and producing wells.

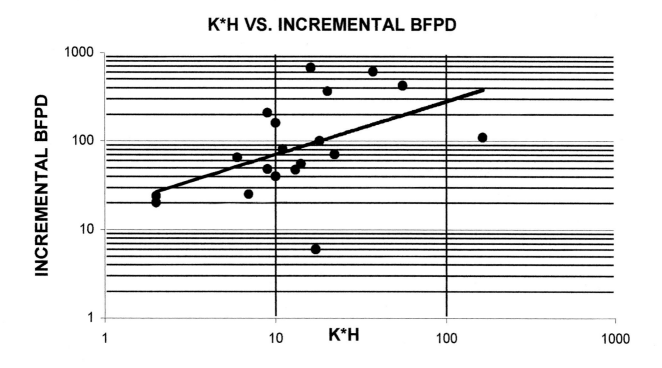

Figure 37. Cross plot between added *kh* from the deepening of 19 wells versus the incremental barrels of fluid per day increase (BFPD).

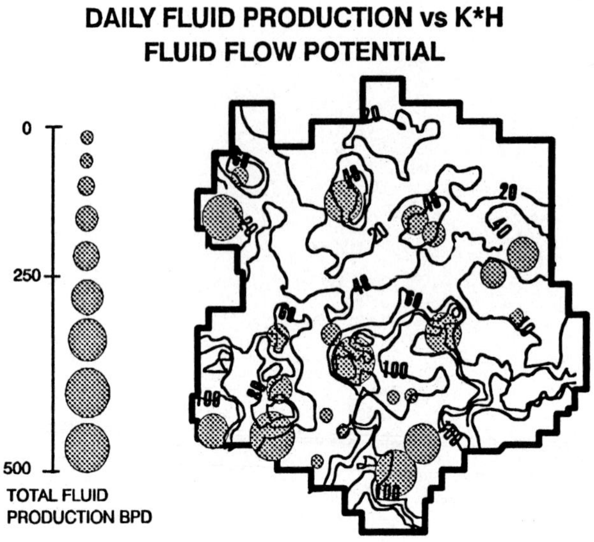

DAILY FLUID PRODUCTION vs K*H
FLUID FLOW POTENTIAL

Figure 38. Daily fluid production in BFPD versus geometric mean of permeability in millidarcy-feet (*kh*). Contour interval = 20 millidarcy-feet.

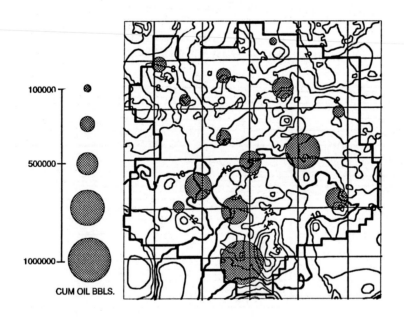

Figure 39. Net porosity-height (Net *phi*h*) map for the Mabee unit showing good agreement with the geostatistically derived map with the cumulative production. Note: Only a few wells are shown for illustration purposes. Contour interval is 2 porosity-feet.

Note that there is a good correlation for most of the wells between high *kh* and high BFPD production. Several wells with low BFPD production and high *kh* are located in the south-central portion of the field. The reasons for this could be that (1) the well is plugging and needs to be cleaned out and stimulated (plugging is a common problem because fresh water is used for injection, which dissolves the anhydrite and carries it in solution to the wellbore, where it precipitates due to the pressure drop), or (2) the well is not deep enough with untapped *kh*. In any case, the *kh* from the three-dimensional model of permeability provides a useful tool not only to qualitatively evaluate a producing or injection well's capacity, but to quantitatively predict the performance. The *kh* map also indicates, in the northern half of the field, that there is significantly less *kh*, indicating a drop in reservoir quality. This is associated with the increasing amounts of supratidal and intertidal sediments that lack significant primary porosity and increasing amounts of pore-filling anhydrite. The increased amount of anhydrite is due to the close proximity of these sediments to the paleoshoreline and increased availability of sulfate-rich fluids. Due to the lower reservoir quality in the northern portion of the field and its associated low flow capacity, four horizontal wells are being drilled. The wells are targeted for the better portions of the reservoir from the three-dimensional model. Since they will be going through 900 ft of pay instead of 75 ft, this will accelerate the movement of fluids through the reservoir because of the added *kh* through which the reservoir fluids can be produced. This is important because it should accelerate the oil production and assist in offsetting the high investment costs of the CO_2 miscible flood.

A comparison of cumulative production to the net reservoir volume (net porosity height or net *phi*h*) was made to verify the geostatistical model and to identify anomalies that may indicate untapped potential. The map was generated from the three-dimensional models of porosity and permeability. This enables the use of a permeability cutoff, rather than a porosity cutoff that is normally done with two-dimensional mapping. Figure 39 is a bubble map showing cumulative oil in barrels with contours of the net *phi*h*. Only a few wells are shown for illustration purposes. The map shows a good correlation with high cumulative production and high net *phi*h*. However, in the southwest portion of the field, well #396 has an anomalously low cumulative production when compared to the net *phi*h*. The three-dimensional model of permeability indicates a high permeability interval between well #396 and its offset injector. This enables the cycling of large volumes of water. Lower permeability zones with high oil saturation are thus unswept, resulting in a lower cumulative production.

SUMMARY

The development of an accurate three-dimensional model of the San Andres at the Mabee Field was highly dependent on the ability to integrate geologic and engineering data. The detailed core descriptions were instrumental in defining the six major facies that control reservoir performance. The development of three zones or genetically related sequences enabled the development of a stratigraphic framework for building the three-dimensional model of porosity and permeability.

The use of geostatistics in the case of the Mabee Field circumvented the problems of a poor correlation of porosity versus permeability and interwell distribution of porosity and permeability. The fractal power-law variogram fit to the "experimental" variograms was essential in improving the interwell estimates of porosity and permeability. The ability of the Gridstats program to integrate three-dimensionally (1) the structure of all three zones, (2) spatial variability (fractal variogram), (3) the "hard data" or core permeability, (4) the "soft data" - the normalized porosity logs transformed to permeability, and (5) the randomness of the porosity-permeability relationship are vital to the development of an accurate reservoir model.

The model for the San Andres Reservoir at Mabee Field not only integrates the data, but preserves the integrity of the input data. The Mabee Field core data extracted from the model showed it to be in excellent agreement from the input core data. In addition, the model demonstrated good agreement between past and present reservoir performance, cumulative production, and current fluid production. The most important aspect of the reservoir characterization and the three-dimensional reservoir modeling is the development of a model accurate enough that reservoir performance can be predicted and distinguished from wellbore problems.

ACKNOWLEDGMENTS

I would like to thank the West Permian Basin Business Unit and Texaco Exploration and Production, Inc., for allowing me to share this work with the geologic profession. Special thanks go to Joe Vogt and An-Ping Yang of the Exploration and Production Technology Department, Texaco, for their assistance in the application of geostatistics and guidance in the use of their geostatistical program, Gridstats. I would also like to thank Frank Wind and William Lawrence of the Exploration and Production Technology Department, Texaco, for their assistance in the core and thin-section photography.

REFERENCES

DULL, D. W., HORVATH, E. A., AND MILLER, K. D., 1991, Geologic reservoir description, San Andres CO_2 Project, Mabee Field: Internal Texaco Report.

DULL, D. W., HORVATH, E. A., AND MILLER, K. D., 1994, Geostatistical method to improve permeability estimates - application to Mabee Field, Andrews and Martin counties, Texas, *in* Gibbs, J. F., ed., Synergy Equals Energy - Teams, Tools, and Techniques: West Texas Geological Society Fall Symposium Publication 94-94, p. 23-33.

FOGG, G. E., AND LUCIA, F. J., 1990, Reservoir modeling of restricted platform carbonates: geologic/geostatistical characterization of interwell-scale reservoir heterogeneity, Dune Field, Crane County, Texas: The University of Texas at Austin, Bureau of Economic Geology Report of Investigation No. 190, 66 p.

FRIEDMAN, G. M., GHOSH, S. K., AND URSCHEL, S., 1990, Petrophysical characteristics related to depositional environments and diagenetic overprint: a case study of the San Andres Formation, Mabee Field, west Texas, *in* Bebout, D. G., and Harris, P. M., eds., Geologic and Engineering Approaches in Evaluation of San Andres/Grayburg Hydrocarbon Reservoirs - Permian Basin: The University of Texas at Austin, Bureau of Economic Geology, p. 125-144.

GHOSH, S. K., AND FRIEDMAN, G. M., 1989, Petrophysics of a dolostone reservoir: San Andres Formation (Permian), west Texas: Carbonates and Evaporites, v. 4, no. 1, p. 45-117.

HORVATH, E. A., AND MILLER, K. D., 1994, Geologic description of the San Andres reservoir facies in the Mabee Field, Andrews and Martin counties, Texas: West Texas Geological Society Bulletin, v. 33, no. 7, p. 5-13.

RAMONDETTA, P. J., 1982, Genesis and emplacement of oil in the San Andres Formation, Northern Shelf of the Midland Basin, Texas: The University of Texas at Austin, Bureau of Economic Geology Report of Investigations No. 116, 39 p.

SILVER, B. A., AND TODD, R. G., 1969, Permian cyclic strata, northern Midland and Delaware basins, west Texas and southeast New Mexico: American Association of Petroleum Geologists Bulletin, v. 53, p. 2223-2251.

TODD, R. G., 1976, Oolite bar progradation, San Andres Formation, Midland Basin, Texas: American Association of Petroleum Geologists Bulletin, v. 60, p. 907-925.

TYLER, N., AND BANTA, N. J., 1989, Oil and gas resources remaining in the Permian Basin: targets for additional recovery: The University of Texas at Austin, Bureau of Economic Geology Geological Circular No. 89-4, 20 p.

GEOSTATISTICAL INTEGRATION OF CROSSWELL DATA FOR CARBONATE RESERVOIR MODELING, MCELROY FIELD, TEXAS

WILLIAM M. BASHORE,[1] ROBERT T. LANGAN,[2,3] KARLA E. TUCKER,[2] AND PAUL J. GRIFFITH[2]

[1]Reservoir Characterization Research and Consulting, Inc.,
2524 Monterey Place, Fullerton, CA 92633;
[2]Chevron Petroleum Technology Company,
1300 Beach Blvd., La Habra, CA 92631;
and [3]Department of Geophysics, Stanford University,
Stanford, CA 94305

ABSTRACT

By establishing a statistical link between a reservoir property of interest (e.g., porosity) and a more extensively available seismic attribute (e.g., acoustic impedance), one can use geostatistical integration algorithms to build improved reservoir models over those generated using well-based information only. This improvement is the result of the additional secondary information regarding interwell heterogeneities of the reservoir property that are known to influence heavily fluid flow in the reservoir (Hewett and Behrens, 1990; Omre, 1991). When these improved models are used in reservoir simulations, the expected value of a suite of simulation outcomes is likely to be closer to the true value than predictions made with only the sparser well control. In addition, the range of outcomes (i.e., uncertainty) should be less with these improved estimates. This reduces the risk associated with using the simulation results when considering various investment options for reservoir management.

In a proof-of-concept exercise, we use the P-wave reflection image from a crosswell seismic survey in the McElroy oil field of west Texas to help build a suite of porosity-permeability models. Because we are dealing with crosswell seismic data, the resolution is more than an order of magnitude better than we can get with surface seismic data. We apply a seismic inversion methodology (Bashore and Araktingi, 1994) to compute an image of acoustic impedance between the wells, which we then correlate at each wellbore with porosities obtained from well logs and core measurements. We expect impedance to be a good predictor of porosity because both velocity and density are variably influenced by differing amounts of porosity.

We generate several equiprobable porosity and permeability models using conditional simulation. A comparison of two reservoir flow simulations, one using well data only and the other using both well data and crosswell seismic data, suggests that models based upon the second set of data provide results more similar to actual field observations.

INTRODUCTION

Crosswell seismic imaging is a relatively new technology that has the potential to enhance the way we manage our petroleum reservoirs. It "plugs" a resolution gap between relatively low resolution, but high coverage, surface seismic data, and relatively high resolution, but low coverage, well-log and core data. By filling this gap we hope that crosswell seismic data will enable us to tie together these various data types. Geostatistics provides one way of doing this. The estimation and conditional simulation techniques used in this study are based on models of spatial continuity and the kriging algorithm (e.g., Isaaks and Srivastava, 1989).

To test this hypothesis, we have at our disposal a high-quality crosswell data set from a producing oil field in a Permian carbonate in west Texas. It was collected by a joint Stanford University-Industry project associated with a CO_2 pilot flood being conducted in the McElroy Field. This data set is intended to provide pre-injection background information and will be followed up at a future date with an identical survey conducted after CO_2 flooding has been in operation for some time. In support of the crosswell data are extensive core data and log data which help establish the "ground truth" of our reflection image.

Our geostatistical analysis is a proof-of-concept study in that we have made numerous simplifying assumptions in order to evaluate the potential impact of this new data type on reservoir management and to provide insight into how crosswell seismic data should be processed in the future with this new application in mind. Our objective in this study is to develop a methodology for integrating crosswell information and assessing its contribution to improved reservoir modeling and subsequent fluid-flow simulations.

SUMMARY OF THE PETROLEUM GEOLOGY

The study area is in the McElroy Field, one of numerous oil fields along the eastern flank of the Central Basin Platform of west Texas (Fig. 1). The reservoir zone is a 100-ft (30 m) interval in the Permian Grayburg Formation. The overall depositional sequence is a series of upwardly shoaling and eastwardly prograding carbonate units, with the lower Grayburg being an open shelf environment, the central reservoir being a shallow shelf environment, and the upper Grayburg being a shallow shelf and intertidal environment. The entire sequence is capped by supratidal and sabkha mudstones, siltstones, and evaporites. Much of the column was dolomitized some time later.

The reservoir interval is primarily peloidal grainstones and packstones, with the porosity being mostly interparticle. Dissolution has increased the porosity in some areas, while interparticle evaporites, like gypsum and anhydrite, have reduced the porosity in other areas. Thin zones of high porosity and high permeability are present which threaten the effectiveness of secondary and tertiary recovery efforts. This heterogeneity is difficult to characterize adequately using conventional geophysical methods. One hope for crosswell seismic imaging is that it will provide high-resolution data for effectively mapping the engineering flow units of a reservoir.

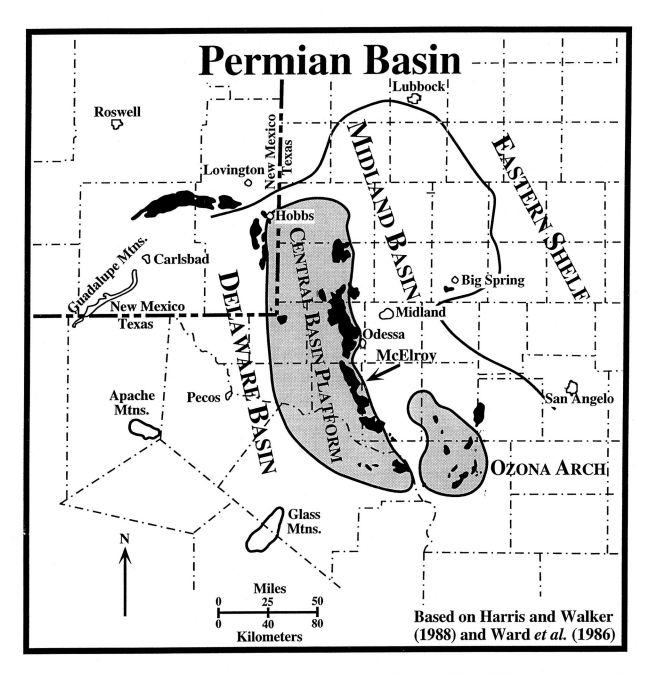

Figure 1. Regional map of a portion of the Permian Basin with highlights showing oil fields which produce from the Grayburg Formation.

CROSSWELL DATA ACQUISITION

The data were collected using a swept-frequency, piezoelectric source and a seven-level hydrophone string that were developed specifically for crosswell seismology. The hydrophone string includes downhole electronics for storage and stacking of the data because of the difficulty in transmitting seven simultaneous samples in real-time. Cross-correlation of the pilot signal with the recorded data is performed on the surface at a later time.

The source operates over a frequency range of 200 to 2000 Hz, which is considerably higher than the frequencies obtained with traditional surface seismic methods (typically <100 Hz). The result is considerably greater resolution with crosswell methods, but over much shorter propagation distances. Therefore, it is currently impractical to try to cover as great an area with crosswell methods as one can with surface seismic methods.

The actual acquisition is done in a manner similar to a logging operation, but two wells are involved: a source well and a receiver well. The zone of interest between these two wells is probed by energy propagating from each source location in one well, to each receiver location in the other well. The hydrophone string is held at a fixed depth in its well. The source is programmed from the surface to fire at preset intervals and is slowly brought up the second well. These intervals determine the number of sweeps that are stacked in a given shot location, the positional "smear" of a given shot location, and the spacing between shot locations. After completing one logging run with the source, the receivers are moved to a new set of locations, the source returns to the bottom of the well, and the process is repeated.

The rate at which the source is brought up the well depends upon the desired source spacing and the number of sweeps to be stacked. The source and receiver spacing must be fine enough to avoid spatial aliasing of the slowest component of the total wavefield (usually the shear wave or tube wave). Closely spaced wells (e.g., 200 m) usually permit small source and receiver spacing, resulting in slow logging speeds (typically around 4 m/min). As the well spacing gets greater, the source spacing is usually increased, which permits faster logging speeds. However, if the signal-to-noise ratio decreases below a tolerable threshold, the number of sweeps that are to be stacked must be increased, and the logging speed must be reduced to prevent smearing of the shot location. Harris and others (1995) cover the details of the acquisition.

The greater the well-to-well spacing, the greater the total range in depth of the source and receiver locations that are necessary to image the reservoir. Obtaining reflections with as high an angle (near to normal) of incidence on the reservoir as possible is desirable. Current practice calls for the most shallow source and receiver locations to be at least as far above the reservoir as the well spacing. Note that this will often be impossible or impractical for widely spaced wells in shallow reservoirs.

The crosswell experiment was conducted between three wells (see Fig. 2). This experiment is at the southern end of the pilot CO_2 flood project outlined by the rectangle, and it ties together an injector (Well 1) and a producer (Well 3). The central well (Well 2) is an observation well drilled specifically for monitoring the progress of the CO_2 flood. A feature of the observation well is that nearly 100 m of continuous core was taken through the reservoir interval and the adjacent strata.

The nominal well spacing for Well Pair 1-2 is 185 ft (56 m) and for Well Pair 2-3 is 600 ft (183 m). Although the spacing for the first profile is considerably less than is typical in this field or most other mature fields, it has provided a high-quality set of data that has permitted us to develop a whole new processing system for crosswell data. Using the knowledge gained from this experiment, the same acquisition and data processing system have now operated successfully at distances of 1500 ft (457 m) and at depths of 10,000 ft.

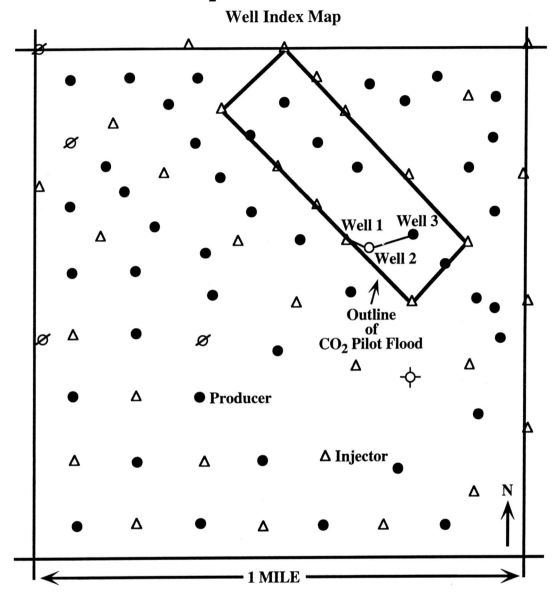

Figure 2. Location diagram of the CO2 Pilot Flood and the wells used for the crosswell experiment. The separation of Wells 1 and 2 is approximately 185 feet and Wells 2 and 3 is approximately 585 feet. Well 2 is an observation well.

Profile 1 (Well Pair 1-2) consists of about 55,000 traces and took about 48 hours of continuous operation to acquire. With Profile 2 (Well Pair 2-3) it took a similar period of time to acquire about 40,000 traces. A very fine source spacing and receiver spacing were necessary to prevent spatial aliasing of various events in the data (2.5 ft for Profile 1 and 5 ft for Profile 2), which makes it easier to separate the components of the wavefield during processing.

In this study we focus on Profile 1 only in order to demonstrate the principles involved while using our highest-quality data set.

SEISMIC IMAGING

There are two aspects to the processing: obtaining a velocity image from the direct or transmitted arrivals (this is often referred to as the velocity tomogram), and obtaining a reflection image from the reflected arrivals. Inverting the travel times of the direct arrivals for the velocity structure is a relatively straightforward procedure, for which various methods are extensively documented in the literature. Computing a reflection image from crosswell data is a relatively new process that has been pioneered on the data set contained in this study.

A tomographic image for the compressional direct arrivals based upon ray tracing was computed by Harris and others (1995) and is shown in Figure 3. The reservoir interval is a prominent low-velocity zone between about 2850 and 2950 ft in depth. Nothing is distinct about this image that might have an impact on how to manage the reservoir. In addition to the ambiguity associated with trying to interpret velocities, the resolution is sufficiently diffuse (probably on the scale of about 20 ft or 6 m) to render the image of limited use for this study. The pinched look in the center of the reservoir interval, and the suggestion of "X"-shaped streaks in a few places, are probably artifacts attributable to the tomographic inversion method.

The sonic log velocities (which have been smoothed) are displayed along with the tomographic velocities obtained near each well. The tomographic velocities follow the sonic velocities quite well, although their smoothness gives an indication of the spatial resolution for the tomographic algorithm employed and the acquisition parameters.

The details associated with obtaining a reflection image are beyond the scope of this paper. See Lazaratos and others (1995) and Rector and others (1995) for details. The imaging algorithm is a variant of the VSP-CDP transform (Wyatt and Wyatt, 1984), which maps each point in the seismic traces to a single point in the resulting depth image. As such, the algorithm may be viewed as a special case of Kirchhoff migration. It has proven to be quite robust for flat-lying geology, but, at this time, it is uncertain as to how well it images lateral variability (e.g., the extent to which there is lateral smearing of the image).

The resulting P- and S-wave reflection images for Profile 1 are shown in Figure 4. The reservoir interval is from 2850 to 2950 ft in depth. The P-wave image has about 8 events within that 100-ft (30 m) reservoir, and the S-wave image has about 12 events in the same interval. The apparent resolution in these two images is quite striking (the wavelengths are on the order of 8-12 ft) when compared with traditional surface seismic data, which usually have an upper frequency limit on the order of 100 Hz, resulting in wavelengths on the order of 150 to 200 ft (45-60 m) in our study area.

One striking feature about these images is the unconformity between the overlying Grayburg Formation and the San Andres Formation at about 3050 ft depth. This unconformity

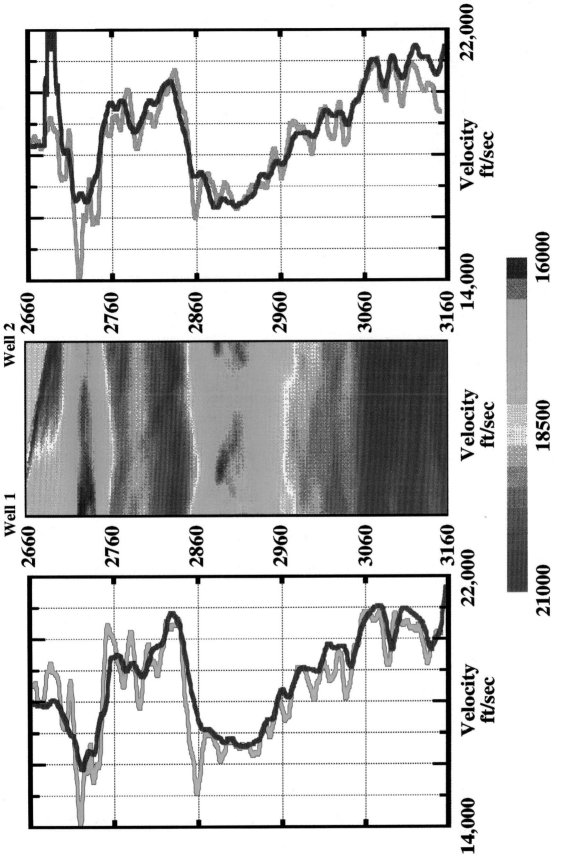

Figure 3. A tomographic image of the P-wave velocities obtained from the crosswell data set. The sonic log velocities (blue) are displayed along with the tomographic velocities obtained near the well (red). The sonic velocities have been smoothed with a 10 foot long averaging function. From Harris et al. (1995).

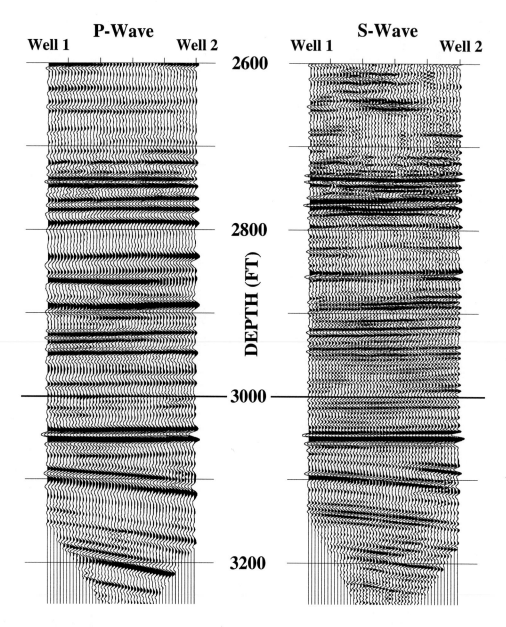

Figure 4. P-Wave and S-Wave reflection images generated from a set of crosswell seismic data collected between wells separated by about 185 feet. The P-wave image was used for the geostatistical analyses in this paper.

is often difficult to identify precisely in the logs, but it is quite clear in the reflection image. In other locations in west Texas, this unconformity often defines the top of a potential reservoir.

In addition to the higher frequencies involved, there are other aspects of the crosswell data that differentiate them from surface seismic data. First of all, there is a much wider range of reflection angles. This means that the apparent amplitudes may be affected by the partitioning of transmitted and reflected energy into both P- and S-waves as a function of incidence angle. This is usually not an issue with the near normal incidence angles associated with surface seismic data. Second, wide-angle reflections tend to distort or "stretch" the wavelet, especially if the wavelet is narrowly banded. Third, we have not accounted for the radiation pattern of the source, nor the sensitivity pattern of the hydrophone receivers. We have ignored these effects when we assume that the convolutional model is valid later on in this study. Although the seismic processing involved in the imaging has not altered the apparent amplitude relationships, no effort has been made to correct for these shortcomings. The resulting correlations we obtain in our geostatistical analyses between seismic impedance and porosity would likely be improved if we were to account for these factors.

We have also not made use of the relatively unique S-wave reflection image, although it should provide an independent set of information with which to characterize the reservoir.

INTERPRETATION OF REFLECTION IMAGE

What is causing the multitude of reflections we see in the reservoir? By comparing the location of P-wave reflections in the reflection image to changes observed and measured in the core from Well 2, we have drawn several preliminary conclusions with regard to the principal factors that give rise to reflections:

- Total porosity has the strongest influence on velocity and acoustic impedance

- Mineralogy, including variations in the amount of siliciclastics and evaporite cement, strongly influences velocity and acoustic impedance

- Variations in amount of porosity, pore type, and pore size influence velocity and acoustic impedance at the core plug scale

These conclusions are based upon petrophysical analyses of core plug samples (see Nolen-Hoeksema and others, 1995), petrographic analyses of numerous thin sections, and petrochemical analyses.

This gives us hope that our reflection image can be used to predict porosity trends within the reservoir. "True seismic amplitudes" from surface seismic data are sometimes used to compute changes in seismic impedance. When these changes are combined with low-frequency background impedances, they are then used to estimate porosities and, possibly, other important lithologic and engineering parameters. Given the very high resolution of the crosswell data, it makes sense to apply this approach.

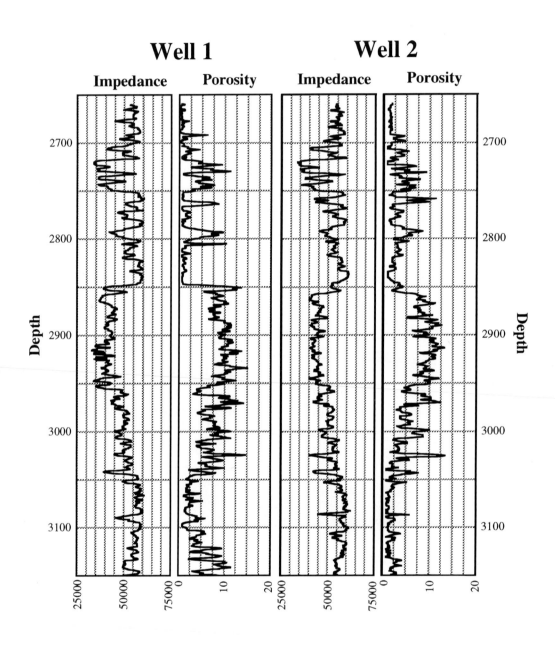

Figure 5. Impedance and porosity logs from Wells 1 and 2, respectively, used in the reservoir modeling.

POROSITY-IMPEDANCE RELATIONSHIP

The integration of multiple data types to improve reservoir description is dependent upon the relationship between the data types. To establish whether information obtained from the crosswell experiment should relate to a reservoir property of interest, we perform a univariate analysis using porosity and acoustic impedance logs. Figure 5 shows the well logs used in this analysis as well as in the reservoir modeling. Presuming that we can derive impedance estimates from the crosswell reflection survey, and that impedance has some correlation to porosity, then the lateral and vertical variations observed in the reflection section may indicate heterogeneities in the porosity distribution.

Quality control of data sets is the foundation for any statistical procedure. Results from a univariate analysis of the impedance and porosity database are displayed in Figure 6. Bimodal distributions are apparent for both impedance and porosity, possibly owing to a mixture of

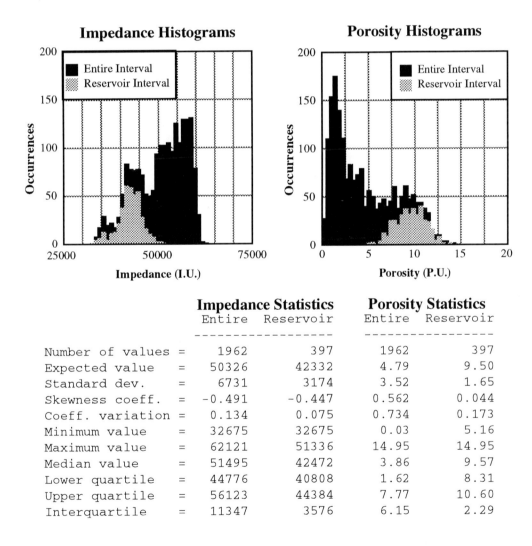

Figure 6. Univariate histograms of impedance and porosity as defined by logs from Wells 1 and 2 (see Figure 5). Histograms for the entire log and the reservoir subset are shown. The corresponding summary univariate statistics are tabulated.

lithofacies populations. Without the necessary discriminating facies information, further subdivision is not possible. However, limiting the population strictly to the reservoir facies does narrow the distributions in both impedance and porosity. These distributions are more closely normal in shape.

Analysis of the summary statistics suggests no aberrant values for either data set. For example, the extreme values (minima and maxima) are well within the expected range for either log property, the coefficients of variation are much smaller than unity, and the interquartile ranges and twice the standard deviations are of similar magnitudes.

Figure 7 contains the results from analyzing the statistical impedance-porosity relationship using the log data from the two wells. The impedance logs are computed from the sonic and bulk-density logs. Two crossplots are given: one for the entire crosswell depth interval and the other over the reservoir interval. Interpretation of the corresponding correlation coefficients (Pearson and Spearman) suggests a strong negative linear relationship between impedance and porosity, especially within the reservoir interval. This relationship suggests that, if impedance estimates can be obtained from the crosswell reflection data, reservoir models generated by integrating the impedances should be more reliable than those based solely on the well-log-derived porosities.

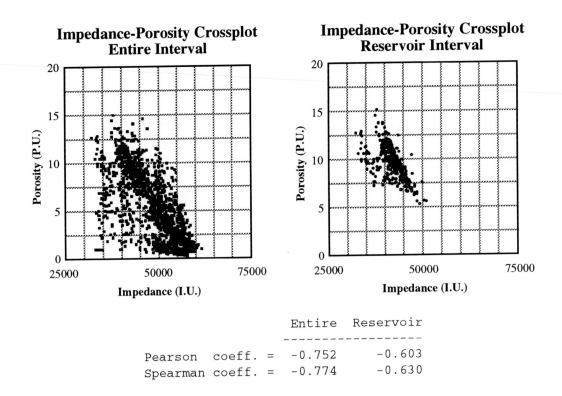

Figure 7. Bivariate crossplots of impedance versus porosity values from Wells 1 and 2 (see Figure 5). The right crossplot covers the entire range of log depths while the left is restricted to the reservoir interval. The corresponding Pearson and Spearman (rank) correlation coefficients suggest a linear relationship between impedance and porosity.

Preliminary laboratory tests on core plugs suggest that changes in rock mineralogy affect rock density that, in turn, affects rock porosity. A major improvement to the bivariate analysis is warranted. The initial univariate and bivariate analyses discussed above do not subdivide the log data into lithofacies populations. We plan to do a lithofacies-based approach using categorical indicators for the next phase of this study. Complete core descriptions were not available at the time of the initial proof-of-concept phase.

This study used only the P-wave reflection survey. We plan to do additional analyses, investigating possible relationships of shear waves to porosity and of Poisson's ratio to fluid properties. We believe, based upon preliminary laboratory results, that when combined with lithofacies discriminators, improved reservoirs may be created.

ACOUSTIC IMPEDANCE INVERSION

Having established that acoustic impedance statistically correlates with porosity in our study area, the incorporation of crosswell reflection data into reservoir porosity models requires converting aggregate seismic amplitudes into point estimates of impedance. The general class of methods that performs this conversion is called seismic inversion (Lavergne and Willm, 1977; Lindseth, 1979; Bamberger and others, 1982). The inversion process decodes each seismic trace into an acoustic impedance trace. These impedance traces may then be integrated directly with well logs, using geostatistical techniques to build the desired reservoir models.

Many approaches are utilized in performing seismic inversion. The method described here is specifically designed to be constrained by the available geologic control, while preserving the inherent degree of confidence in how well the resulting pseudo-logs (impedance) predict the reservoir property of interest. The power of many geostatistical algorithms resides in their ability to temper the contribution of secondary data according to its level of prediction confidence. The inversion process should not impose too much of the well information into the inversion so as to lose the measure of this prediction calibration.

The inversion of crosswell data is more straightforward than inversion of surface seismic data, because the data are recorded and processed in depth, not in time. Therefore, intensive effort to determine sample-by-sample depth-time pairs is not required. The inverted impedance traces are already in the same domain as the well logs. Additionally, the well logs are exactly located with the seismic data, because these data were recorded in the wells.

In order to apply the inversion methodology outlined in Figure 8 (Bashore and Araktingi, 1994), we assume a one-dimensional convolution of the earth reflectivity with a filter. Results obtained by match-filtering the log reflectivity with the corresponding seismic trace indicate that the processed waveform contained in the reflection data is approximately zero-phase. Modeling by Araktingi and Bashore (1992) has shown that flow results are relatively insensitive to minor phase misestimation, and no further dephasing was applied. In this regard, each trace is assumed to be a band-limited (although not properly scaled) reflectivity trace.

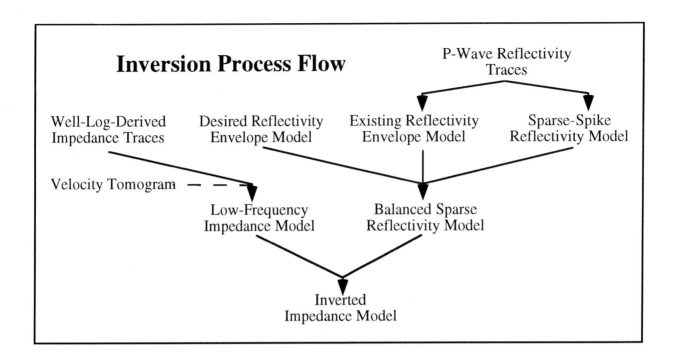

Figure 8. Process flow for performing a geologically constrained seismic inversion which preserves the inherent uncertainty in the calibration between the resultant inverted traces and the well logs.

Although the processing of the crosswell image produces a zero-phase waveform, it does not produce absolute-value reflectivity estimates. To achieve balanced estimates necessary for the inversion to produce absolute-value impedances, a two-step process is employed (Figs. 9 and 10). First, a sparse-spike deconvolution algorithm is applied to sharpen the band-limited reflectivity into "spikes" (Walker and Ulrych, 1982: Oldenburg and others, 1983). This high-frequency spectral extrapolation presumes a "blocky-shaped" inherent impedance log. The unbalanced sparse-spike result is shown in Figure 9b next to its input, the original crosswell section, Figure 9a. While the sparse reflectivity section definitely has higher frequency content, the vertical resolution is still the same as the original data.

The second step balances the spikes to appropriate reflection coefficients, but the balancing must be done to preserve the fine-scale relative variability in the crosswell reflectivity. The balance functions are conditioned by the well reflectivity in a low-frequency sense so as not to bias unduly the crosswell reflectivity. We computed well-reflectivity traces from the impedance logs, and then we derived low-frequency envelopes from these traces via the Hilbert transform and a low-pass filter. The two reflectivity envelopes are linearly interpolated along dip to produce a model of the desired low-frequency reflectivity bias (Fig. 9d). We could have used kriging instead, but the wells are so close together and the envelopes are so smooth that no perceptible difference would result.

Similar low-frequency envelopes can be computed from the crosswell reflectivity and assumed to be the existing bias. Figure 9c contains these existing envelopes. Comparison with the desired envelopes reveals that the crosswell processing produced a uniformly balanced section,

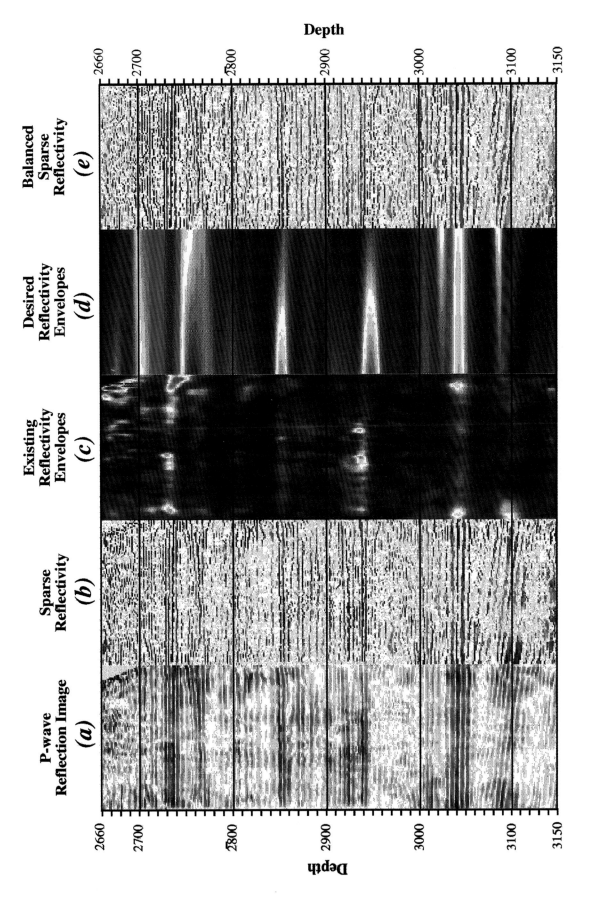

Figure 9. Processing images for generating a balanced reflectivity section: a) original crosswell reflection image, b) sparse-spike reflectivity, c) filtered Hilbert envelopes of sparse reflectivity, d) linear interpolation along dip of sparse reflectivity, d) filtered Hilbert envelopes of well log synthetic seismograms, and e) balanced sparse reflectivity formed by dividing sparse reflectivity by existing envelopes and then multiplying result by the desired envelopes.

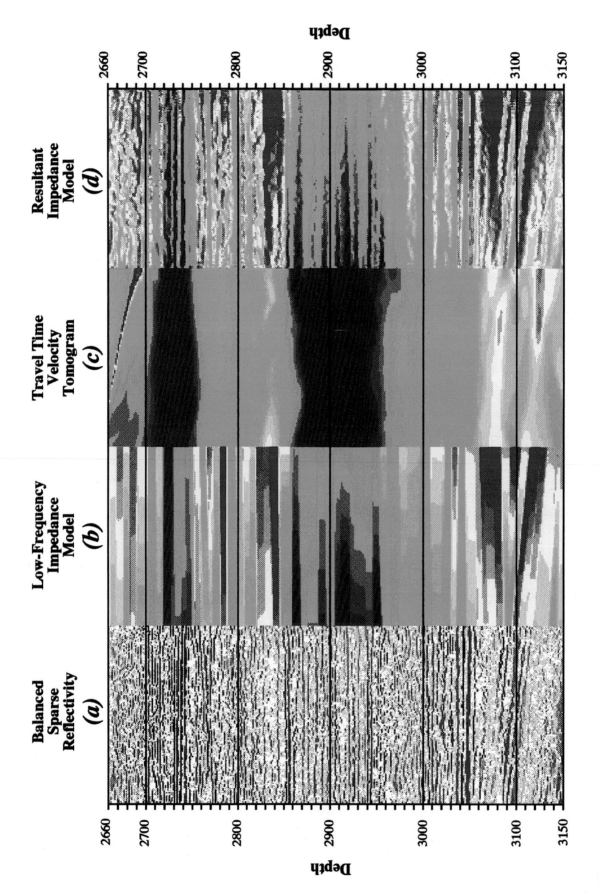

Figure 10. Processing images for generating an acoustic impedance inversion section: a) balanced sparse reflectivity image (see Figure 9e), b) linear interpolation along dip of low-frequency component of impedance well logs, c) velocity tomogram (see Figure 3), and d) the resultant acoustic impedance inversion section by combining the balanced sparse reflectivity and the low-frequency impedance model.

as might be achieved by an AGC operation. The desired section more clearly shows the major impedance changes, as is found at the top and bottom of the reservoir. To remove the existing bias and impose the well-derived bias, each crosswell trace is divided by its existing bias (envelope) and then multiplied by the corresponding interpolated bias trace. In this way lateral and vertical balancing is performed which preserves the high-frequency relative variations, but dynamically scales them to proper magnitudes defined by the well control. The balanced sparse reflectivity is shown in Figure 9e; this image is inverted to impedance.

The spectral extrapolation operated only on the high-frequency end of the reflectivity spectrum. The generation of absolute impedance values requires an estimation of the missing low-frequency reflectivity components, including the d.c. component. The primary source of this information again comes from the well-log-derived impedances (Araktingi and others, 1990). Other low-frequency data may be available, and, in the case of crosswell seismic data, the velocity tomogram contains information on the low-frequency variability of impedance.

Figure 10b shows a linear interpolation of the low-frequency components that were derived from the well logs. Figure 10c is the preliminary travel-time tomographic solution from the crosswell data (shown in Fig. 3). Unfortunately the structure within the tomogram does not tie sufficiently to the reflection structures in order to be useful for the inversion process. The inversion is performed in the frequency domain, which requires the low-frequency model and the reflectivity to tie exactly in the depth domain. In future work we will incorporate the structural control, as defined by the reflectivity, to constrain the tomographic inversion.

The balanced reflectivity (Fig. 10a) and the low-frequency impedance model (Fig. 10b) are input to the frequency-domain inversion algorithm (Santosa and others, 1986) to produce the impedance model (Fig. 10d). This impedance model, after calibration with the porosity logs, is used in the geostatistical integration for generating the reservoir models.

INVERSION-POROSITY CALIBRATION

As previously mentioned, a powerful attribute of certain geostatistical procedures is their ability to limit the influence of the secondary variable to the degree of predictability of its primary variability. In our case the question of how well the inverted impedance traces predict porosity must be answered. Because porosities are known at the well locations, a crossplot of the corresponding inverted impedance values with the log porosities will provide this measure of prediction confidence.

Collocated cokriging is the selected technique for integrating the inversion with the porosity logs (Xu and others, 1992). This method utilizes the assumption that, where primary and secondary data are jointly located, the primary data dominate. With this assumption it is possible to derive the theoretical cross-variance model of the primary variable to the secondary variable from the cross-correlation coefficient and the primary data variogram. In essence, the correlation coefficient acts as a dial for increasing or decreasing the contribution of the secondary data in the modeling process.

Figure 11 displays the porosity logs for Wells 1 and 2, and the respective end traces from the inversion that correspond to the well locations. Visual comparison of the porosity logs and the inverted traces, and the corresponding crossplot, indicates a strong negative correlation coefficient of approximately the same magnitude as that obtained in the bivariate analysis of the well porosity and impedance logs. Because of the manner in which the inversion is constrained by the low-frequency log information, the trend in the inversion traces should correlate nicely with the porosity logs and does add substantially to the correlation coefficient. For the same reason, the high-resolution variation is obtained only from the seismic traces. Thus, the calibration between the logs and inverted traces should reflect the correlation between the inverted traces and porosity variations away from the wells.

GEOSTATISTICAL RESERVOIR MODELING

We conducted two modeling and flow-simulation studies to assess the difference made by the incorporation of any interwell heterogeneity information obtained from the crosswell data. One study utilizes only well data to produce reservoir porosity models, and the other study geostatistically integrates the inverted impedance estimates to constrain interwell predictions of porosity.

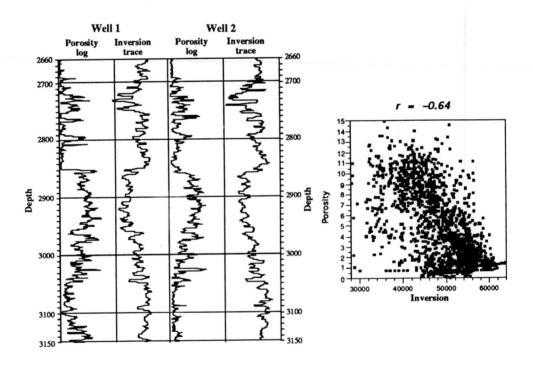

Figure 11 - The inversion-porosity calibration traces and crossplot from Wells 1 and 2. The correlation coefficient (r=0.64) is used in the collocated cokriging procedure for integrating the two data sets when forming the porosity models.

The process flow combining the porosity logs with the inverted traces is shown in Figure 12. The flow consists of three primary elements: the generation of porosity models using kriging (well logs only) or collocated cokriging (well logs and crosswell inversion), the generation of permeability models using a stochastic transform, and the subsequent fluid-flow simulations.

To perform geostatistical model building with kriging or collocated cokriging requires a theoretical model of spatial continuity. This theoretical model is called a variogram. A pair of horizontal variograms were computed over the entire crosswell interval and over only the reservoir interval. The former has a correlation range of 36 ft (10 m), while the latter is greater at 75 ft (23 m). This difference underscores our earlier concerns from the univariate and bivariate analyses about not subdividing the data set into more than one population. Because we are limiting the output models to only the reservoir interval, an indicator approach was not used. A categorical indicator approach, even for the reservoir interval, will be employed for the next phase of the study when the lithofacies descriptions are available.

At the same time, however, these variograms reinforce the value of integrating seismic data with the well-log data. The well spacing is much too great to obtain any statistical information on interwell property variations on this scale from only the well logs. Moreover, these ranges

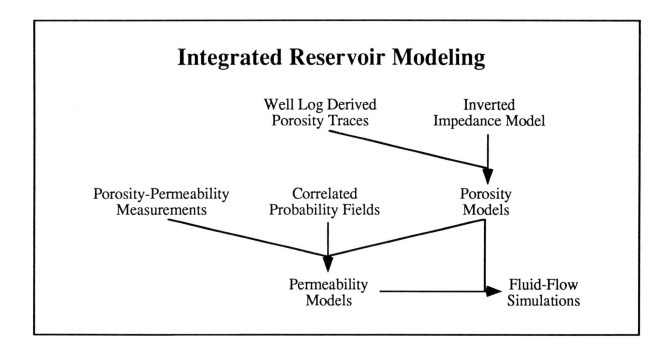

Figure 12 - The integrated process flow for generating porosity and permeability models for use in a fluid-flow simulator.

are slightly larger than those found from horizontal air-permeability transects made on outcrop faces of a similar depositional system (Grant and others, 1994). The fact that the range obtained from the reflection image is greater than that obtained from outcrop studies is consistent with our expectation of lateral smearing associated with the selected imaging algorithm. In theory, migration algorithms should have greater lateral resolution than the VSP-CDP method applied. The resultant models to be submitted to the flow simulator are analogous to performing an outcrop flow simulation. This scale of flow model helps to bridge the long-standing gap in resolution between core models and full-field models.

Porosity models based upon the well-log information only and based upon information from both the logs and the seismic data can now be generated. For comparative purposes Figure 13 contains the kriging solution and the cokriging solution. The variogram used in the kriged model is the one traditionally used for building models in this field. The correlation range from well logs only is about 1200 ft (365 m), and it produces a smoothly interpolated model (Fig. 13a) between the two wells. Conversely, the incorporation of an impedance trace every 2 ft at the $r=0.64$ confidence level and a correlation range of 75 ft produces a model (Fig. 13b) with significantly more short-scale variation. Figure 13b is an estimation, not a conditional simulation.

One of the primary benefits of using kriging-based interpolators is the determination of uncertainty associated with the predictions. To obtain the uncertainty profile multiple, equiprobable models are generated. Each of these "realizations" honors the input primary data and is constrained by the same variograms and calibration parameters. This study uses a sequential Gaussian simulation technique with both the kriging and collocated cokriging (Deutsch and Journel, 1991). Displayed in Figure 13c and 13d are two realizations of both models respectively. The two realizations based on the well-log and crosswell data are far more similar than the two based solely on the well logs. The differences in similarity are especially evident in the mid portion of the models. This increased similarity translates directly to reduced uncertainty in the porosity predictions.

Because insufficient permeability data exist to simulate this property directly at each grid node, permeability models can be generated from the porosity model in a number of ways. The most common approach is to fit an analytic function through laboratory-derived measurements of porosity and permeability obtained from core samples. Linear regression on porosity versus the logarithm of permeability is one such approach. Figure 14 is a crossplot of porosity and permeability core measurements from Wells 1 and 2. There exists a broad range of permeability values for any given porosity value. The linear regression solution, however, will produce only one permeability despite the scatter or uncertainty. To include this inherent uncertainty, a stochastic transform is used to generate many permeability models from a single porosity model. Each of these permeability models, when crossplotted against the porosity model, will reproduce the shape and scatter in the original crossplot of porosity and permeability. In this way the uncertainty in flow results due to the uncertainty in permeability prediction may be quantified.

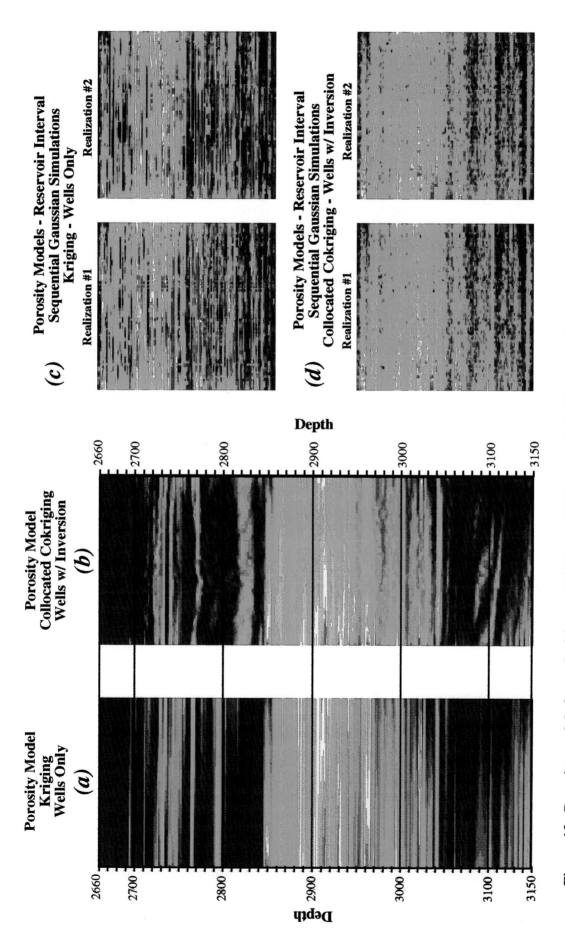

Figure 13. Porosity models formed with geostatistical techniques: a) kriged model using well logs only, b) collocated cokriged model using well logs and crosswell inversion, c) two conditional simulation models using sGs with well logs only, and d) two conditional simulation models using sGs with well logs and crosswell inversion.

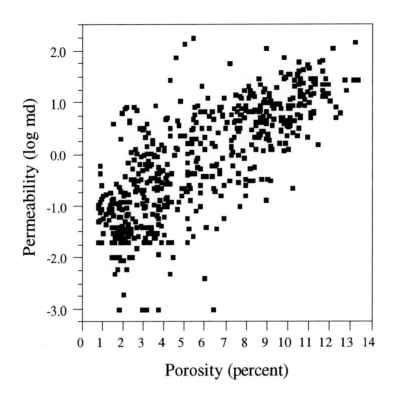

Figure 14. Crossplot of porosity and the logarithm of permeability. A linear relationship is seen but with considerable scatter from a straight line. These data are used to generate the stochastic transform of porosity to permeability.

FLUID-FLOW SIMULATIONS

The final step in this proof-of-concept exercise is fluid-flow simulation. Only one model pair of porosity and permeability for each case (well logs only and well logs with crosswell data) was used, because these models are not the final ones upon which decisions will be based. Ideally, as many models as are necessary to capture the distribution and associated probabilities of flow responses of interest should be subjected to simulation. Displacement images in Figure 15 show water displacing oil as a function of the amount of water (measured as a fraction of the total pore volume) that has been injected in the left well. Comparison of the displacement patterns yields some interesting observations.

The simulation on the "wells plus seismic" model could only be accomplished by using the full finite-difference solution with the smallest possible time steps. This implies that the flow paths through this model are extremely complicated. Surprisingly this model actually predicts breakthrough of water into the right-hand well earlier than the "wells only" model. This means that the seismic data were not only predicting shorter-scale variations, but also more continuous high-permeability zones. The more diffuse flooding in the "wells plus seismic" versus the fine-fingering of flow through the "wells only" model indicates better sweep efficiency in the former because of its finer-scale variations.

Fluid Displacements

a) Model from wells only

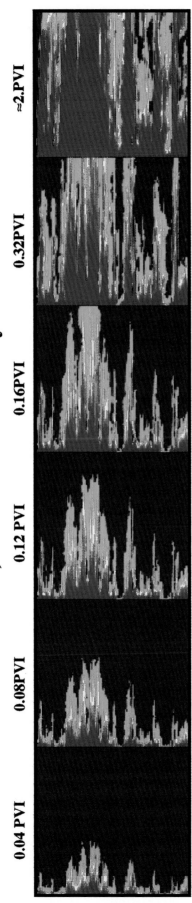

b) Model from wells and crosswell

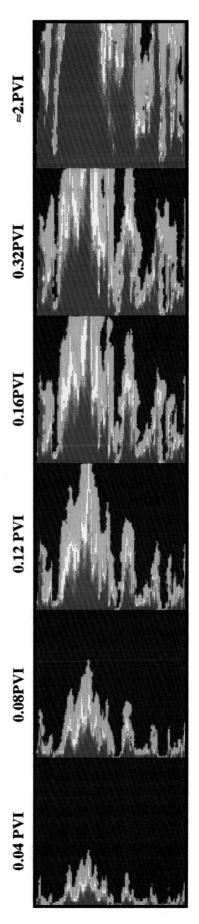

Figure 15. Fluid displacement images for selected time steps (in pore volumes injected, PVI) from flow simulations on: a) realization 1 from Figure 13c and b) realization 1 from Figure 13d. Blue represents original oil saturation, red represents high water saturation.

While these observations are not statistically significant, as they are based on just a single realization, they are thought-provoking. Historically two observations are commonly seen when comparing field results with model predictions: breakthrough occurs sooner, but recovery rates are generally higher. The "wells plus seismic" model is more coincident with both of these than the "wells only" model.

We have ended our simulation study here, but one could run a series of realizations through flow simulation to build histograms of flow results, both with and without crosswell seismic data. Economic scenarios could be attached to each simulation to see what kind of return would be predicted for a particular investment and to obtain a measure of uncertainty associated with the predictions. Our hope is that the combination of geostatistics and crosswell imaging will reduce the uncertainty in such investment scenarios. Demonstration of this idea is the objective of the next phase where more geologic control is incorporated. Also we have the opportunity to take advantage of the higher-resolution shear-wave image and post-CO_2 images for both P- and S-waves. Combined use of these data may yield detailed images of flood fronts and viscous fingering on a very fine scale.

CONCLUSIONS

As part of a proof-of-concept exercise, we have presented a methodology for integrating crosswell-based information with well data to produce reservoir models. Comparative fluid-flow simulations on these models, both with and without the crosswell data, show that shorter- and longer-scale porosity patterns are captured from the crosswell data. This improvement in heterogeneity characterization translates to prediction of earlier breakthrough times and to predictions of improved sweep efficiency. These results are common observations during actual field trials. Models generated from well logs only tend to be more optimistic in breakthrough and less optimistic in recovery.

Geostatistics provides a tremendous variety of tools with which to integrate additional types of data in the estimation and conditional simulation processes. We have not utilized many of these tools to constrain our models further with core descriptions and lithofacies interpretations. Also we have not considered the integration of surface seismic data, while of lesser vertical resolution, to provide extensive lateral sampling of the reservoir. Because of these additional considerations, and given the opportunity to continue this exercise, several things will be addressed:

- What impact does the kind of processing (VSP-CDP mapping, Kirchhoff migration, reverse time migration) have on the key geostatistical measures (e.g., range of spatial correlation)?

- Do images processed with relative amplitudes have an impact on the geostatistical measures and resultant impedance-porosity calibration?

- Do S-wave reflection images make different predictions of porosity than P-wave images? If so, how might the simultaneous use of both images improve the reservoir models (e.g., porosity, permeability, fluid content, and saturations)?

- Does using the tomogram for the low-frequency impedance, in addition to the low-frequency log traces, alter the inverted impedances? What effect does the low-frequency component have on the flow simulations?

- How does incorporating lithofacies dependencies alter spatial measures, calibrations, and univariate distributions and alter the ultimate models and flow simulations?

- By running a more complete set of realizations, both with and without the crosswell information, do the range and associated probabilities of outcomes from the flow simulator support reduced uncertainty conclusions from the simulator results (e.g., breakthrough times, recovery rates)?

- What is the contribution of crosswell data to improving the utilization of surface 3D seismic data for building 3D reservoir models? What geostatistical methodologies should be followed to integrate well logs, crosswell data, and surface data, while honoring their inherent abilities to predict reservoir distributions?

The final point may be the most important consideration, given that it is possible to survey a large volume of the reservoir using relatively low-resolution 3D surface seismic data. Because it is currently too expensive or impractical to "shoot" across every interwell distance with crosswell seismology, high-resolution crosswell data should be considered as a means "to calibrate" the surface data and the detailed geologic descriptions.

REFERENCES

ARAKTINGI, U. G., BASHORE, W. M., HEWETT, T. A., AND TRAN, T. T. B., 1990, Integration of seismic and well-log data in reservoir modeling: Proceedings from the Third Annual NIPER Conference on Reservoir Characterization, Tulsa, October 5-7.

ARAKTINGI, U. G., AND BASHORE, W. M., 1992, Effects of properties in seismic data on reservoir characterization and consequent fluid-flow predictions when integrated with well logs: Proceedings of the Society of Petroleum Engineers Technical Conference, Washington, D.C., October 4-7, SPE 24572, p. 913-926.

BASHORE, W. M., AND ARAKTINGI, U. G., 1994, Seismic inversion methodology for reservoir modeling: presented at the Middle East Geosciences Conference, GEO94, Manama, Bahrain, April 25-27.

BAMBERGER, A., CHAVENT, G., HEMON, C., AND LAILLY, P., 1982, Inversion of normal incident seismograms: Geophysics, v. 47, p. 757-770.

GRANT, C. W., GOGGIN, D. J., AND HARRIS, P. M., 1994, Outcrop analog for cyclic-shelf reservoirs, San Andres Formation of Permian Basin: stratigraphic framework, permeability distribution, geostatistics, and fluid-flow modeling: American Association of Petroleum Geologists Bulletin, v. 78, no. 1, p. 23-54.

HARRIS, J. M., NOLEN-HOEKSEMA, R. C., LANGAN, R. T., VAN SCHAACK, M., LAZARATOS, S. K., AND RECTOR, J. W., III, 1995, High resolution imaging of a west Texas carbonate reservoir: part 1 - project summary and interpretation: to appear in Geophysics, v. 60, no. 3.

HARRIS, P. M., AND WALKER, S. D., 1988, McElroy Field, Central Basin Platform, U.S. Permian Basin, in Beaumont, E. A., and Foster, N. H., eds., American Association of Petroleum Geologists Treatise of Petroleum Geology: Atlas of Oil and Gas Fields, no. 1, 32 p.

HEWETT, T. A., AND BEHRENS, R. A., 1990, Considerations affecting the scaling of displacements in heterogeneous porous media: Proceedings of the Society of Petroleum Engineers Technical Conference, New Orleans, SPE 20739, September 23-26.

ISAAKS, E. H., AND SRIVASTAVA, R. M., 1989, An introduction to applied geostatistics: New York, Oxford University Press, 561 p.

LAVERGNE, M., AND WILLIAMS, C., 1977, Inversion of seismograms and pseudo velocity logs: Geophysical Prospecting, v. 11, p. 231-250.

LAZARATOS, S. K., RECTOR, J. W., III, HARRIS, J. M., AND VAN SCHAACK, M., 1995, High resolution imaging of a west Texas carbonate reservoir: part 4 - reflection imaging: to appear in Geophysics, v. 60, no. 3.

LINDSETH, R. O., 1979, Synthetic sonic logs - a process for stratigraphic interpretation: Geophysics, v. 44, p. 3-26.

NOLEN-HOEKSEMA, R. C., WANG, Z., HARRIS, J. M., AND LANGAN, R. T., 1995, High resolution imaging of a west Texas carbonate reservoir: part 5 - core analysis: to appear in Geophysics, v. 60, no. 3.

OLDENBURG, D. W., SCHEUER, T., AND LEVY, S., 1983, Recovery of the acoustic impedance from reflection seismograms: Geophysics, v. 48, p. 1318-1337.

OMRE, H., 1991, Stochastic models for reservoir characterization: Norwegian Computing Center.

RECTOR, J. W., III, LAZARATOS, S. K., AND HARRIS, J. M., 1995, High resolution imaging of a west Texas carbonate reservoir: part 3 - wavefield separation: to appear in Geophysics, v. 60, no. 3.

SANTOSA, F., SYMES, W. W., AND RAGGIO, G., 1986, Inversion of band-limited reflection seismograms using stacking velocities as constraints: Inverse Problems, v. 3, p. 477-499.

WALKER, C., AND ULRYCH, T. J., 1982, Autoregressive recovery of the acoustic impedance: Geophysics, v. 47, p. 1160-1173.

WARD, R. F., KENDALL, C. G. ST. C., AND HARRIS, P. M., 1986, Upper Permian (Guadalupian) facies and their association with gydrocarbons - Permian Basin, west Texas and New Mexico: American Association of Petroleum Geologists Bulletin, v. 70, p. 239-262.

WYATT, K. D., AND WYATT, S. B., 1984, Determining subsurface structure using vertical seismic profiling, *in* Toksoz, M. N., and Stewart, R. R., eds., Vertical Seismic Profiling: Advanced Concepts: Geophysical Press, p. 148-176.

XU, W., TRAN, T. T. B., SRIVASTAVA, R. M., AND JOURNEL, A. G., 1992, Integrating seismic data in reservoir modeling: the collocated cokriging alternative: Proceedings of the Society of Petroleum Engineers Technical Conference, Washington, D.C., October 4-7, SPE 24571, p. 833-842.

RESERVOIR CHARACTERIZATION AND MODELING OF THE JURASSIC SMACKOVER AND NORPHLET FORMATIONS, HATTER'S POND UNIT, MOBILE COUNTY, ALABAMA

ELLIOTT P. GINGER, ANDREW R. THOMAS,
W. DAVID GEORGE, AND EMILY L. STOUDT
Texaco E & P Technology Department, Houston, TX 77042

ABSTRACT

Hatter's Pond Field, discovered in 1974, has produced over 241 billion cubic feet of wet gas from the Upper Jurassic Smackover and Norphlet Formations. From 1974 until 1988, the field was produced by pressure depletion. In 1985, the field was unitized to facilitate a gas injection program that began in 1988, aimed at retarding pressure depletion and thus delaying the onset of liquid condensation in the reservoir. This paper describes the geological and reservoir engineering studies involved in building an integrated reservoir model for Hatter's Pond Unit, a model that has been used to evaluate, monitor, and predict reservoir performance under a variety of reservoir management scenarios.

The Norphlet Formation at Hatter's Pond Unit consists of feldspathic lithic arenites deposited mostly in an eolian environment as dunes and interdune facies. Stratigraphically, dune successions are essentially flat lying and correlate across the Unit, and are terminated at the top and bottom by fairly sharp dune succession boundaries. There was essentially no relief on the Hatter's Pond structure at the time of Norphlet Formation deposition. This influenced the way that the reservoir model and subsequent simulation units were constructed. The current distribution of gross reservoir properties in the Norphlet Formation at Hatter's Pond Unit is controlled by this sedimentary template. However, that distribution has been severely modified by diagenetic processes. The quantity of authigenic illite decreases upward away from the gas/water contact, and the process is governed by the water saturation of the sandstone. Illite grows at the expense of feldspar dissolution. Norphlet intergranular volume (IGV) ranges from 16% to 27%. Zones of high IGV usually show low intergranular cement volume, low frequencies of stylolites/foot, and packing heterogeneities suggestive of secondary porosity. A recently dissolved intergranular cement (dolomite or anhydrite) is interpreted to have inhibited compactional processes in high porosity, high IGV zones.

The overlying Smackover Formation carbonates were deposited during the first major marine transgression after the opening of the Gulf of Mexico Basin. At Hatter's Pond Unit, the Smackover Formation consists of shallow marine carbonate deposits; however, texturally destructive dolomitization has obliterated most original rock fabrics and this strong diagenetic overprint makes recognition of original depositional environments difficult. Seven distinctive lithofacies units are recognized, and they occur in three styles of vertical successions. The lower lithofacies are present in all wells in the Unit. Porosity and permeability in the lower units are largely the result of dolomitization and dissolution, and although reservoir quality is somewhat variable, it is significantly improved on the crest of the structure. Tremendous lateral variations occur within the upper Smackover lithofacies, mostly due to early diagenetic events controlled by paleotopography. Reservoir quality in these upper Smackover lithofacies is largely a function of fabric selective, meteoric dissolution of carbonate grains. This dissolution was controlled by the occurrence of localized islands and therefore is unrelated to present–day structure.

A quantitative model of reservoir properties for Hatter's Pond Unit was constructed using Stratamodel's SGM™. Core, log, and production data were used from 27 wells within the Unit to build a stratigraphic framework, a framework that had a robust geological characterization as its foundation. A template was necessary in the upper Smackover Formation in order to bias the attribute interpolation to best match the conceptual geologic model and production data. Four reservoir properties were exported from SGM: porosity, permeability, net pay thickness, and initial water saturation. They were imported into Western Atlas' GeoLink® and vertically averaged into thirteen simulation layers: three for the Smackover Formation and 10 for the Norphlet Formation.

The Hatter's Pond Unit reservoir simulation was conducted using Western Atlas' VIP™ Compositional Model. The system can simulate gas condensate flow behavior, taking into account the fact that fluid properties and phase behavior vary strongly with fluid temperature, pressure, and composition. Fluid properties and phase equilibrium are governed by a generalized cubic equation of state in which reservoir fluid is treated as a mixture containing an arbitrary number of hydrocarbon and non–hydrocarbon components.

Accuracy and usefulness of any reservoir model is heavily dependent on the quality of the underlying geologic model, both qualitative and quantitative. Furthermore, it is critical to retain the character from the geologic model when scaling–up to the chosen simulation grid. Development of a geologically–based reservoir model formed a solid foundation for the Hatter's Pond reservoir simulation model.

INTRODUCTION

Hatter's Pond Unit, located in southwestern Alabama (Fig. 1) was discovered in 1974 with Getty Oil Company's successful completion of the Peter Klein 3-14#1 well. As of 9/1/94, the field had produced 241 billion cu. ft. of wet gas which has yielded 58 million barrels of condensate, 15 million barrels of natural gas liquids, and 168 billion cu. ft. of dry gas from two Upper Jurassic units, the Norphlet Formation (eolian and reworked shallow marine sandstone) and the overlying Smackover Formation (shallow shelf carbonate). The top seal overlying these two reservoirs is the Buckner Formation, which consists of thick evaporites and carbonates deposited in intertidal and supratidal environments. The combined net pay section of the Smackover and Norphlet Formations is a maximum of 415 feet (127 meters) thick at subsea elevations ranging from approximately 18,000 feet (5490 meters) to 18,400 feet (5612 meters).

The field consists of a north–south trending anticline that is approximately seven miles long and three miles wide (12.9 by 4.8 kilometers; Fig. 2). It is bounded on the east by a down–to–the–east fault that is part of the Mobile Graben fault system (see Fig. 1). This fault forms the updip seal for the reservoir, bringing the porous Smackover and Norphlet Formations in contact with the impermeable Buckner Formation evaporites and evaporitic dolostones.

The other significant faulting in the field is an east–west trending graben block located in the southern part of the field (see Fig. 2). The top of the Norphlet Formation is offset by 341 feet (104 meters) in the two wells that border the fault on the north flank of the graben (the HPU 16-9#1 and 21-7#1 wells). The HPU 21-7#1 well is cut by this fault in the middle section of the Norphlet, and it contains an abundance of high–angle fractures (as many as seven/foot), both open and cemented by dolomite, anhydrite and silica. The fault has been modeled as a total flow barrier between the main part of the field to the north and the graben and south fault block. However, pressure data collected during recent drilling of the HPU 21-7#2 well in the central part of the graben indicate there is partial communication across the fault. Offset along the southern margin of the graben is slight less, approximately 200 to 250 feet (61 to 76 meters).

Since its discovery, approximately 49 wells have been drilled in and around Hatter's Pond Unit. Most of these wells were cored extensively. Geologic reservoir characterization has been a continuous process in support of engineering modeling/simulation and infill drilling. The availability and utilization of actual core data has been an invaluable aid in understanding and modeling this complex reservoir.

GEOLOGIC SETTING

Opening of the Gulf of Mexico Basin in early Mesozoic time resulted in the deposition of widespread Triassic and Jurassic age formations in the Gulf Coast (Fig. 3 and Braunstein et al., 1988). In ascending order, the synrift sequence begins with graben–fill sandstones and redbeds of the Triassic Eagle Mills Formation. These, in turn, are unconformably overlain by Middle Jurassic evaporites of the Werner Formation (basal clastics and anhydrite) and Louann Formation (primarily halite). The Louann salt is an extremely important unit in the Gulf Coast because post–depositional loading caused it to develop many low relief to piercement salt structures that are associated with hydrocarbon production.

Unconformably overlying the Louann salt are clastics of the Late Jurassic (Oxfordian) age Norphlet Formation. Typically, the Norphlet consists of reddish to clean, brownish–gray sandstones. Updip feldspathic conglomerates and black, anoxic, basal shales can be local variations. Over much of the western and central Gulf Basin, Norphlet deposits are of fluvial to marine origin with total thicknesses measured in tens of feet, but in southeastern Mississippi, Alabama and Florida, the Norphlet contains a prominent upper unit of eolian dune sands and

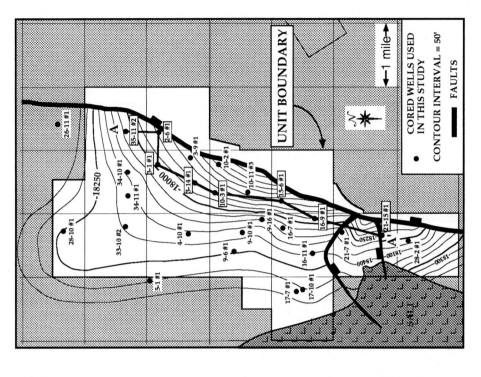

Figure 1. Regional map of southern Alabama showing the location of Hatter's Pond Unit with respect to other regional tectonic features.

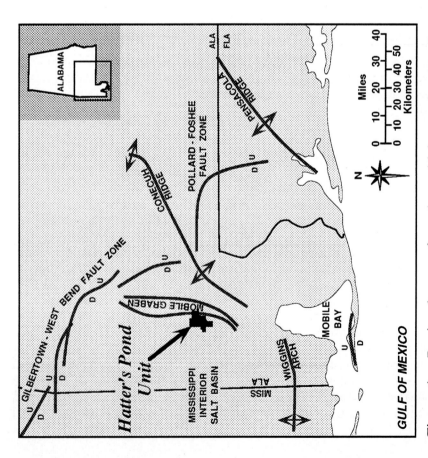

Figure 2. Structure map of Hatter's Pond Unit on top of the Smackover Formation. The black circles show the wells, all of which are coreholes, used to build the reservoir model discussed later in this paper. Note the north-south trending fault (heavy black line) that defines the eastern margin of the field, and the east-west trending faults (heavy black lines toward the bottom) that bound the graben fault block. The cross-section line labeled A–A' is shown in Figure 22.

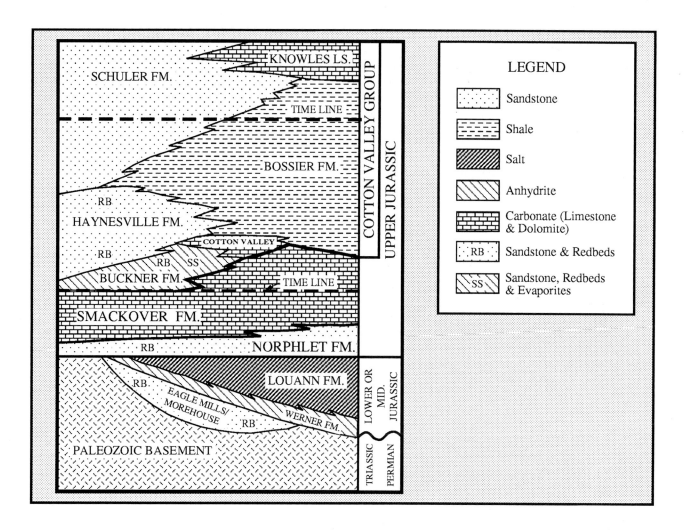

Figure 3. Gulf Coast Permian–Jurassic geologic column (based on Braunstein et al., 1988).

reworked, transgressive marine deposits with thicknesses in excess of 500 feet (153 meters). Norphlet dune sands have good reservoir potential in fields where they are structurally high enough to lie above the gas/ or oil/water contact. Examples of such fields include Hatter's Pond, Flomaton Field in western Florida and several offshore fields in the Mobile Bay area.

The Smackover Formation overlies the Norphlet Formation and was deposited during the first major marine transgression after the opening of the Gulf of Mexico Basin. It consists of transgressive marine deposits, mostly carbonate mudstones and wackestones containing a variety of fossil fragments and scattered argillaceous and/or sandy beds, overlain by shoaling upward (regressive) carbonate phases that includes oolitic, oncolitic and pelletal packstones and grainstones of widespread distribution. These Smackover grainstones are a major hydrocarbon reservoir throughout the upper Gulf Coast.

Over most of the central and western upper Gulf Basin, the total Smackover section averages in excess of 1,000 feet (305 m) thick. In contrast, the Smackover Formation in the eastern Gulf ranges from less than 100 feet (31 m) to a maximum of no more than 500 feet (152 m) thick. During Norphlet and Smackover deposition, the Alabama–Florida shelf seems to have been more restricted and shallower than the remainder of the Gulf rim. Norphlet dunes developed only in this area; Smackover deposition was thinner, contained a more restricted fauna (virtually no corals, for example), and consisted mostly of pellet (rather than oolite) grainstones and packstones.

Smackover carbonates are overlain by, and interfinger updip with, Buckner Formation clastics, evaporites and carbonates. The lower part of the Buckner is an effective seal facies, characterized by nodular mosaic to massive anhydrite with thin interbeds of limestone, dolomite and sandstone. Basinward, Buckner evaporites disappear and Smackover carbonates grade without a distinct break into overlying Cotton Valley deposits of similar lithology (see Fig. 3).

RESERVOIR DESCRIPTION—SMACKOVER FORMATION

Lithofacies and Reservoir Quality

The Smackover Formation at Hatter's Pond Unit (hereafter referred to as HPU) consists of shallow marine carbonate deposits; however, texturally destructive dolomitization has obliterated most original rock fabrics and this strong diagenetic overprint makes recognition of original depositional environments difficult. Seven distinctive lithofacies units are evident in the Smackover in HPU. They occur in 3 types of vertical successions (type 1, type 2, and type 3), that are illustrated in Figure 4. The areal distribution of these succession types is shown in Figure 5. The following paragraphs describe the seven lithofacies in ascending order as observed within the three vertical successions in the Smackover. The lower units (medium crystalline dolomite, finely crystalline dolomite) are present in all wells in the unit. Variations occur in the upper Smackover lithofacies units.

Type 1 vertical succession

Medium crystalline dolomite.—This unit is a distinctly mottled (medium gray to white), anhydritic, stylolitic, medium to coarsely crystalline, anhedral to euhedral dolomite with scattered, irregular shaped, organic–rich patches and rare elliptical "ghosts" that may be relict carbonate grains (Figs. 6A, D and F). The mottled appearance of this unit makes it strikingly different from any other Smackover lithofacies at HPU. The mottling is strongly suggestive of thrombolitic algal sediment binding. Anhydrite occurs as nodules or clusters of felted crystals and/or blades within the dolomite matrix (Fig. 6D). It can also develop along stylolites. The basal few feet of the medium crystalline dolomite contain reworked siliciclastic sand grains and crenulated, subhorizontal organic–rich laminations that are probably algal mats (Figs. 6B and C).

Porosities in the medium crystalline dolomite range from 3–18 percent; they average about 10–14 percent. Pore geometries indicate vuggy and intercrystalline origins; none of the pore shapes suggest molds of preexisting grains (Fig. 6E). Permeabilities are 0.01 md up to 100 md, averaging about 5–15 md.

Finely crystalline dolomite.—As its name implies, this rock consists of a meshwork of finely crystalline, mostly subhedral, stylolitic dolomite with no preserved relicts of original depositional texture other than very rare peloid ghosts (Figs. 7A–D). Typically it displays a uniform dark brownish black color with no porosity visible to the naked eye or under low magnification. Under higher magnification (> 40x) a network of tiny, intercrystalline dolomite pores is sometimes seen (Fig. 7D). The unit often contains scattered anhydrite nodules (Fig. 7B), but anhydrite rarely plugs individual intercrystalline pores

The finely crystalline dolomite can have porous streaks; however, it is not consistently porous and permeable either vertically or laterally. Porosities range from less than three percent to as much as 21 percent, but average 9 to 12 percent. Permeabilities range from 0.01 md to only 50 md and tend to average around 1–5 md. The low permeability of this unit is due to the fact that most

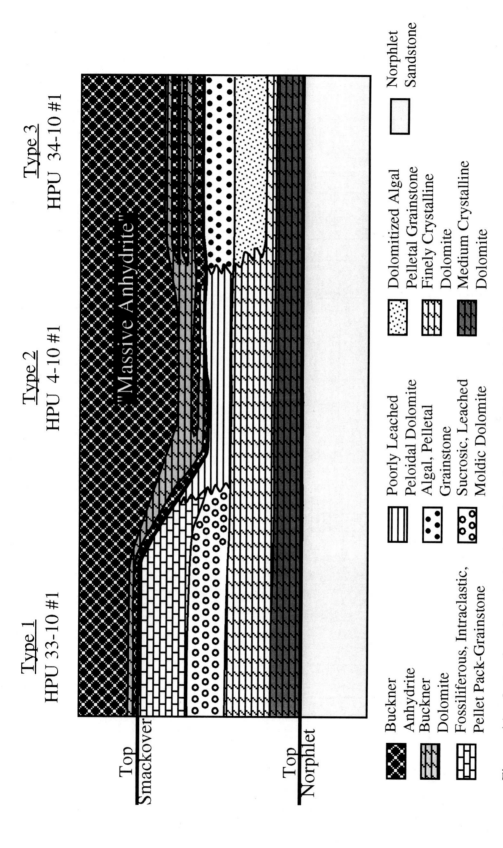

Figure 4A. Lateral and vertical relationships between lithofacies within the Smackover Formation at Hatter's Pond Unit. The contact between the Smackover and underlying Norphlet Formation is very sharp whereas the contact between the Smackover and overlying Buckner Formation is gradational and interfingered. For vertical scale, the sucrosic, leached moldic dolomite in the 33-10#1 well is 35 feet (11 meters) thick.

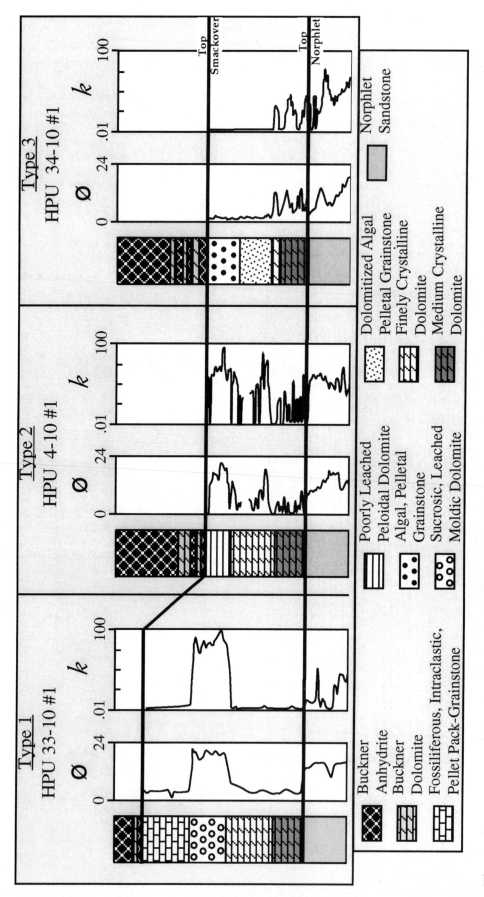

Figure 4B. Relationship of Smackover Formation lithofacies to core-measured porosity and permeability. There is often a significant increase in reservoir quality in the uppermost part of the Norphlet Formation relative to the lower Smackover Formation (i.e., the medium-crystalline dolomite). For vertical scale, the sucrosic, leached moldic dolomite in the 33-10#1 well is 35 feet (11 meters) thick.

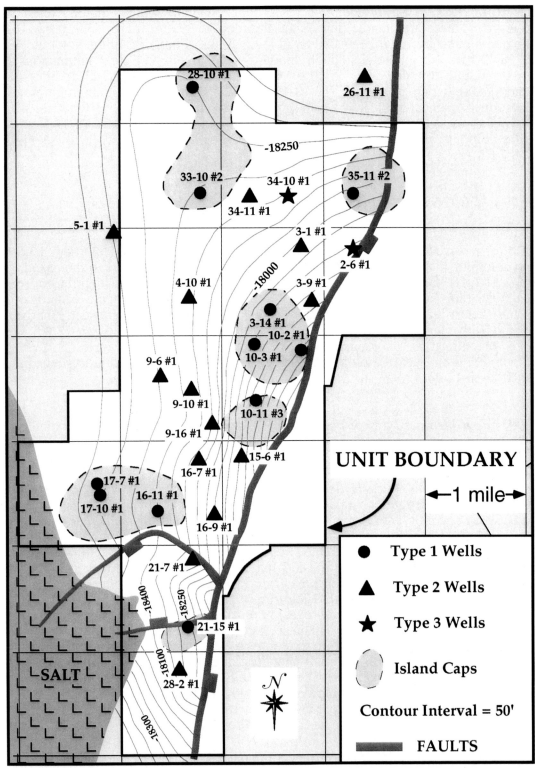

Figure 5. Structure contour map on top of the Smackover Formation at Hatter's Pond Unit. The various well symbols show the areal distribution of the vertical succession types as described in the text. The shaded gray regions with the dashed outline schematically illustrate the distribution of the fossiliferous, intraclastic, pellet pack–grainstone lithofacies which is interpreted to represent an island cap facies that developed over pellet–ooid grain shoals. Their distribution was used to help construct the geologic template that was used in SGM™ to model the upper Smackover Formation (see Figure 38).

Figure 6. *Medium Crystalline Dolomite*. **(A)** slab photo of typical mottled texture in this lithofacies, dark areas are organic rich, ?algal thrombolitic textures, HPU 33-10 #2, 18,435 ft.; **(B)** slab photo of sandy, algal laminated basal interval in medium crystalline dolomite, HPU 33-10 #2, 18,450 ft.; **(C)** photomicrograph showing sand grains and algal laminations in basal medium crystalline dolomite, HPU 33-10 #1, 18,435 ft.; **(D)** anhydrite crystals (white) associated with dark, organic rich patches, HPU 10-11 #3, 17,987 ft.; **(E)** photomicrograph of vuggy porosity (arrows) typical of medium crystalline dolomite in crestal wells, HPU 10-11 #3, 17,997 ft.; **(F)** photomicrograph of nonporous, interlocking dolomite crystals typical of medium crystalline dolomite in downdip well locations, note possible pellet ghosts (arrows), HPU 33-10 #1, 18,402 ft.

Figure 7. *Finely Crystalline Dolomite*. See photographic plate for figure caption.

Figure 8. *Sucrosic, Leached Moldic Dolomite*. **(A & B)** slab photos of light gray, porous, sucrosic dolomite, blackened, hydrocarbon–lined moldic pores are visible to the naked eye, A = HPU 33-10 #2, 18,354 ft., B = HPU 33-10 #2, 18,370 ft.; **(C-E)** photomicrographs of rounded to elliptical pores (arrows) characteristic of this lithofacies, pore shapes indicate leaching of pellets, oolites and skeletal grains; some irregularly shaped pores (p) have experienced enlargement by nonfabric selective leaching, C & D = HPU 10-11 #3, 17,923 ft., E = HPU 10-11 #3, 17,921 ft.; **(F)** high magnification photomicrograph of intercrystalline pores (arrows) that provide connectivity between leached moldic pores in this lithofacies, note euhedral to subhedral crystal shapes, HPU 10-11 #3, 17,908 ft.

Figure 9. *Fossiliferous, Intraclastic, Pelletal Pack/Grainstone*. See photographic plate for figure caption.

Figure 10. *Fossiliferous, Intraclastic, Pelletal Packstone-Grainstone With Exposure Textures*. Photomicrographs of the fabrics, grains and textures observed in this lithofacies; **(A)** stylolitized contact between oncolitic, intraclastic, pelletal packstone with calcite cemented fenestral fabric and reverse graded bedded oncolitic, intraclastic pelletal grainstone, HPU 33-10 #1, 18,301 ft.; **(B)** magnified view of thin section depicted in Figure 10A showing pellets, intraclasts, oncolites including one with bivalve nucleus (O), and foraminifer (arrow) in cemented grainstone; **(C)** fossiliferous, pelletal, intraclastic grainstone from the base of this lithofacies (see slab in Fig. 9D), note all grains are cemented with a thin rim of fibrous calcite, HPU 17-10 #1, 18,368 ft.; **(D)** cement–filled fenestral pores (bird's–eyes) in oncolitic, algal laminated packstone, HPU 33-10 #1, 18,299 ft.; **(E)** magnified view of thin section shown in Figure 10D illustrating bladed rim cement and equant calcite spar fill in fenestral pores; **(F)** magnified view of thin section depicted in Figure 10A illustrating details of small oncolites, large oncolite encrusting bivalve and void filling calcite cements in shelter porosity beneath the large oncolite; **(G)** magnified view of thin section depicted in Figure 10A showing details of void filling cements, including dolomite rim cements (arrows) and blocky calcite spar void fill.

Figure 11. Annotated photographs of continuous slabbed core through the Smackover Formation in a type 1 vertical sequence. Abbreviations used are as follows: Buck = Buckner Formation, Smck = Smackover Formation, Norph = Norphlet Formation, FIPG = Fossiliferous, Intraclastic, Pelletal Packstone-Grainstone with Exposure Textures, SLMD = Sucrosic, Leached Moldic Dolomite, FCD = Finely Crystalline Dolomite, MCD = Medium Crystalline Dolomite.

MEDIUM CRYSTALLINE DOLOMITE LITHOFACIES

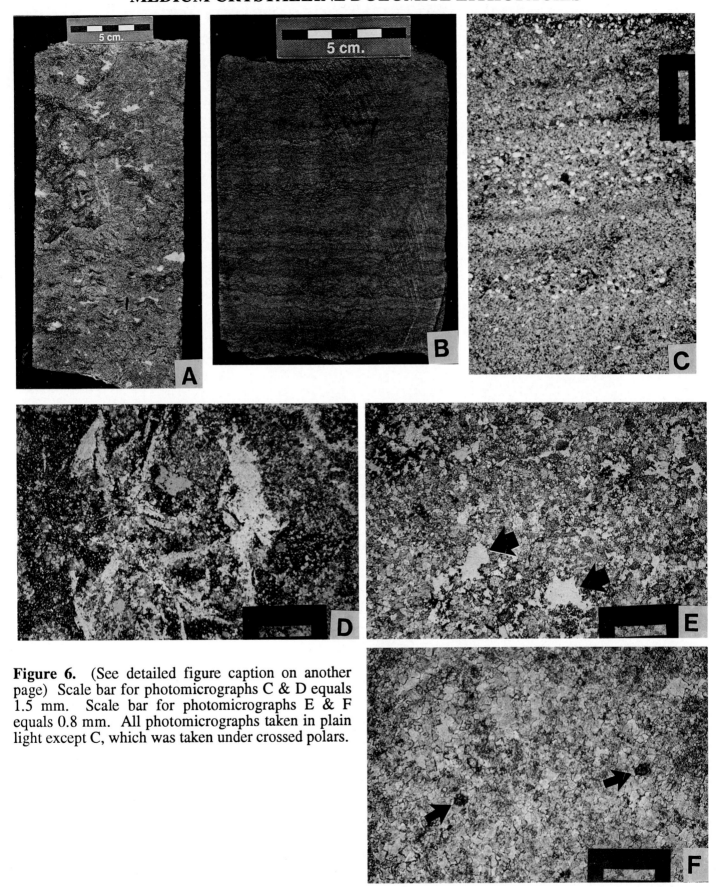

Figure 6. (See detailed figure caption on another page) Scale bar for photomicrographs C & D equals 1.5 mm. Scale bar for photomicrographs E & F equals 0.8 mm. All photomicrographs taken in plain light except C, which was taken under crossed polars.

FINELY CRYSTALLINE DOLOMITE LITHOFACIES

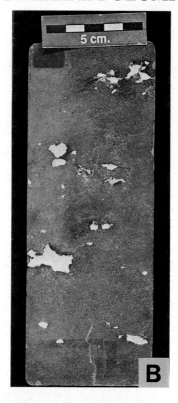
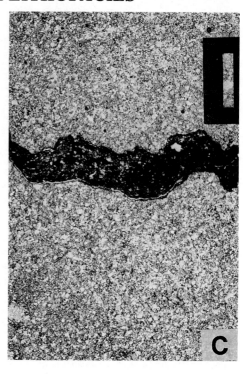
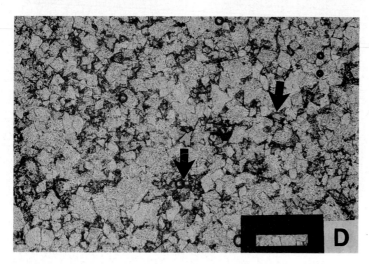

Scale bar for photomicrograph C equals 0.8 mm. Scale bar for photomicrograph D equals 0.2 mm. All photomicrographs taken in plain light.

Figure 7. *Finely Crystalline Dolomite.* (**A**) Slab photo of brownish black, textureless dolomite typical of this lithofacies, HPU 33-10 #1, 18,377 ft.; (**B**) slab photo of white anhydrite nodules in finely crystalline dolomite, HPU 33-10 #1, 18,387 ft.; (**C**) photomicrograph of crystals in finely crystalline dolomite, compare crystal size and shape to medium crystalline dolomite (Fig. 6E & F) taken at the same magnification, HPU 10-11 #3, 17,958 ft.; (**D**) photomicrograph of intercrystalline pores (arrows), note high magnification required to illustrate these pore types, HPU 10-11 #3, 17,935 ft.

SUCROSIC, LEACHED MOLDIC DOLOMITE LITHOFACIES

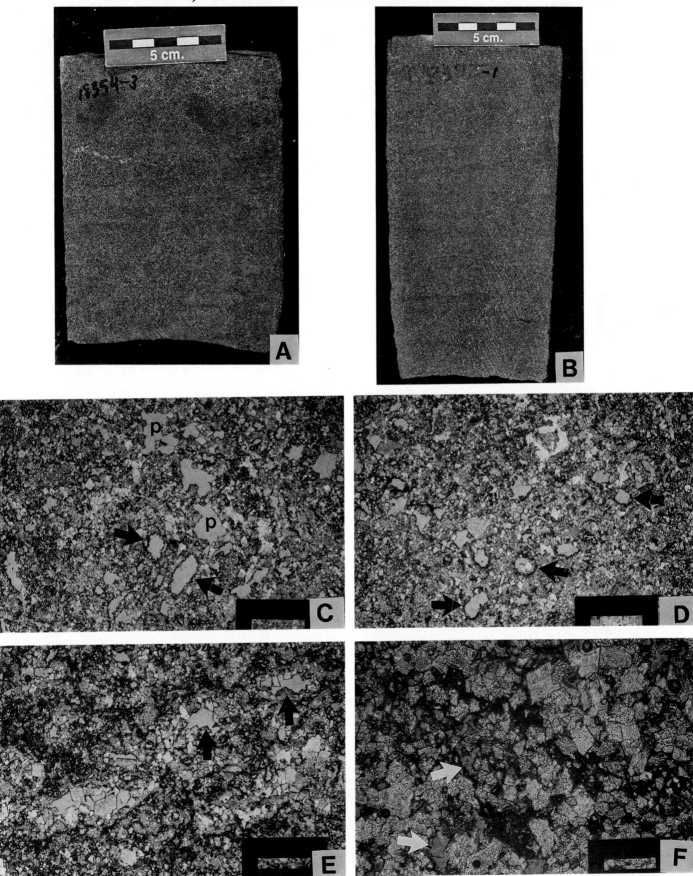

Figure 8. (See detailed figure caption on another page) Scale bar for photomicrographs C-E equals 0.8 mm. Scale bar for photomicrograph F equals 0.2 mm. All photomicrographs taken in plain light.

FOSSILIFEROUS, INTRACLASTIC, PELLETAL GRAINSTONE LITHOFACIES

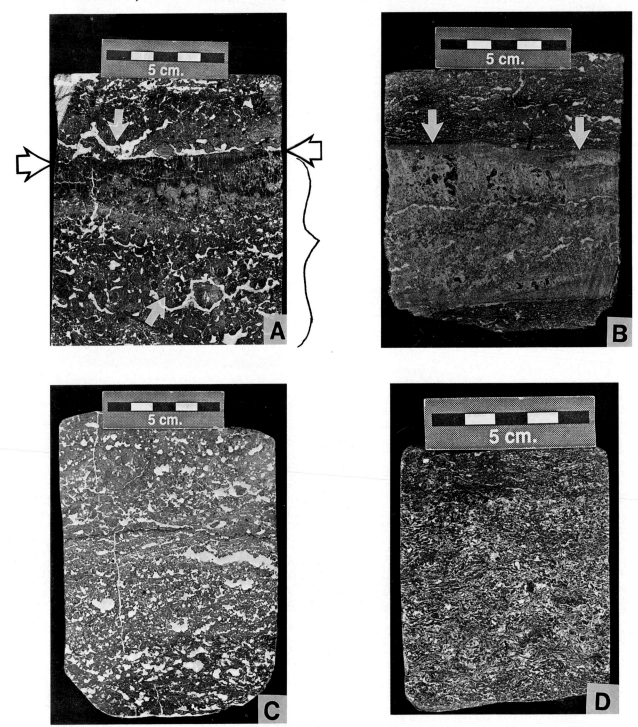

Figure 9. *Fossiliferous, Intraclastic, Pelletal Packstone-Grainstone With Exposure Textures.* (**A-D**) slab photos of typical fabrics & grains in this lithofacies; (**A**) oncolites and pisolites with dripstone cements (arrows) and cement filled fenestral fabrics, surface marked with large arrows is sharp and contains vertical, hairline shrinkage cracks. Bed marked by '}' displays normal (coarse–to–fine) grading of pisolites to pellets, HPU 3-14 #1, 18,026 ft.; (**B**) "hardground" of tightly cemented oncolitic, intraclastic, pelletal grainstone (upper surface marked by arrows), overlain by pelletal, oncolitic packstone with fenestral textures, HPU 33-10 #2 18,340 ft.; (**C**) bird's–eye (fenestral) pores filled with white, equant calcite spar cement, matrix is composed of intraclastic, oncolitic, pelletal packstone-grainstone, HPU 28-10 #1, 18,469 ft.; (**D**) fossiliferous, intraclastic, pelletal grainstone from near the base of this lithofacies, HPU 17-10 #1, 18,361 ft.

FOSSILIFEROUS, INTRACLASTIC, PELLETAL GRAINSTONE LITHOFACIES

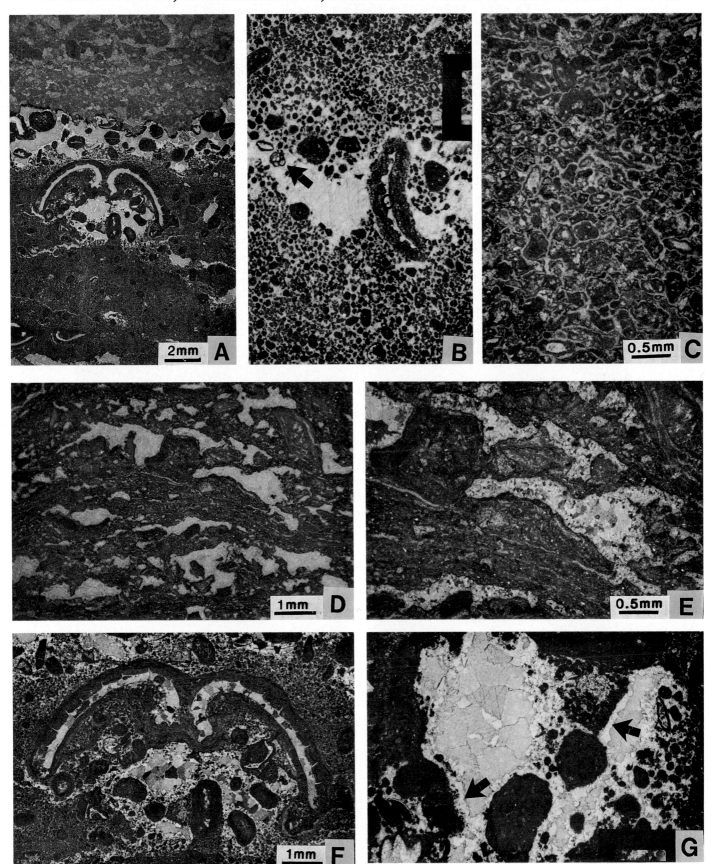

Figure 10. (See detailed figure caption on another page) Scale bar for photomicrographs B & G equals 0.8 mm. Photomicrographs E & F were taken under crossed polars, all other photomicrographs were taken in plain light.

HATTER'S POND UNIT 33-10 #2
MOBILE COUNTY, ALABAMA

Figure 11A

HATTER'S POND UNIT 33-10 #2
MOBILE COUNTY, ALABAMA

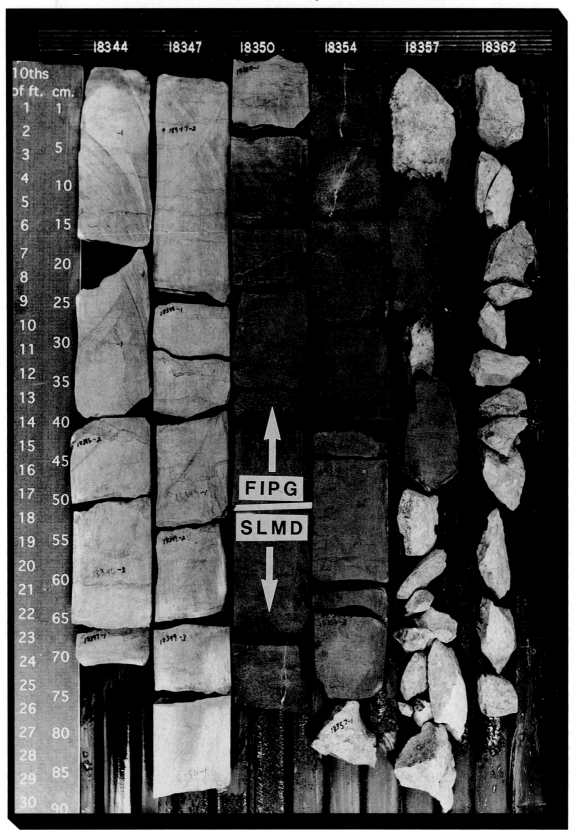

Figure 11B

HATTER'S POND UNIT 33-10 #2
MOBILE COUNTY, ALABAMA

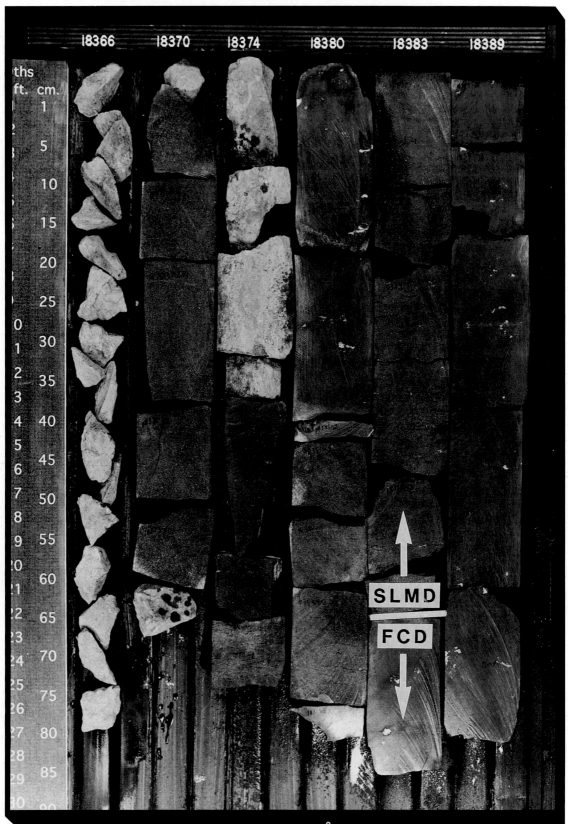

Figure 11C

HATTER'S POND UNIT 33-10 #2
MOBILE COUNTY, ALABAMA

Figure 11D

HATTER'S POND UNIT 33-10 #2
MOBILE COUNTY, ALABAMA

Figure 11E

HATTER'S POND UNIT 33-10 #2
MOBILE COUNTY, ALABAMA

Figure 11F

porosity is intercrystalline, and the small crystal size of the finely crystalline dolomite results in very small pore throats (Fig. 7D).

Sucrosic, leached moldic dolomite.—This unit consists of medium gray, medium to coarsely crystalline, euhedral to subhedral dolomite with numerous rounded to elliptical or elongate pore spaces that are large enough (> 0.2mm) to be clearly visible to the naked eye. These pores appear to be the leached molds of pellets and possibly scattered oolites and fossil fragments (Figs. 8A–E). Abundant intercrystalline pores between euhedral dolomite crystals provide connectivity pathways between the moldic pores (Fig. 8F). The rock is virtually 100 percent dolomite, with only scattered traces of blocky calcite and anhydrite filling moldic pores. Porosity in this unit ranges from 9–21 percent (averaging 15 percent); permeability ranges from 1–100 md and averages 10–50 md. *The sucrosic, leached moldic dolomite constitutes the best reservoir rock in the Smackover Formation in HPU.*

Fossiliferous, intraclastic, pelletal packstone/grainstone with exposure textures.—This lithofacies is one of only two in the Smackover formation at HPU that is largely undolomitized limestone. It consists of medium dark gray, pelletal, oolitic, pisolitic, oncolitic, intraclastic, fossiliferous packstone–grainstone (Figs. 9 and 10). Nonskeletal grains predominate; skeletal grains include scattered gastropods, pelecypods, ostracodes, echinoderm plates, green algae and foraminifera. Blocky, equant calcite spar is the most common cement. It fills sheet cracks, bird's-eye pores and leached moldic porosity in fossil fragments (Figs. 9A and C, 10A, B, D–G). Under cathodoluminescence, some of the blocky spars display numerous zonations. In places, particularly in the lower part of this unit, fibrous to bladed, isopachous, iron–free calcite rim cements are developed on most grains (Figs. 9D and 10C). Scattered (10–20 percent), fine to medium crystalline, euhedral to subhedral dolomite rhombs and traces of blocky, coarsely crystalline anhydrite occur, but this unit is principally limestone. Porosity and permeability are 1–3 percent and 0.01 to 0.1 millidarcies, respectively.

The most distinctive aspect of this unit, regardless of its original packstone to grainstone fabric, is the strong diagenetic overprint of exposure textures. These include: 1) fenestral (bird's-eye) porosity that is infilled by blocky calcite spar cement (Figs. 9A and C; 10D and E); 2) possible "vadose" pisolites; 3) bearded, dripstone cements on many large grains (Fig 9A); and 4) micritic layers with shrinkage and desiccation cracks (Fig. 9A).

Wells that display the type 1 vertical succession are shown in Figure 5. They include the HPU 28-10#1, 33-10#2, 35-11#2, 3-14#1, 10-2#1, 10-3#1, 10-11#3, 17-7#1, 17-10#1, 16-11#1, and 21-15#1 wells. Figure 11 shows slabbed core photographs through the entire type 1 vertical section in the HPU 33-10#2 well.

Type 2 vertical succession

The lower two units of this vertical succession are the medium crystalline and finely crystalline dolomite. Their characteristics were discussed under the type 1 vertical succession.

Poorly leached, peloidal dolomite.—This unit represents the upper Smackover in wells that display a type 2 vertical succession. Lithologically it is a medium dark gray, finely to coarsely crystalline dolomite with anhedral to euhedral crystals (Figs. 12A–F). There are scattered ghosts of pellets and other grains; but in most samples the original depositional texture has been obliterated. Porosity and permeability in the poorly leached, peloidal dolomite are highly variable. As implied by the name, most of this unit contains very little leached moldic porosity. In many places, the unit consists of a tightly interlocking meshwork of dolomite crystals that is virtually nonporous and impermeable. However, there are stringers or zones in this unit in which leached moldic and intercrystalline porosity do occur (Figs. 12E and F). The porosity in these streaks can be as high

Figure 12. *Poorly Leached, Peloidal Dolomite.* **(A & B)** slab photos of dark gray to black dolomite with scattered visible pores, except for larger crystal size, this lithofacies is not significantly different in appearance from the finely crystalline dolomite lithofacies that underlies it, A = HPU 16-9 #1, 18,039 ft.; B = HPU 16-9 #1, 18042 ft.; **(C & D)** photomicrographs of subhedral to anhedral dolomite in this lithofacies, peloid ghosts are rarely evident (arrow), but very few leached moldic pores are observed, compare to Figures 8C-E, taken at the same magnification, and note lack of open pores, C = HPU 15-6 #1, 17,931 ft., D = HPU 15-6 #1, 17,935 ft.; **(E & F)** magnified views of C & D respectively, showing numerous intercrystalline pores (arrows), but no moldic porosity.

Figure 13. Annotated photographs of continuous slabbed core through the Smackover Formation in a type 2 vertical sequence. Abbreviations used are as follows: Buck = Buckner Formation, Smck = Smackover Formation, Norph = Norphlet Formation, PLPD = Poorly Leached, Peloidal Dolomite, FCD = Finely Crystalline Dolomite, MCD = Medium Crystalline Dolomite.

Figure 14. *Dolomitic, Algal, Pelletal Grainstone To Boundstone.* See photographic plate for figure caption.

Figure 15. *Algal, Pelletal Grainstone To Boundstone.* **(A & B)** slab photos of this lithofacies, algally bound areas indicated by arrows, A = HPU 2-6 #1, 17,957 ft., B = HPU 2-6 #1, 17,965 ft.; **(C)** photomicrograph of grain types in a non–algally bound grainstone area, grains include intraclasts, oncolites, pellets, and an algally coated bivalve fragment (arrow), note thin rims of fibrous calcite surrounding each grain, scattered dolomite rhombs and anhydrite crystals appear white, HPU 2-6 #1, 17,958 ft.; **(D)** photomicrograph of margin of algal head (arrows) and non–algally bound, pelletal grainstone sediment, HPU 34-10 #1, 18,246 ft.; **(E)** magnified view of algal head depicted in Figure 15D, showing clotted pelletal texture in algal material; **(F)** photomicrograph of intraclasts and algal clasts, each surrounded by thin rim cements of fibrous calcite, HPU 2-6 #1, 17,960 ft.; **(G)** high magnification photomicrograph of several generations of cements in grainstone, note bivalve in center of photo, it contains both an early generation of fibrous rim cement and void filling calcite spar in the center. Similar spar fills any voids remaining after precipitation of the fibrous rim cements. Clear rhombs are dolomite, HPU 2-6 #1, 17,961 ft..

Figure 16. *Buckner Formation Seal Facies.* (A-D) slab photos of typical Buckner lithofacies, **(A & B)** = nodular, chickenwire anhydrite with dark colored, terrigenous carbonate between anhydrite nodules, A = HPU 4-10 #1, 18,277 ft.; B = HPU 5-1 #1. 18,427 ft.; **(C)** bedded to massive anhydrite with very little carbonate "matrix", HPU 16-11 #1, 18,196 ft.; **(D)** reddish colored, argillaceous siltstone to silty claystone, tiny white patches are anhydrite nodules, HPU 4-10 #1, 18,249 ft.; **(E)** photomicrograph of silty claystone similar to slab shown in Figure 16D, white specs are quartz silt & sand, HPU 33-10 #1, 18,263 ft.; **(F)** photomicrograph of chickenwire anhydrite nodules similar to those in Figures 16A and B, note finely felted texture to anhydrite in nodules, HPU 33-10 #1, 18,279 ft.; **(G)** photomicrograph of pellets in Buckner carbonate bed, note excellent preservation of original depositional textures in these dolomitized grainstones. Some pellets are filled with anhydrite cement. HPU 33-10 #1, 18,297 ft..

POORLY LEACHED, PELOIDAL DOLOMITE LITHOFACIES

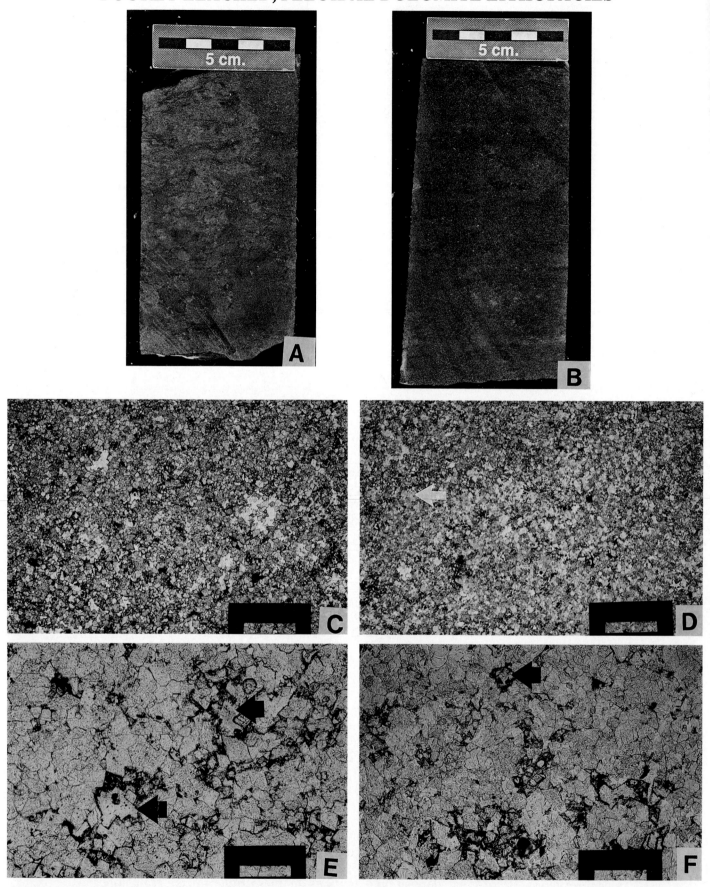

Figure 12. (See detailed figure caption on another page) Scale bar for photomicrographs C & D equals 0.8 mm. Scale bar for photomicrographs E & F equals 0.2 mm. All photomicrographs taken in plain light.

HATTER'S POND UNIT 16-9 #1
MOBILE COUNTY, ALABAMA

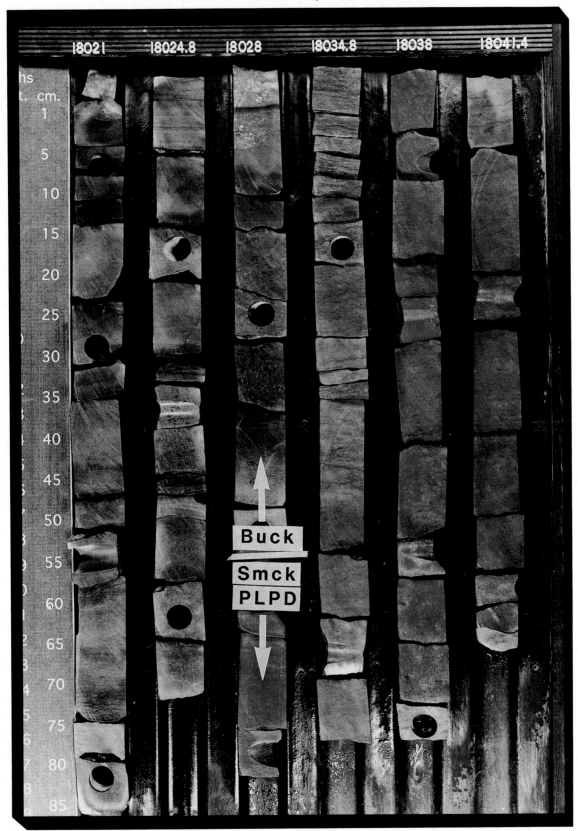

Figure 13A

HATTER'S POND UNIT 16-9 #1
MOBILE COUNTY, ALABAMA

Figure 13B

HATTER'S POND UNIT 16-9 #1
MOBILE COUNTY, ALABAMA

Figure 13C

HATTER'S POND UNIT 16-9 #1
MOBILE COUNTY, ALABAMA

Figure 13D

HATTER'S POND UNIT 16-9 #1
MOBILE COUNTY, ALABAMA

Figure 13E

DOLOMITIC, ALGAL PELLETAL GRAINSTONE LITHOFACIES

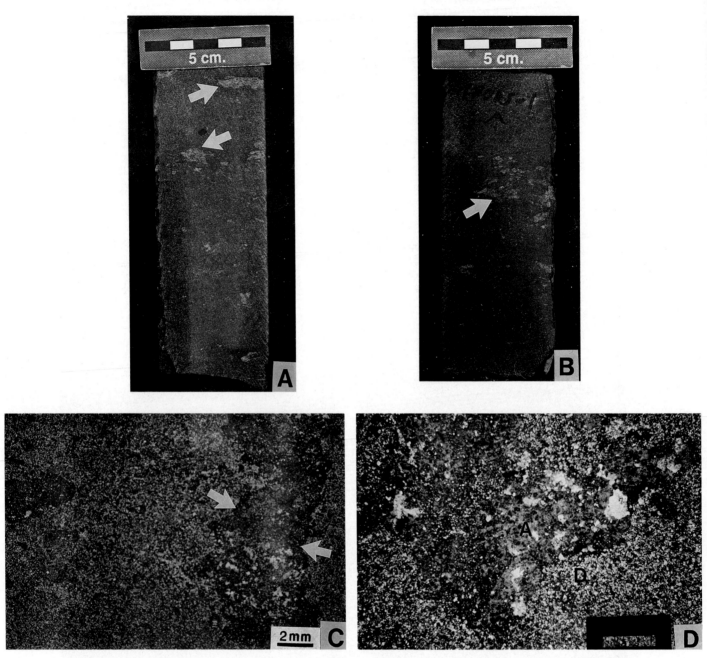

Scale bar for photomicrograph D equals 1.5 mm. All photomicrographs taken in plain light.

Figure 14. *Dolomitic, Algal Pelletal Grainstone.* **(A & B)** slab photos of this lithofacies, note only the scattered calcareous patches (arrows) display relict depositional textures, A = HPU 2-6 #1, 17,985 ft., B = HPU 2-6 #1, 18,001 ft.; **(C)** low magnification photomicrograph of relict algally bound, pelletal packstone (boundaries marked by arrows) in otherwise dolomitized material, HPU 34-10 #1, 18,266 ft.; **(D)** photomicrograph of dolomitized (d) fabric with patch of relict algally bound packstone (a), this algally bound material is identical to algal textures in the overlying algal, pelletal grainstone (Figs. 15C and D), HPU 2-6 #1, 17,985 ft.

ALGAL PELLETAL GRAINSTONE LITHOFACIES

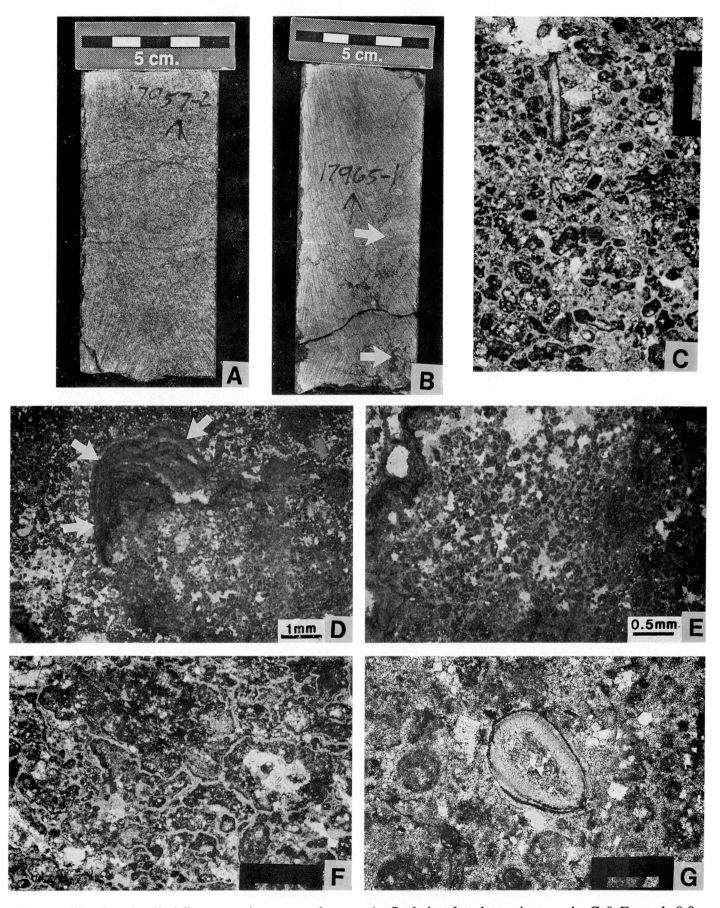

Figure 15. (See detailed figure caption on another page), Scale bar for photomicrographs C & F equals 0.8 mm. Scale bar for photomicrograph G equals 0.2 mm. All photomicrographs were taken in plain light.

BUCKNER FORMATION LITHOFACIES

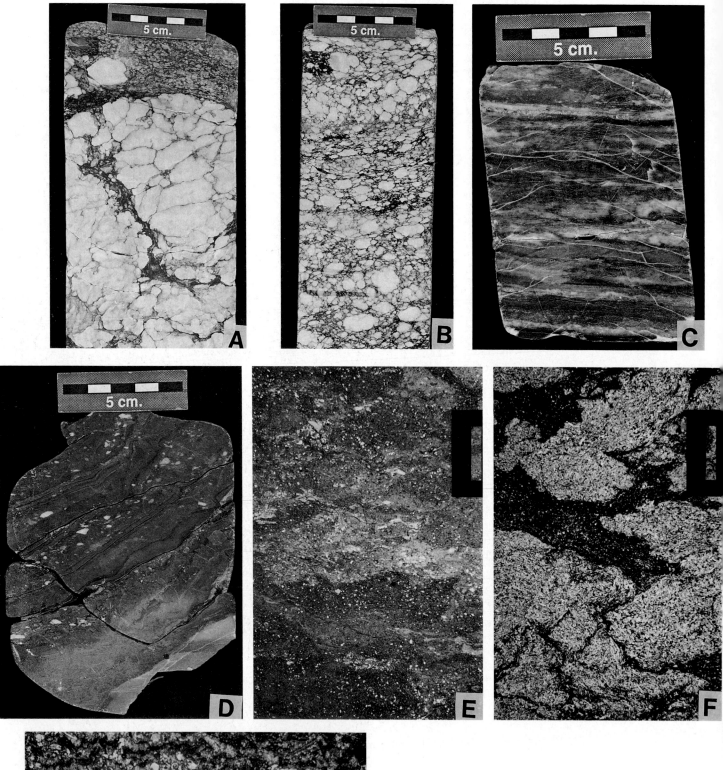

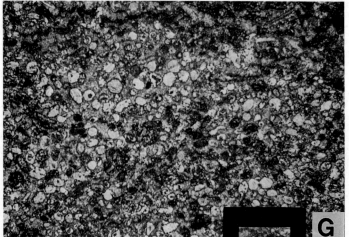

Figure 16. (See detailed figure caption on another page) Scale bar for photomicrographs E-G is 1.5 mm. Photomicrograph F was taken under crossed polars, photomicrographs E & G were taken in plain light.

as 15–18 percent, but the permeability rarely exceeds 10 md and averages about 1–2 md.

As illustrated in Figure 5, the type 2 vertical succession is the most common in the unit. It occurs in the HPU 34-11#1, 5-1#1, 4-10#1, 3-1#1, 3-9#1, 9-6#1, 9-10#1, 9-16#1, 16-7#1, 16-9#1, 15-6#1, 21-7#1, and 28-2#1 wells. Figure 13 shows slabbed core photographs through a typical type 2 vertical section from the HPU 16-9#1 well.

Type 3 vertical succession

As in the type 1 and type 2 examples, the lower two units of this vertical succession are the medium crystalline and finely crystalline dolomite. They are overlain by two units, distinguished from each other on the basis of extensive dolomitization in the deeper unit and sparse dolomitization in the shallower unit.

Dolomitic, algal pelletal grainstone.—This unit consists of very dolomitic, medium to dark gray, sparsely fossiliferous pellet–peloid packstone to grainstone (Figs. 14A–D). It is texturally very similar to the overlying algal pelletal grainstone, but it contains much more dolomite (average 70–90 percent). Even though dolomitization is nearly complete in this unit, porosity (1–3 percent) and permeability (< 0.01 md) remain very low.

Algal pelletal grainstone.—This is the other lithofacies in the Smackover Formation at HPU that is largely limestone. It consists of medium to dark gray, wispy, stylolitic, sparsely fossiliferous pellet–peloid grainstone to packstone with zones of oncolites and intraclasts (Figs. 15A–G). Many of the intraclasts are algally bound sediment aggregates. Fossils include ostracodes, echinoderm plates and mollusk fragments (Figs. 15C and G).

There are two, and in some places three, generations of calcium carbonate cements in this unit. All grains are surrounded by well developed isopachous rims of iron–free calcite cement (Figs. 15C, F and G). After rim cementation, pores that remained as intergranular voids or internal fossil molds were filled by a slightly ferroan, equant sparry calcite (Fig. 15G). In a few places, small fractures or remaining open porosity were filled with blocky, iron–free calcite that appears to postdate the slightly ferroan calcite.

The algal, pelletal grainstone unit contains dolomite in amounts varying from about 20 percent to as much as 70 percent. Dolomite content does not increase consistently downward. Instead, there are several heavily dolomitized intervals, separated by others that are less dolomitic. Dolomite can be divided into several types: 1) a finely crystalline to scattered medium crystalline, euhedral to subhedral dolomite that replaces grains and matrix and represents the bulk of the existing dolomite; 2) a medium to coarsely crystalline, euhedral dolomite, often with a final zone of ferroan dolomite along its outer crystal margins, that replaces scattered patches of anhydrite and/or grows into pores; and 3) scattered medium to coarsely crystalline, subhedral to anhedral, baroque dolomite that replaces limestone and develops near stylolites. Dolomitization has not produced or enhanced porosity, and this unit is virtually nonporous and impermeable (1–3 percent porosity; < 0.01 md permeability).

The type 3 vertical succession occurs in only two wells in HPU. They are the HPU 34-10#1 and 2-6#1 wells.

Relationships with Smackover Bounding Units

At HPU, the Smackover Formation is underlain by reworked marine transgressive sandstones of the Norphlet Formation (described below). The contact between the Norphlet and overlying

Smackover is sharp. Thin stringers and scattered, isolated grains of Norphlet–type siliciclastics are present in the basal few feet (1 m) of the Smackover.

The Smackover Formation is overlain by carbonates and evaporites of the Buckner Formation (Figs. 16A–G). The contact is essentially conformable, although local exposure and brief hiatuses undoubtedly occurred because of the regressive nature of the upper Smackover and overlying Buckner. Since formation tops are often picked from electric logs, some distinctive log break is typically used to pick the Smackover–Buckner contact. Some geologists use the top of significant carbonate porosity as the top of the Smackover, others place it at the base of the first distinctive anhydrite bed.

A slightly different method for picking the Smackover–Buckner contact is suggested for Hatter's Pond. Clearly units above the base of the "massive anhydrite" (Figs. 4A, 16A–C, F) would be included in the Buckner Formation and the porous sucrosic leached moldic dolomites and underlying units are part of the Smackover (Fig. 4). Lithologies subject to question are the limestones, dolomites and scattered thin anhydrite beds between these two markers.

A pronounced difference in the style of dolomitization was found between "typical" Smackover units (sucrosic, leached moldic dolomite, and underlying rocks) and interbedded dolomites and anhydrites just below the Buckner "massive anhydrite" (Fig. 4). Smackover dolomites are generally medium to coarsely crystalline, typically with euhedral to subhedral crystals. Commonly, original depositional textures have been obliterated and very few ghosts of grains are easily distinguished. In addition, anhydrite is relatively rare in the Smackover dolomite, except as scattered small nodules. It does not develop as poikilotopic cement and/or replacement crystals. In marked contrast, dolomites just below the Buckner "massive anhydrite" are finely to very finely crystalline with subhedral to anhedral crystals. They display excellent preservation of original textures, containing many allochem ghosts and bedding features (Fig. 16G). Anhydrite is very abundant (as much as 50 percent) in these dolomites and it frequently forms as poikilotopic cement and replacement crystals as well as nodules (Fig. 16G).

It is suspected that the differences in dolomite and anhydrite style reflect different diagenetic conditions and modes of origin of the dolomites. Vinet (1984) saw similar differences at Jay–Big Escambia Creek Fields (southern Alabama – western Florida). He attributed the finely crystalline, texturally–preserving dolomites to supratidal (sabkha) dolomitization and the medium–coarsely crystalline, texturally–destroying dolomites to mixing zone dolomitization. Geochemical data support his hypotheses.

Since the finely crystalline dolomites probably have a sabkha origin, they are placed in the Buckner Formation to distinguish them from the Smackover dolomites. Thick limestone units, with or without exposure textures, that are not interbedded with and do not contain "sabkha–type" dolomites, are assigned to the Smackover Formation. Such units are present in the HPU 33-10#2 well and the HPU 34-10#1 well (Fig. 4).

Depositional Environment of the Smackover Formation

Most of the Smackover Formation at Hatter's Pond is completely dolomitized. This makes recognition of original depositional textures and environments difficult. The following hypotheses concerning depositional environment are based on varied amounts of evidence.

The lowermost Smackover lithofacies, the medium crystalline dolomite, is present everywhere in HPU. It contains reworked or wind deposited Norphlet clastics in its basal few feet, some horizontal algal laminations, and scattered peloid ghosts. It displays an overall "mottled" texture that is very similar in appearance to better preserved algal wackestones to boundstones that have been observed in other Smackover cores from the eastern Gulf area (Baria et al, 1982). This unit probably represents shallow, somewhat abnormally saline, tidal flat to lagoonal conditions that developed as the Smackover seas deepened over the marine–reworked Norphlet sandstones. Blue–green algal activity was common, some peloids (pellets, oncolites, algally coated clasts) were present (Fig. 17).

The finely crystalline dolomite overlies the medium crystalline dolomite. Thickness variations in this lithofacies are present across the unit. It is identified from diagenetic characteristics, such as crystal size, color, lack of allochem ghosts, lack of significant porosity and permeability. The aforementioned diagenetic characteristics probably reflect a change in the original sediment texture. Lacking any clear original depositional fabrics, it is suggested that the finely crystalline dolomite was deposited as a lime mudstone or wackestone with very few allochems. The fine dolomite crystal size may reflect the fine grain size of the original mud. It probably represents the maximum water depth during the Smackover transgression, as overlying units reflect a shoaling upward (regressive) depositional environment (Fig. 17).

Overlying the finely crystalline dolomites are several possible vertical sequences (Fig. 4). Regardless of the vertical sequence, all overlying units reflect a shoaling upward, higher energy, less muddy, more allochem–rich environment.

Best preservation of overlying lithofacies occurs in units like the algal pelletal grainstones of the HPU 34-10#1 well, where dolomitization is incomplete. These units are packstones and grainstones with primarily nonskeletal allochems (pellets, peloids, oncolites, algally–bound intraclasts) and well developed, fibrous to acicular rim cements. Algal and algally–coated clasts are very common. These units probably reflect very shallow, high energy, submarine deposits. They lack exposure textures, contain apparent marine cements, and have a preponderance of slightly deeper water allochems such as algal balls. In general, they constitute a thinner depositional sequence (Figs. 4 and 14) than adjacent pellet shoals. Extensive early marine cementation plugged intergranular porosity, preventing dolomitizing fluids from entering, and reacting with, this lithofacies.

Units that are apparently laterally equivalent to the algal pelletal grainstones are the poorly leached, peloidal dolomite (HPU 4-10#1 well) and the sucrosic, leached moldic dolomite (HPU 33-10#2 well). The poorly leached, peloidal dolomite retains scattered pellet and peloid ghosts and some leached molds of the same size and shape as the pellets and peloids. In places, the density of preserved allochem ghosts suggests an original packstone–grainstone texture. This unit represents a poorly leached, flank–equivalent of the pellet bars represented by the sucrosic, leached moldic dolomite (Figs. 4 and 17).

The sucrosic, leached moldic dolomite displays very few allochems ghosts because most of the grains have been leached away. Size and shape of the resulting molds (average range 0.3 – 0.8 mm) suggests that the grains were originally mostly pellets and not oolites. There are scattered molds of fossil fragments, and a few moldic pores that are large enough to have been oncolites or algally coated intraclasts similar to those seen in the algal, pelletal grainstone. Packing density of the molds indicates a grain supported texture; the presence or absence of mud or early rim cements is uncertain due to complete dolomitization. Cross–bedding is only sporadically observed in this unit, probably because of the very uniform grain size. The sucrosic, leached moldic dolomite has a patchy, discontinuous distribution within HPU.

The distribution, packing density, mold size and shape, and relationship to the overlying pellet pack–grainstone with exposure textures indicates that the original depositional environment of the sucrosic, leached moldic dolomite was as pellet bars or shoals, similar to modern pellet–ooid bar complexes widely distributed in the Bahamas and Persian Gulf.

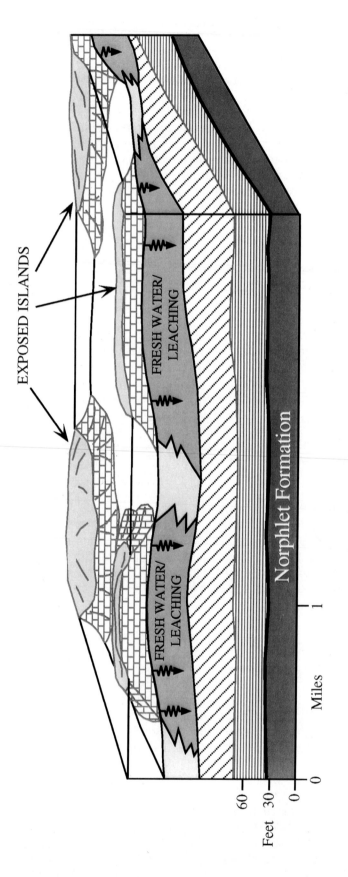

Figure 17. Schematic diagram showing the areal distribution of Smackover Formation lithofacies and the depositional environment for the upper Smackover Formation at Hatter's Pond Unit.

Fossiliferous, intraclastic, pellet pack–grainstone caps the sucrosic, leached moldic dolomite. The two units occur as a couplet. Where one is present, the other also occurs. Textures are well preserved in the pellet pack/grainstone because it contains very little dolomite. The predominant grains are pellets, but there are also abundant intraclasts, oncolites, algal–coated grains, and fossils (mostly pelecypods, gastropods, ostracodes, and echinoderm plates). The fauna and intraclasts suggest very shallow, somewhat salinity–restricted conditions in the shallow subtidal to intertidal range. Superimposed on the rock are a number of features such as inorganic pisolites, bird's-eye textures, micritic or coarse calcitic "crusts", desiccation shrinkage cracks, and brecciation that strongly suggest subaerial exposure. These sediments appear to represent island caps that formed on top of the pellet bar shoals (Figure 17).

Intertidal and supratidal (arid sabkha) conditions, exemplified by Buckner carbonates and evaporites, prograded over the upper Smackover regressive carbonate deposits.

Diagenesis of the Smackover Formation

Porosity and permeability in the Smackover Formation are entirely secondary, all primary pore space has been filled with calcite or anhydrite cement. Secondary pores result from varying combinations of fabric selective and non–fabric selective dissolution, dolomitization, and fracturing. The most important diagenetic phenomena responsible for these pore networks are dissolution and dolomitization.

Dissolution

Both fabric selective and non–fabric selective pores exist. Fabric selective pores were created by an early diagenetic leaching event. Dissolution resulted from flushing non–marine (fresh) water through limestones prior to carbonate mineral stabilization, when metastable minerals such as aragonite were selectively dissolved. Prominent fabric selective pores are evident in the sucrosic, leached moldic dolomite, where virtually all carbonate grains have been dissolved (see Fig. 8C and D). Fabric selective porosity is mostly confined to areas under, within, and immediately adjacent to island caps, suggesting that the leaching fluids that produced such pores came from fresh water lenses developed within the islands. Non–fabric selective pores occur locally in most of the Smackover dolomites, most prominently in parts of the medium crystalline dolomite along the eastern side (crest) of the Hatter's Pond structure (see Fig. 6E). These pores cut across dolomite crystal fabric and pre–existing voids. Their mode of origin appears to be leaching under deep burial conditions, perhaps by basinal brines that moved upward along the Mobile Graben fault system.

Dolomitization

Dolomite has completely replaced the original Smackover Formation limestone in all units except the island cap and the algal, pelletal grainstone. Intercrystalline pores between dolomite rhombs contribute to both porosity and particularly permeability in the reservoir (see Figs. 7D, 8F, and 12D–F). Geochemical data (Prather, 1992) indicate that many Smackover Formation dolomites are the result of mixing zone and/or burial dolomitization. Leaching of any remaining calcite between dolomite rhombs produced the existing pore network.

RESERVOIR DESCRIPTION—NORPHLET FORMATION

Depositional Environment of the Norphlet Formation

The Norphlet Formation in the HPU area is almost entirely eolian dune and interdune facies with rare representations of non–eolian facies. The Norphlet Formation trends northwest–southeast across southern Mississippi, Alabama, and Florida (Mink and others, 1990; Marzano and others, 1988). A regional dip profile through the Norphlet Formation depositional system exhibits alluvial fan and braided stream sediments updip grading into eolian sheet and dune sands mid–dip in the vicinity of HPU and offshore Alabama (Fig. 18). Compositionally, the Hatter's Pond Norphlet Formation is composed of feldspathic lithic arenites as shown in Figure 19. The Norphlet Formation exists only in the subsurface and the eolian dune facies is the subject of continued hydrocarbon exploration (Tew and others, 1991).

The Norphlet Formation eolian dunes consist of stacked sand dune foreset strata. Grain flow, grain fall, and wind ripple drift laminae can be seen in conventional core. These are commonly represented by a succession of low angle alternating very coarse and very fine thin laminae (traction and grain fall deposits) overlain by laminae of increasing thickness, increasing depositional dip, and decreasing textural heterogeneity within one dune section. These higher dip laminae are representative of grain flow (avalanche) processes and wind ripple formation. As dips increase to maximums of 30 degrees preserved in the core, grain flow strata dominate and wind ripple drift laminae are difficult to discern. Typical facies can be seen in Figures 20 and 21. The dune succession boundaries are fairly sharp and are represented by flat lying beds which truncate the subjacent steeply dipping dune foresets (Fig. 21). Sometimes these boundaries correspond to drastic changes in reservoir properties. Typical reservoir property variations can be seen in Figure 22.

Dune successions such as those found at approximately 120 to 160 feet (37 to 49 meters) below the top of the Norphlet Formation (Fig. 22) are flat lying throughout the unit, correlate across the interfield graben, and are bounded top and bottom by typical dune boundaries like those described above. Since the pattern can be correlated across the interfield graben, there was little to no relief on the Hatter's Pond structure at the time of Norphlet Formation deposition. This influences the way that the reservoir simulation units are constructed, as will be demonstrated later. The current distribution of reservoir properties found in Hatter's Pond are controlled by this sedimentary template, but are severely modified by diagenetic processes.

Diagenesis of the Norphlet Formation

Diagenesis in the Norphlet Formation has been aggressive, and is quite variable throughout the unit. Quartz cementation associated with pressure solution and intergranular volume (IGV) reduction, illite cementation, feldspar dissolution, and dissolution of a precursor cement presumed to be anhydrite or carbonate are the main diagenetic events altering Norphlet Formation reservoir quality (Fig. 23). The key to understanding Norphlet Formation diagenetic modification lies in integrating the petrographic data with structural data from the unit. Simply, the diagenesis varies with structural elevation above the free water level.

Diagenetic illite

Stoudt and others (1992) show that there is a difference in the porosity/permeability relationship between crestal wells and flank wells in the Norphlet Formation at HPU due to the quantity of authigenic illite present near the water level. The amount of authigenic illite present in the HPU 16-9#1, an extensively studied well, decreases systematically upsection, as shown in Figure 24 and

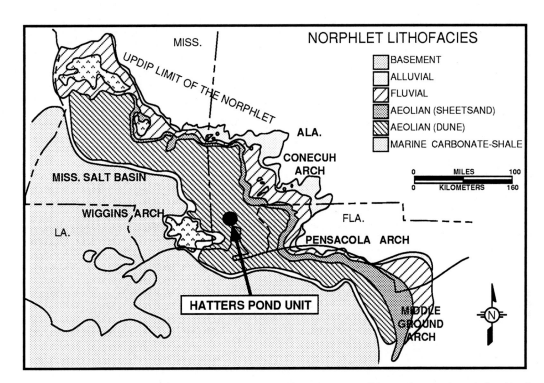

Figure 18. Map of the Norphlet Formation depositional system illustrating regional distribution of facies. Hatter's Pond Unit is located within the eolian dune sand facies. Figure modified from Marzano and others, 1988.

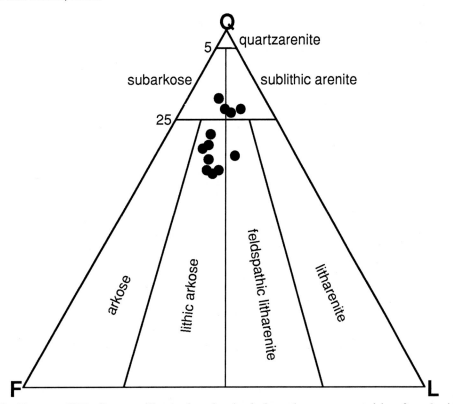

Figure 19. Ternary QFL diagram illustrating the detrital sandstone composition found within the Norphlet Formation of Hatter's Pond Unit. Ternary classification of Folk (1968).

HATTER'S POND UNIT 16-9 #1
MOBILE COUNTY, ALABAMA

Figure 20 - This photograph illustrates selected depths and fabrics representative of the HPU 16-9#1 Well Norphlet Formation. Steep foreset bed dips (up to 30 degrees) can be found within the core photo. In addition, stylolites and pressure solution seams (stylolites with no transform offsets) can be seen within the field of view. Pressure solution features can be seen conforming to the sedimentary template identified here as foreset bed dip preserved along grainflow laminae. Some of the pressure solution seams have insoluble residues thicker than others, presumably indicating larger quantities of pressolved material at that site. White spots seen at lower depths are nodules of anhydrite, quartz, and/or dolomite which cemented the sandstone prior to pyrobitumen deposition.

Figure 21 - A) 18,335':Flat lying strata at the base of a dune section which contain white cemented spots. White spots are nodules of anhydrite, quartz, and/or dolomite which cemented the sandstone prior to pyrobitumen deposition. B) 18,183':Dune boundary showing low angle grainfall and coarse traction load sandstone laminae truncating higher angle dip foreset laminae. Intersecting stylolitized surfaces exxagerate the dip contrast between the two facies and demonstrate how pressure solution can be focused by the sedimentary template. C) 18,184':A bundle of stylolites formed within some moderately dipping foreset laminae which contain some textural heterogeneity. One thicker solution seam can be seen near the top of the core piece. D) 18,279': Several poorly discernable stylolites accentuate the left-to-right dipping foreset strata.

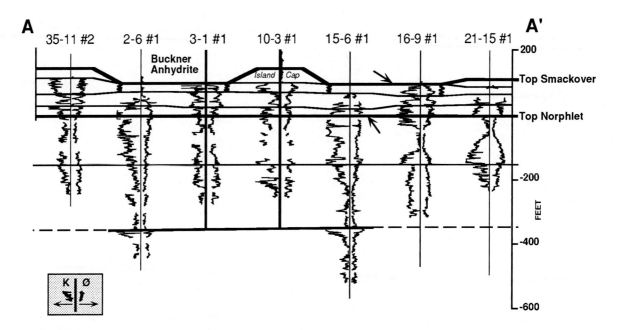

Figure 22. Cross section A-A' illustrates the geologic criteria used to develop the simulation layers for Hatter's Pond Unit. The upper Smackover Formation lithofacies are strongly controlled by the position of the island caps that focused early, fabric–selective dissolution within the unit. Thicknesses within Smackover lithofacies are somewhat variable. The Norphlet Formation appears to have been deposited in an erg uninfluenced by basement structure, and two reservoir units marked illustrate the correlatability of flat lying Norphlet Formation units. Note that the southern end of the cross section jumps across the interfield graben and a good correlation still exists. The location of cross–section A-A' is marked on Figure 2.

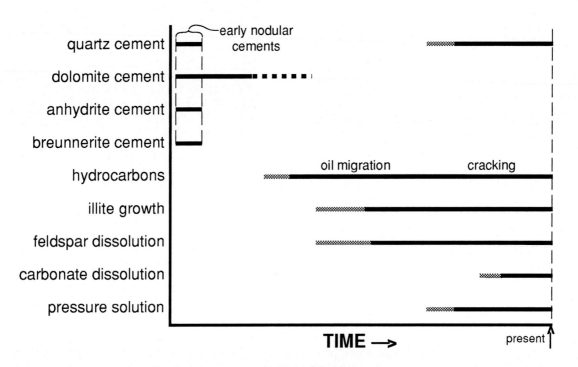

Figure 23. The generalized paragenetic sequence illustrates the sequence of events related to Norphlet Formation diagenetic processes in Hatter's Pond Unit.

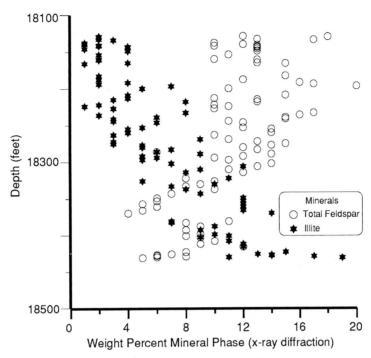

Figure 24. Illite and feldspar weight percentage variability shown with respect to depth in the HPU 16-9#1. All data is gathered by x-ray diffraction of whole rock powders. Illite growth is interpreted to take place at the expense of feldspar dissolution. K-spar is thought to be the main source of potassium necessary for illite precipitation.

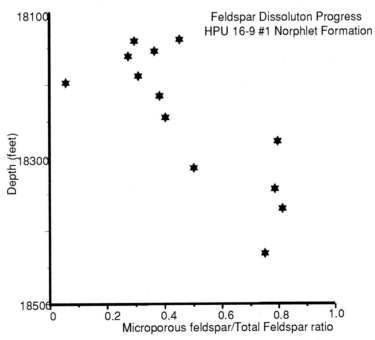

Figure 25. Petrographic data illustrates that the ratio of microporous feldspar/total feldspar increases with depth in the HPU 16-9#1 Well, as might be expected if feldspar was being lost because of dissolution.

Appendix A. K–spar is found to be 50 percent to 70 percent of the detrital feldspar population in the Hatter's Pond area, and its dissolution provides the potassium necessary for illite growth. The volumetric ratio of microporous feldspar/total feldspar increases dramatically with depth as seen in thin section (Fig. 25) and is an expression of the feldspar dissolution process. The variability in weight percent feldspar is due solely to feldspar dissolution and does not express any change in provenance. Both illite growth and the development of an intragranular micropore system increase the tortuosity of the pore network and increase the porosity/permeability ratio with depth in the reservoir.

McBride and others (1987) dated the authigenic illite in the Norphlet Formation at HPU at 55 Ma, timing its growth after hydrocarbon emplacement. Their isotopic data also suggest that illite growth occurred at burial temperatures of 130 to 155 degrees Centigrade. Current bottom hole temperatures in the unit are near 160 degrees Centigrade, illustrating that illite growth is a relatively late diagenetic process. We also agree that oil migration, hydrocarbon cracking, and illite growth overlapped in time (see Fig. 23). The vertical variability in illite percentage and feldspar dissolution is interpreted to be due to the quantity of water available to do the diagenetic work. Capillary pressure studies show that the ambient water saturation is quite variable through the formation (Fig. 26), and we interpret that the quantity of available water directly influenced the rate at which diagenetic work could be accomplished. In fact, the abundance of feldspar dissolution and illite growth found near the gas/water contact suggests that these reactions may still be occurring, and a vertical age profile developed for Norphlet Formation authigenic illite may not be uniform.

Clay coatings on quartz grains have been called upon to inhibit quartz cementation in certain diagenetic settings (Thomson, 1979; Dixon and others, 1989), but usually the clay coatings have been authigenic chlorites. The data found in the HPU 16-9#1 illustrate that grain coating authigenic illite has had a similar effect on quartz cementation. Figure 27 illustrates that a maximum measured reservoir porosity is associated with between 5 percent and 10 percent illite by weight. Grain coating illite percentages of less than 5 percent allow quartz cementation to proceed relatively unimpeded. Where illite coatings are discontinuous, quartz crystals can be seen growing from exposed grain surfaces (Fig. 28). Rocks containing greater than 10 percent illite tend to have diminishing reservoir quality. However, breunnerite cement, a rare ferroan magnesite also found in the Norphlet Formation by Kugler and McHugh (1990), is also shown to be more abundant with depth (Appendix A) and probably contributes to the porosity reduction associated with higher illite percentages (see also Fig. 28). Since grain coating clays are present in lower percentages near the top of the Norphlet Formation, areas which are tightly cemented by quartz tend to concentrate there. One of these areas is present in the HPU 16-9#1 approximately 40 to 140 feet (13 to 46 meters) below the top of the Norphlet Formation (see Fig. 22). These quartz cemented Norphlet zones are the most tightly cemented areas found in the Norphlet Formation.

Pressure solution and evidence for secondary porosity

Two styles of pressure solution, stylolitization and intergranular pressure solution, are both present in HPU. Stylolitization, observable in the slabbed conventional core face, was recorded throughout the Norphlet Formation within the HPU 16-9#1 type well (see Fig. 21). Some zones contain over 10 stylolites/foot (Fig. 29). Stylolites liberate significant quantities of silica which can be reprecipitated as quartz cement in the formation, lowering reservoir porosity and permeability. Stylolites are late diagenetic features which generate significant quantities of quartz cement in the Norphlet Formation, offshore Alabama (Thomas and others, 1993) and appear very similar to those found in this onshore setting. Much is unknown about the controls governing the lateral extent of stylolite formation patterns, and vertical stylolite frequency. Some areas of the HPU 16-9#1 core show many closely spaced stylolites with paper–thin seams of insoluble residue, some isolated stylolites show single thick seams of insoluble residue, some zones contain bundles

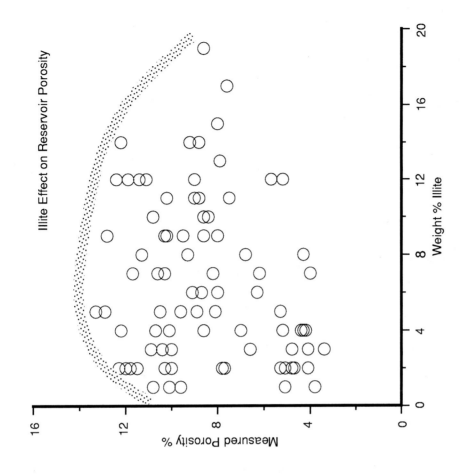

Figure 27. The percentage of 5% to 10% illite grain coatings (by weight) seems to inhibit quartz cement in the Norphlet Formation in the Hatters Pond 16-9#1 Well. Lower percentages of illite correspond to lower porosities and zones of abundant quartz cement. Higher percentages of illite (greater than 10%) also correspond to poor reservoir quality.

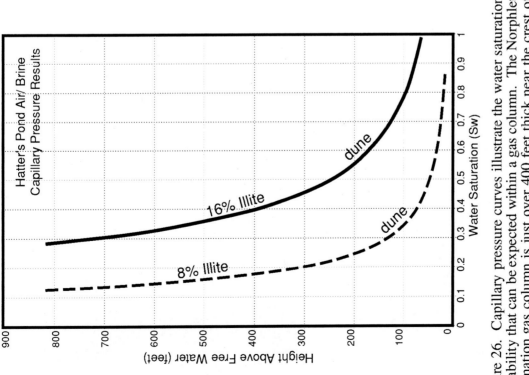

Figure 26. Capillary pressure curves illustrate the water saturation variability that can be expected within a gas column. The Norphlet Formation gas column is just over 400 feet thick near the crest of the field. Depending on textural parameters, the water saturation can be expected to vary by 80 saturation units within the gas column. This difference in water availability is thought to be responsible for the varying degree of illite growth within the reservoir.

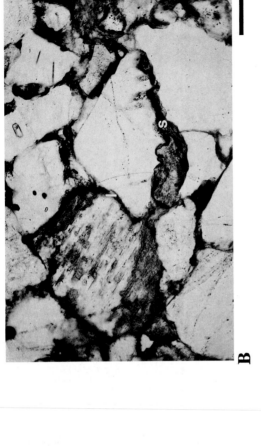
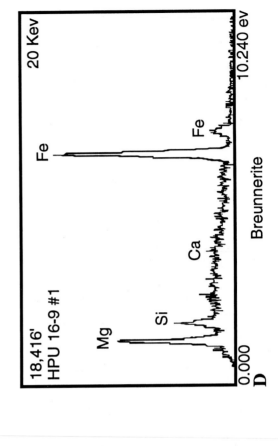
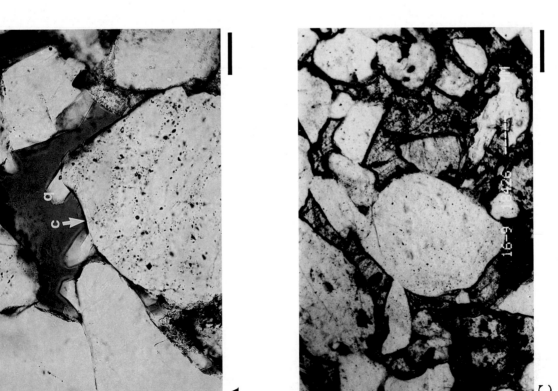

Figure 28. A) 18,144': Quartz cement crystals (q) can be seen growing out of a host grain where illite grain coatings (c) are incomplete. Scale bar is 0.04 mm. B) 18,431': Intergranular pressure solution is common within this field of view. Sandstone grain contacts are long (l) and sutured (s). Scale bar is 0.1 mm. C) 18,426': Breunnerite cement, easily distinguished by high relief in the field of view, fills intergranular pore space in this sandstone. Scale bar is 0.1 mm. D) Breunnerite EDS spectra collected at 20KV on a Tracor Northern 5500 without a light element detector.

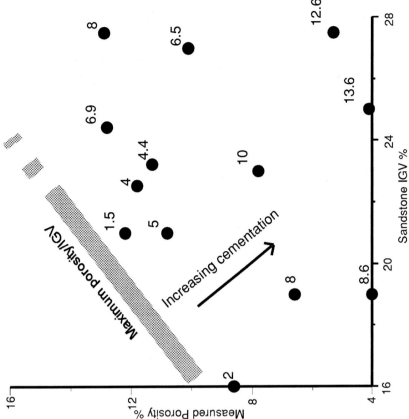

Figure 29. Measured core porosity and numbers of stylolites and solution seams per foot (as determined by visual inspection of the slabbed conventional core face) are plotted for a portion of the HPU 16-9#1 Well. Sandstone IGV's, as determined from point counts of 200 points, are also plotted. Zones of high porosity correlate to zones of diminished stylolitization and high IGV.

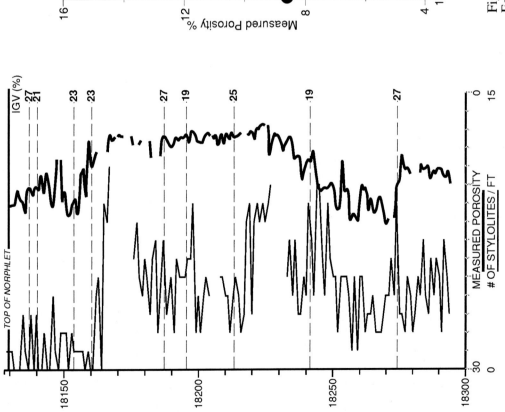

Figure 30. Sandstone IGV is plotted with measured porosity. Each point is labelled with the volume of quartz cement (the main volumentric mineral cement in the HPU 16-9#1 Well) measured by point count. A maximum porosity per IGV trend is described by the samples with less mineral cement, while the samples with greater quantities of quartz cement fall lower on the measured porosity axis. Ideally the quartz cement percentages would contour linearly parallel to the maximum porosity/IGV trend.

of diffuse wispy stylolites. Stylolites always follow some predesigned sedimentary template when they occur. Frequently stylolites nucleate in finer–grained laminae, but assessing this relationship quantitatively is difficult since the nucleation site strata are the first to be pressolved. Stylolites were found to be more commonly developed in rocks of finer average grain size in an offshore Alabama well (Thomas and others, 1993). Sometimes stylolite bundles form in sands showing maximum textural heterogeneity, such as alternating grainfall silts and traction load coarse sands associated with flat lying eolian foreset dune bases. Although numbers of stylolites/foot may not be a true accurate descriptor of pressure solution processes or intensity, it is one piece of information which is easily gathered from a slabbed core face.

Intergranular pressure solution has been examined by measurements of overlap quartz (Houseknecht, 1984) and point counts with cathodoluminescence (Sibley and Blatt, 1976). Simple petrographic calculations of sandstone IGV, or intergranular volume, will be shown to be an effective measure of intergranular pressure solution when examined along with cement volume and measured porosity. Intergranular volume is defined as the quantity of intergranular porosity, mineral cement, and matrix present in a sandstone, and was first used by Weller (1959). In sandstones where dust rims, illite coatings, and pyrobitumen coatings are visible to help determine original grain boundaries, IGV can be measured reasonably well with normal petrographic point counts.

The Norphlet Formation section in HPU contains evidence that compactional processes have reduced sandstone porosity to varying degrees. Sandstone IGV's in the HPU 16-9#1 range from 16 percent to 27 percent, and high reservoir quality obviously is associated with the samples containing high ratios of IGV/mineral cement (Fig. 30). Zones containing significantly lower quantities of stylolites, high sandstone porosity, and high IGV appear to be anomalous in terms of normal compactional processes, yet zones of this type are present within Hatter's Pond and are interpreted to be zones where a pre–existing mineral cement has dissolved. This form of secondary porosity is the most difficult to distinguish since the complete dissolution of the intergranular mineral may leave very few hints of its previous existence. In zones where intergranular pressure solution and stylolitization combine to reduce porosity and sandstone IGV, textural anomalies provide valuable clues concerning zones where intergranular secondary porosity has developed.

The uppermost portion of the Norphlet Formation commonly contains several zones of extremely high reservoir quality (see Fig. 22). These zones are of varying thickness, ranging from 0 feet to nearly 100 feet (0 to 31 meters) within the unit. The HPU 16-9#1 contains two zones of this type displayed in Figure 29, one from 18,130 feet (5530 m) to 18,160 feet (5539 m) core depth, and another from 18,240 feet (5563m) to 18,275 feet (5574 m) core depth. IGV is consistently greater than 20 percent and as high as 27 percent in these zones. IGV greater than 26 percent is especially significant within sandstones with little mineral cement because compaction and grain reorientation have been shown to be incomplete (Szabo and Paxton, 1991). This suggests that the high porosity/high IGV areas are zones of decementation or secondary porosity generation within the unit. A pre–existing mineral cement, probably dolomite or anhydrite, is inferred to have been present during pressure solution, thus preventing overcompaction of the zones. Recent anhydrite and/or carbonate dissolution during thermochemical sulfate reduction (Siebert, 1985) is a likely scenario for the generation of secondary porosity in this setting. Stylolitization is nearly absent in the top zone shown in Figure 29, while represented to a lesser degree in the lower zone. The variation in degree of stylolitization could be due to differences in the timing of decementation, or more likely, differences in the degree of original cementation. It is remarkable to note that the stylolitization pattern mimics the measured porosity pattern in the core in many places (note in detail the zone from 18,240 feet to 18,275 feet), indicating again that stylolitization is a major influence on porosity reduction in the Norphlet Formation. Petrographic fabric elements also indicate that these are zones of secondary porosity development (Fig. 31), as oversized pores and packing heterogeneity are common.

Figure 31 - A) 18,137': Abundant intergranular porosity (p) is shown immediately adjacent to a zone containing long grain contacts (c). Such packing heterogeneity is suggestive of secondary porosity after intergranular cement dissolution. Scale bar is 0.1 mm. B) 18,277': Low volumes of mineral cement and oversize pores characterize the field of view. This sandstone has an IGV of 27%. Scale bar is 0.1 mm. C) 18,137': Abundant intergranular porosity (p) is shown immediately adjacent to a zone containing long grain contacts (c). Such packing heterogeneity is suggestive of secondary porosity after intergranular cement dissolution. Scale bar is 0.1 mm. D) 18,194': Low IGV (19%) and abundant intergranular pressure solution characterize this sandstone. Scale bar is 0.1 mm. All photos are taken in the HPU 16-9#1 well.

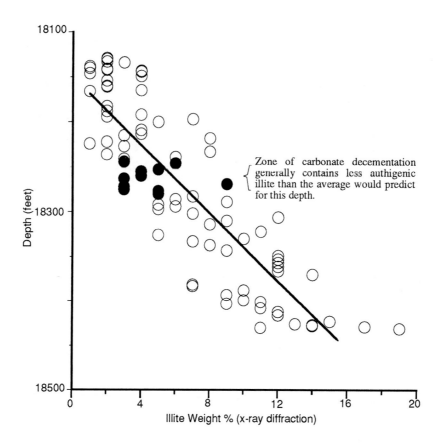

Figure 32. The illite versus depth trend is replotted to highlight the zone of probable secondary porosity found at 18,245' to 18,275' core depth in HPU 16-9#1. The zone of interest contains less authigenic illite than the trend would predict, suggesting that illite growth has lagged other adjacent zones. A precursor mineral cement may have kept the illite cement from growing with the rest of the well trend.

Additionally, the lower zone seems to have developed less authigenic illite than the Norphlet trend would predict (Fig. 32). This would appear to be due to the fact that the formation was only recently decemented such that illite growth in this zone lags other zones. Many of the low IGV thin sections contain a great deal of pyrobitumen, but many of the high IGV/high porosity sections contain little. High in the reservoir where oil saturations should have been the highest (prior to cracking), the reservoir should be uniformly high in pyrobitumen unless a precursor cement prevented migration in some zones. In other areas of the core, zones of high IGV and low porosity are commonly quartz cemented (Fig. 33A), and zones of low porosity and low IGV commonly show accelerated intergranular pressure solution (Fig. 31D). Although sandstone decementation is interpreted to have occurred during thermochemical sulfate reduction and within the hydrocarbon column, the volume of water necessary to do that dissolution seems problematic.

It is clear that there are nodular zones of quartz cement which were deposited prior to hydrocarbon emplacement. White nodules seen in Figure 20 represent zones of anhydrite, dolomite, and quartz cementation which took place prior to hydrocarbon emplacement. The majority of the Norphlet Formation core is gray to dark gray due to the presence of pyrobitumen, the cracked residual of a precursor Norphlet Formation oil. In fact, some of the offshore upper Norphlet Formation reservoir sands contain total organic carbon contents approaching 1 percent by

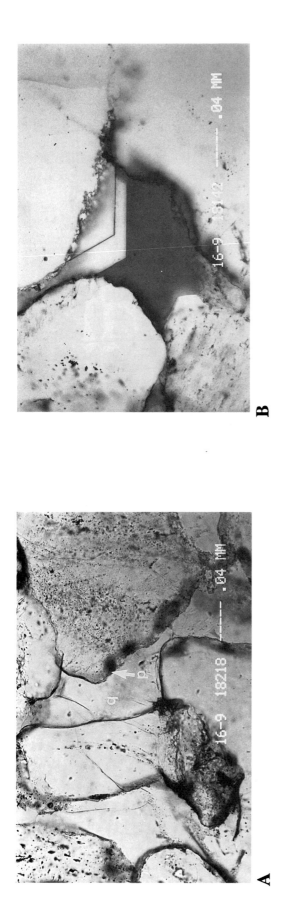
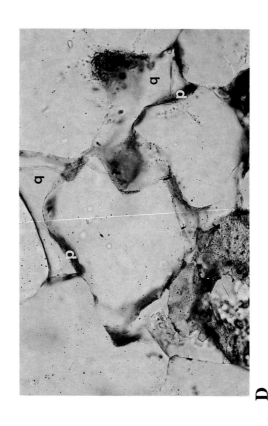
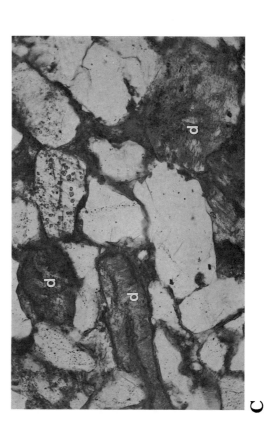

Figure 33 - A) 18,218': Pyrobitumen (p) coatings exist on areas of the framework grain underneath quartz cement. This indicates that quartz cement post-dated hydrocarbon cracking and took place within the hydrocarbon column. B) 18,142': Quartz overgrowths are sometimes found with birefringent illite layers sandwiched within, indicating that they overlapped in time. C) 18,431': Abundant feldspar dissolution (d) has resulted in large quantities of intragranular pores which are usually filled in part by fibrous illite cement. Scale bar is 0.1 mm. D) 18,215': Quartz cement is seen growing over the top of pyrobitumen coatings (p) on illite coated sand grains. Scale bar is 0.04 mm.

weight due to pyrobitumen deposition, but it is interesting to note that residual water saturations in the oil column apparently influenced the quantity of pyrobitumen deposited in the upper versus the lower Norphlet Formation (Thomas and others, 1993). Quartz cement is commonly deposited on top of this discontinuous organic layer (Fig. 33), and serves as additional evidence that diagenesis in the Norphlet Formation did not stop when hydrocarbons migrated into the reservoirs. In water wet sandstones undergoing pressure solution, it is probable that liberated silica could be expelled out of the sandstone or precipitate within the sandstone depending on local hydraulic and chemical conditions. In hydrocarbon charged systems, pressure solution can continue as shown in Thomas and others (1993), Saigal and others (1992), and Walderhaug (1990), but the ability of the liberated silica to be exported is much diminished. The quartz liberated during Norphlet Formation stylolitization and intergranular pressure solution is interpreted to have precipitated in the immediately adjacent sandstone because of steep concentration gradients built up in the residual water saturation of the gas sand. The linkage between low resident water saturation and resultant limited silica movement is reinforced by the observation of quartz cemented zones that correlate to zones of high stylolite frequency seen in the slabbed core face.

GEOLOGIC RESERVOIR MODEL

The lithofacies as established and discussed above for the Smackover and Norphlet Formations served as the framework for a reservoir model that was constructed for HPU in Stratamodel's SGM™ (an acronym for **S**tratigraphic **G**eocellular **M**odeling). The primary goal was to provide a geologically–constrained distribution of reservoir properties (e.g., porosity, permeability) to:

- better understand the three–dimensional distribution of reservoir properties and how their distribution relates to hydrocarbon production,
- aid operations in short–term reservoir management decisions (e.g., locations for replacement wells, infill wells, perforation strategies, etc.), and
- aid operations in long–term reservoir management decisions (e.g., a methane–injection program) by providing a reservoir model for production forecasting.

In order to build a model in SGM, two types of input data are required: structural surface grids and well data. The structure grids provide the architecture or framework for the model—they guide the layering within the model and therefore influence the correlation of properties from well to well. Thus, these surfaces have tremendous impact on the continuity (or lack thereof) of properties such as lithology, porosity, permeability, etc.

The well data, which include any properties that can be represented in X–Y–Z space (e.g., core–measured properties, raw or calculated log properties, perforations, seismic attributes, etc.), are used to populate the model. The interpolation of properties is based on a distance–weighted algorithm, but there are several ways within SGM to significantly bias the attribute distributions. Core and log data from 27 wells were used to build a reservoir model of HPU in SGM (see Fig. 2).

Surface Grids—The Stratigraphic Framework

To build the stratigraphic framework for HPU, the lithofacies in the Smackover Formation were grouped into three packages (Fig. 34):

- The upper unit included three laterally equivalent lithofacies—the sucrosic,

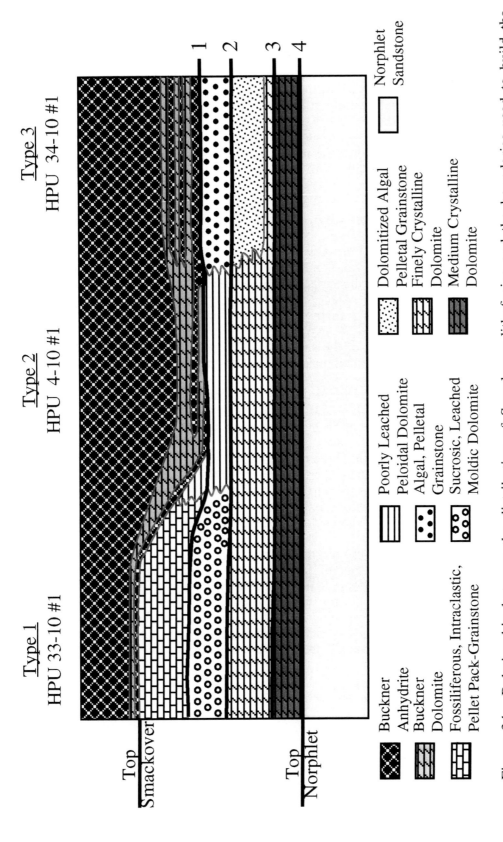

Figure 34. Relationship between the distribution of Smackover lithofacies and the boundaries used to build the stratigraphic framework for the reservoir model in SGM™. The boundaries are shown as heavy black lines and are numbered along the right margin of the display. For vertical scale, the sucrosic, leached moldic dolomite in the 33-10#1 well is 35 feet (11 meters) thick.

leached moldic dolomite, the poorly leached peloidal dolomite and the algal pelletal grainstone.
- The middle unit was mostly finely crystalline dolomite, but also contained the dolomitized algal pelletal grainstone.
- The lower unit consisted entirely of medium crystalline dolomite.

The uppermost Smackover lithofacies in the Type 1 wells (i.e., the fossiliferous, intraclastic, pellet pack–grainstone) is completely nonporous and impermeable and was therefore *not* included as part of the HPU reservoir model.

The top of the Norphlet Formation was the only sequence boundary used to model the Norphlet portion of the reservoir. All layers within the Norphlet Formation are two feet (0.6 meters) thick and parallel to its top.

The tops for the lithofacies packages were picked in each of the 27 wells and used in Z–MAP *Plus*™ to construct structure grids. All of the grids were built with the same dimensions (i.e., X and Y limits and grid node spacing). The well spacing and distribution, and the area of interest to be modeled (in SGM *and* in the reservoir simulator) were critical factors in determining the limits and node spacing in Z–MAP *Plus* for the surface grids because:

- the boundaries of the surface grids are used in SGM as the model boundaries, and
- the grid node spacing of the surface grids is used as the length and width dimensions of the cells that make up the stratigraphic framework.

The map surfaces were projected to the grid margins so there were no null values in the grids using the editor within Z–MAP *Plus*. It was critical that the input grids used to build a stratigraphic framework did *not* contain nulls, particularly since the model was to include faults.

Structure grids were also made of the three main faults: the northern margin of the graben, the southern margin of the graben, and the eastern boundary of the unit. A 3–d seismic survey shot over the southern part of the unit was used to locate and characterize the fault planes. The fault grids were built with the same dimensions and grid node spacing as the surface grids. Similarly, the fault grids were extrapolated above and below the reservoir interval to eliminate null values.

Prior to exporting the grids, they were resampled to a finer grid node spacing more suitable for use in SGM. The appropriate node spacing for creating 2–dimensional grids is *not* necessarily ideal for building 3–dimensional models in SGM, particularly in a faulted model. A general rule of thumb for grid node spacing when building surface grids is one–half the distance of the closest well pair. However, a finer grid node spacing may be necessary in SGM due to the way grid cells are terminated against faults. Nevertheless, grid resampling must be weighed against model size—the greater the number of cells in the model, the larger the size of the model.

Well Data and Reservoir Model Attributes

Porosity and permeability

Core– and log–derived porosity and permeability data from 27 wells were used to build a reservoir model for HPU. An abundance of evaporites and brittle dolomites as well as the reservoir depth (> 18,000') contributed to borehole instability and made it difficult to run logging tools and obtain reliable measurements. Thus, most of the wells were cored through the reservoir interval in order to adequately characterize the distribution of porosity and permeability. This core–

measured porosity and permeability data formed the foundation for the reservoir model.

Several adjustments, enhancements and corrections had to be made to the well data prior to being used in SGM:

- The core data was depth–shifted to match the resistivity curve depth. Due to the depth of the reservoir, this depth–shift was often significant (ten feet or more).
- The raw porosity logs were converted to lithology–adjusted porosity data.
- The lithology–adjusted porosity data was bridged where appropriate.
- All porosity data was regressed against the bridged porosity data to get the best regressed porosity.
- The best regressed porosity was substituted for the gaps in the core data to obtain the best continuous porosity curve.
- The core permeability was bridged where appropriate.
- Best continuous porosity was regressed against bridged core permeability to get a continuous permeability curve.
- The continuous permeability curve was substituted for gaps in the core permeability in order to obtain a continuous permeability curve.
- A 5000 pound overburden correction was made to adjust the porosities and permeabilities to reservoir conditions, since the values were obtained from plugs under surface conditions. The correction had a minor affect on the porosities, but permeabilities were often reduced by more than one order of magnitude.

The well data, once corrected and adjusted, was imported as an ASCII file into SGM, placed within the stratigraphic framework built from surface grids, and used to populate the reservoir model with values of porosity and permeability. In addition, the well perforation data was imported so that perforated intervals could be displayed and compared to the distribution of reservoir properties.

Net pay thickness and initial fluid saturations

Net pay thickness[1] and initial water saturation attributes were calculated using the "model operations" utility in SGM. Net pay thickness was calculated by summing the thickness of individual cells whose values for porosity and permeability exceeded the cutoffs *and* whose elevation was above the gas/water contact. The net pay thickness attribute was used to create isopach grids (discussed and displayed below) as well as exported from SGM to be used for reservoir simulation.

The functions used in SGM for calculating the initial water saturation (Sw_i) were derived from capillary pressure studies done in the late 1970's. Core data from the Smackover and Norphlet Formations were used to establish a relationship between interstitial water saturation and permeability. A separate function was established for each of the two formations:

Smackover $Sw_i = 0.1847 - 0.06079 * \log k^{5000}$
Norphlet $Sw_i = 0.2590 - 0.07064 * \log k^{5000}$

---

[1] At HPU the reservoir "pay" has been established at 6% porosity and 0.1md permeability *at surface conditions*. When corrected for overburden pressure, the Smackover Formation pay cutoffs are 5.4% porosity and 0.022md permeability; the Norphlet Formation pay cutoffs are 5.2% porosity and 0.01md permeability.

where k^{5000} is the permeability under reservoir conditions. Using these functions, a model operation was designed to calculate the initial water saturation in each cell throughout the model.

The initial water saturation for each cell was then converted to initial gas saturation (Sg_i) using the simple operation:

$$Sg_i = 1 - Sw_i$$

The initial gas saturation was used to calculate the original gas in place (OGIP) for each cell using the function

$$\begin{aligned}OGIP &= (\text{Cell length} * \text{Cell width} * \text{Cell thickness}) * \text{Porosity} * Sg_i \\ &= \text{Total cell volume} * \text{porosity} * Sg_i\end{aligned}$$

The original gas in place for the individual cells was summed together to obtain an estimate of original gas in place for the reservoir. The gas volume derived from SGM was somewhat larger than previous modeling studies where more traditional methods were used for calculating original gas in place. Most of the additional gas is attributable to the use of a flat gas/water contact in this model. Future studies will attempt to account for this difference by using a variable gas/water contact.

Although many other attributes were created and calculated within the geologic model, only four reservoir properties were exported from SGM for use in reservoir simulation: porosity, permeability, net pay thickness, and original water saturation (discussed further below).

The Reservoir Model—Smackover Formation

Understanding the depositional and diagenetic controls on the distribution of porosity and permeability is *critical* to predicting and modeling reservoir quality, establishing hydraulic flow units and maximizing recovery from the unit. Reservoir temperatures and pressures at HPU result in hydrocarbons that are primarily in a gaseous state, hence the cutoffs for effective porosities and permeabilities are quite low. These low reservoir cutoffs allow much of the Smackover and Norphlet Formations above the gas/water contact to be effective reservoir. Nevertheless, it is the distribution of the *extremes*—the high and low permeability zones—that must be understood and modeled in order to maximize sweep efficiency because it is these extremes that control flow within the reservoir.

The Smackover carbonates were modeled as three lithofacies packages as previously discussed (see *Surface Grids—The Stratigraphic Framework* and Fig. 4). The lithofacies packages were not only used for modeling the Smackover Formation reservoir properties in SGM, but they also served as simulation layers in the subsequent engineering studies. Reservoir quality and distribution are significantly different for the uppermost Smackover unit relative to the lower two units. The underlying reasons relate back to the depositional and diagenetic history for the Smackover Formation.

Upper unit

The upper unit in the Smackover Formation is composed of three distinct lithofacies: the sucrosic, leached moldic dolomite, the poorly leached peloidal dolomite, and the algal pelletal grainstone. The overall thickness of the upper Smackover package ranges from 11 to 47 feet (3 to 14 meters); net pay thickness ranges from 0 to 39 feet (0 to 12 meters).

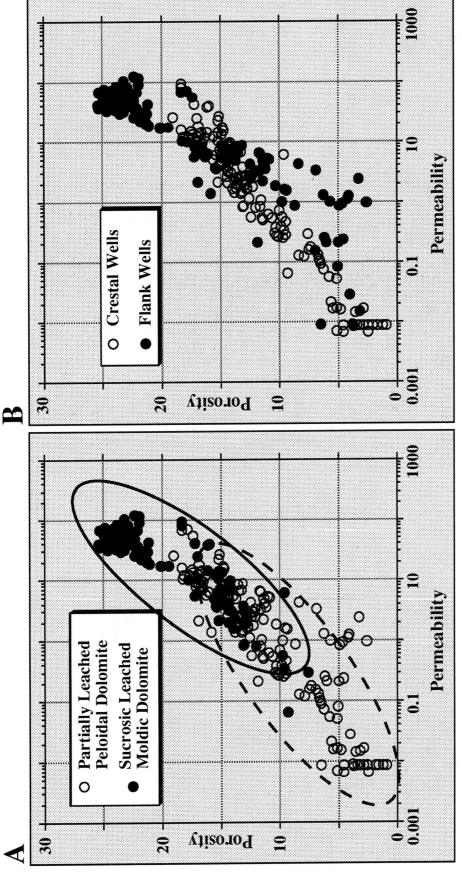

Figure 35. Porosity–permeability crossplots for the upper Smackover lithofacies package. Data in plot A are separated on the basis of lithofacies; data in plot B on the basis of well position on the structure (crest versus flank). Plot A shows that the data associated with the sucrosic, leached moldic dolomite consistently has better reservoir character and tends to cluster towards the "high" end of the crossplot (solid circle). In contrast, most of the data associated with the laterally equivalent lithofacies plot toward the lower, less porous and permeable end of the graph (dashed circle). Plot B shows significant scatter of reservoir quality based on structural position, demonstrating that the distribution of reservoir quality in the upper Smackover flow unit is independent of structural position and is instead a function of lithofacies. The data in this plot are only from four crestal wells and two flank wells in order to minimize overposting and cluttering of the data. Figure 36 shows the distribution of net pay thickness for the upper Smackover based on data from all wells.

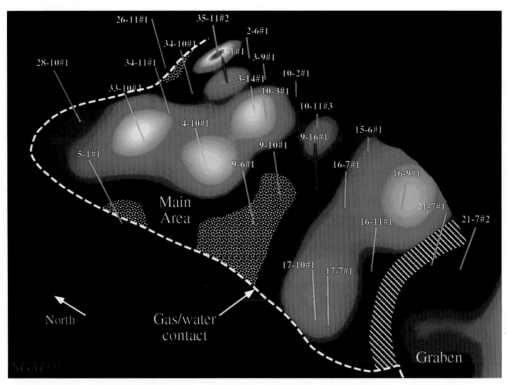

Figure 36. Net pay thickness of the upper Smackover Formation draped over its structural surface. This display demonstrates that good reservoir quality (green, yellow and red), which is mostly within the porous sucrosic dolomite lithofacies, is independent of structural position (crest vs. flank; compare to Figure 41 below). Red areas are approximately 39 feet (12 meters); green about 20 feet (8 meters) thick. Regions of the upper Smackover unit that are completely nonpay (zero feet net pay) are shown in a white stippled pattern. The white striped area toward the lower left corner of the figure represents the north graben fault escarpment. Inactive wells are shown in gray, producers in green and injectors in red.

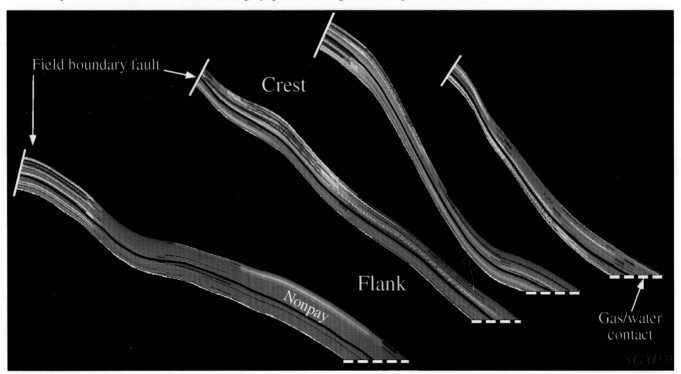

Figure 41. East–west structural cross–sections through the main part of HPU showing significantly higher reservoir quality on the crest in the middle and lower Smackover Formation. The middle and lower Smackover units are the only two that are displayed and the contact between them is shown as a solid black line. The colors represent porosity from 6 percent (dark blue) to ≥ 15 percent (red). *All regions of the Smackover that are below the porosity and/or permeability cutoffs (i.e., the nonpay) are shown in gray.* The view is from the north toward the south, so the cross–section panel on the right side of the display is the most distant and is just north of the graben. The total thickness of the section in the northern–most cross–section is 70 feet (21 meters).

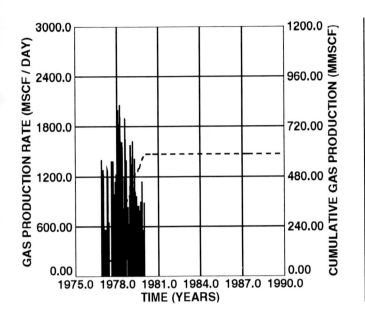
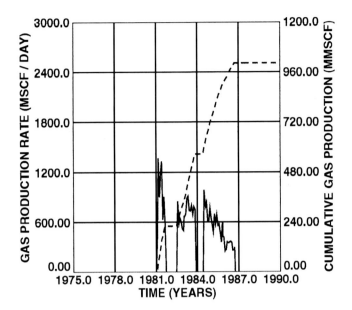

Figure 37. Production histories for two wells that produced exclusively from the upper Smackover sucrosic, leached moldic dolomite lithofacies. Gas production rate is shown as a solid line and cumulative gas production is represented by the dashed line. Their rapid depletion and relatively short lives (e.g., compare to wells in Fig. 53) suggest that the sucrosic, leached moldic dolomite lithofacies has a very limited lateral extent. It was necessary to use a geologic template in SGM™ (see Figure 38) in order to limit the distribution of the high porosity and permeability associated with the sucrosic, leached moldic dolomite lithofacies to discrete "pods".

Porosity formation in these uppermost lithofacies is very closely related to depositional fabric andrelatively early, near–surface diagenetic events. Thus, reservoir quality in the upper Smackover is *not* related to location on the structure (i.e., crestal versus flank) but instead is a function of lithofacies (Figs. 35 and 36).

Depositional and diagenetic events in the sucrosic, leached moldic dolomite have resulted in some of the best reservoir rock at HPU—porosities average 19.5 percent and permeabilities 17.5 millidarcies (see Fig. 4B).

Wells that perforated and produced solely from the sucrosic, leached moldic dolomite lithofacies typically had high production rates but were relatively short lived (Fig. 37), suggesting that the lateral extent of the porous sucrosic dolomite lithofacies is rather limited. This is supported by the fact that most sucrosic dolomite pods were penetrated by only one or two wells (see Fig 5).

The porous and permeable sucrosic, leached moldic dolomite is top–sealed by the fossiliferous, intraclastic, pellet pack/grainstone (i.e., island cap) lithofacies. This lithofacies is completely nonporous and impermeable, thus it was not explicitly included as part of the reservoir model in SGM. However, its development controlled the formation of freshwater lenses and the resultant distribution of porosity in the underlying sucrosic dolomite. Thus, understanding the relationship between the occurrence of the island cap lithofacies and the underlying sucrosic dolomite significantly influenced the modeling of upper Smackover Formation reservoir pay in SGM (discussed further below).

Laterally the sucrosic, leached moldic dolomite is flanked by the poorly leached peloidal dolomite and the algal pelletal grainstone. Porosities and permeabilities within the poorly leached peloidal dolomite, although not as high as the sucrosic dolomite, average 9.2 percent and 0.41 millidarcies, respectively. The algal pelletal grainstone is only present in two wells (HPU 2-6#1 and HPU 34-10#1) and is essentially non–reservoir, with an average porosity of 2.0 percent and an average permeability of less than 0.01 millidarcies.

Use of a template to model the upper Smackover in SGM.—In order to properly model the discrete lithologic (and reservoir) boundaries between the porous sucrosic dolomite "pods" and the surrounding, (relatively) tight lithofacies, a geologic template was constructed in SGM (Fig. 38). A template allows the application of geological interpretations and concepts in order to bias the attribute interpolations.

Modeling without using a template in the upper Smackover resulted in a very gradual and smooth lateral transition from low to high porosity and permeability and therefore too much pay (Fig. 39). This model was used in a previous simulation study, and artificial vertical barriers around many of the wells with sucrosic dolomite pods were necessary in order to properly match predicted model production with real historical well production.

The use of a template in the upper Smackover unit produced very rapid changes in reservoir properties over a relatively short distance. This more accurately reflected the conceptual geologic model as well as the available production data. The individual "pods" were resized in the template several times until a close match between the initial gas volume in the model and a calculated initial gas volume for each pod was achieved. A calculated initial gas volume was derived for each pod from p/z decline curves (average reservoir pressure divided by gas compressibility plotted against culmulative gas production). Assuming that the pods behaved as isolated reservoirs, the decline curves were extrapolated to zero, which provided an estimate for the original gas in place.

Middle unit

The middle Smackover is primarily composed of finely crystalline dolomite, but in two wells (HPU 2-6#1 and HPU 34-10#1) it also contains the dolomitized algal pelletal grainstone lithofacies. Total thickness ranges from 29 to 65 feet (9 to 20 meters) and net pay thickness ranges from 3 to 48 feet (1 to 15 meters).

The middle unit is composed of dolostones with very little preserved depositional texture and, relative to the overlying lithofacies, the diagenetic fabric is far more uniform. As a result, variations in reservoir quality within the middle unit are relatively gradational. Therefore, a geologic template was not required to bias or discretize the distribution of reservoir properties within this unit.

The *overall* average porosity of the middle unit is 8.0 percent and average permeability is 3.0 millidarcies. However, there is a striking and significant difference in reservoir quality between wells proximal to the structural crest (average porosity: 12.6 percent; average permeability: 1.0 md) versus those toward the flank (average porosity: 6.1 percent; average permeability 0.2 md). This is reflected in Figures 40 and 41, which show the distribution of reservoir for the middle Smackover with respect to structural position.

Reservoir quality in the middle Smackover is largely a function of the amount of non–fabric

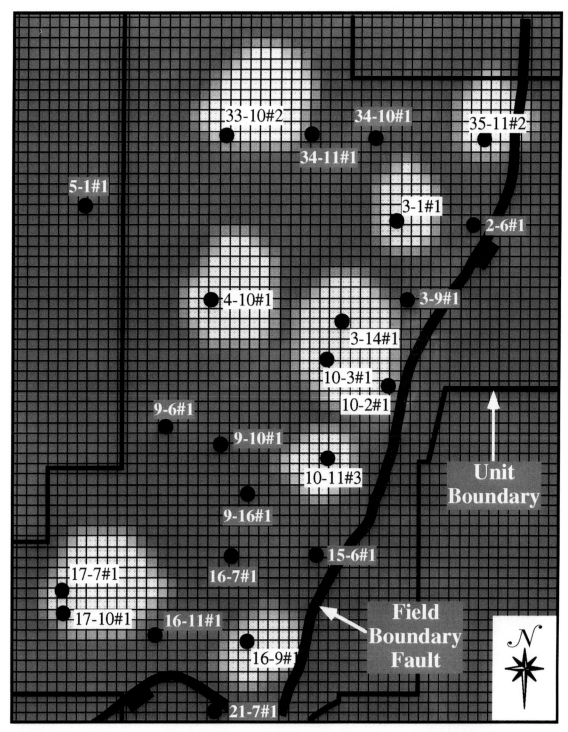

Figure 38. Geologic template used to model the upper Smackover in SGM™. The template was only applied to the north (main) fault block. When the template was applied to attribute interpolation in the upper Smackover Fm., only the wells within the white regions were used to calculate values for cells within the white region; similarly, only wells within the dark gray area were used to calculate values for cells in the dark gray area. The use of this template created very discrete boundaries in the attributes within the upper Smackover, which more accurately reflects the geologic conceptual model (compare Figs. 36 and 39). A combination of core information (i.e., lithofacies characterization as in Figure 5) and pressure data were used to design the size and shape of the pods. Each cell is 100 meters (328 ft.) on a side.

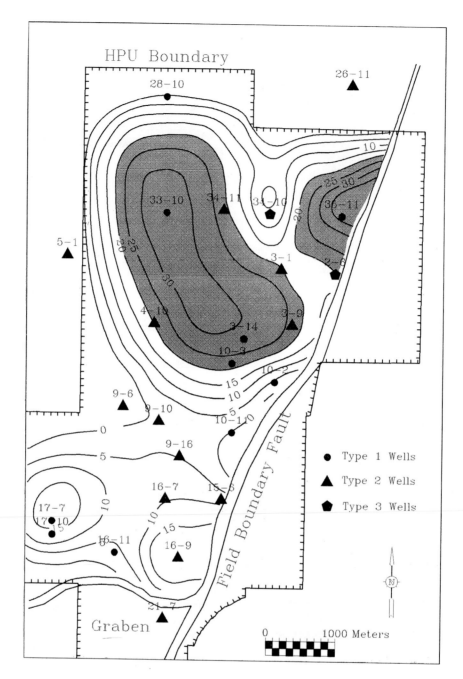

Figure 39. Isopach map of net pay thickness in the upper Smackover lithofacies package as modeled in SGM *without* a geologic template. The map centers on the main fault block because it contains the majority of well data. The contour interval is 5 feet, and all pay thickness greater than 20 feet is shaded gray. The well symbols represent the various Smackover vertical successions as discussed in the text. Notice the thick, continuous pay zone that extends from northwest of the HPU 33-10#2 well to the HPU 10-3#1 and 3-14#1 wells. The majority of the wells surrounding this region are Type 1 wells that contain the sucrosic dolomite lithofacies with some of the best reservoir quality in HPU. SGM used the high porosity and permeability values from these surrounding wells to create a "sweet spot" in the upper Smackover Formation that is more than two miles long. However, core studies and production data indicate that the sucrosic dolomite pods are relatively limited in their extent (Figure 37). In order to match the geologic conceptual model and production data, a template (Figure 38) was necessary in order to bias porosity and permeability distribution in the upper Smackover Formation (Figure 36).

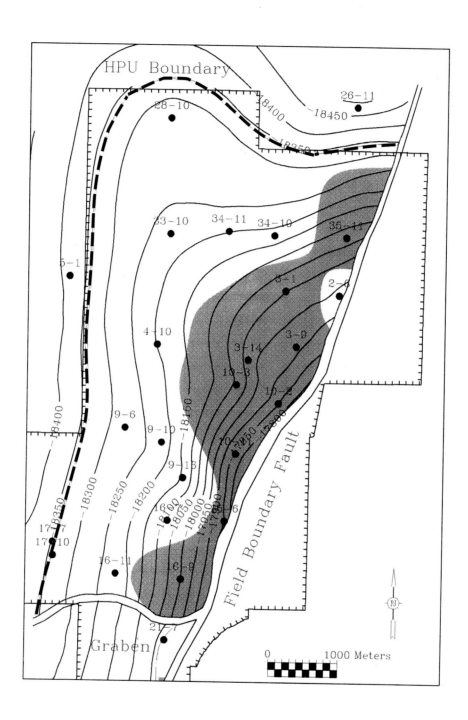

Figure 40. Combined average total porosity (pay *and* nonpay) and structure map of the middle Smackover lithofacies package as modeled in SGM. The map centers on the main fault block because it contains the majority of well data. The solid black contours represent the structural top of the middle Smackover Formation and the contour interval is 50 feet. The black dashed contour is the gas/water contact. The gray shaded area is that portion of the middle Smackover Formation that has an average porosity greater than 8 percent. This figure demonstrates that the best reservoir in the middle Smackover Formation is focused on the crest of the structure.

selective dissolution that has taken place within the dolomites. The finely crystalline dolomite shows progressively less dissolution toward the flanks of the structure. This crest–versus–flank reservoir quality relationship suggests that the diagenetic event(s) responsible for porosity development was a relatively late–stage, deep burial episode that occurred concurrent with, and/or subsequent to, the tectonics responsible for the present day structure at HPU.

A possible source path for the diagenetic fluids responsible for porosity development in the middle Smackover was the north–south trending fault system that runs along the crest of the structure and forms the eastern margin of the unit. Fluids undersaturated with respect to calcium carbonate may have moved up from below along the fault system and out into the Smackover Formation, dissolving the carbonates proximal to the crest. However, as the fluids moved down structure, they reached equilibrium with the dolomites and were unable to further dissolve the carbonates.

Lower unit

The lowermost unit in the Smackover Formation is composed entirely of the medium crystalline dolomite. Its total thickness ranges from 21 to 42 feet (6 to 13 meters) and the net pay thickness ranges from 0 to 42 feet (0 to 13 meters).

Similar to the middle Smackover unit, the character of the lower Smackover is relatively uniform—it is composed of a single lithofacies. Variation in reservoir properties tends to be somewhat gradational, and a geologic template was not necessary to model this unit.

The porosity and permeability trends in the medium crystalline dolomite are similar to the finely crystalline dolomite (i.e., the middle Smackover): for wells along the crest of the structure, average porosity and permeability is 11.9 percent and 3.61 millidarcies, respectively; average porosity is 4.4 percent and average permeability is 1.26 millidarcies for flank wells (Figs. 41, 42 and 43).

The major pore type in the medium crystalline dolomite is non–fabric selective vugs and molds. The principle difference in reservoir quality from the overlying finely crystalline dolomite is the coarser crystal size in the medium crystalline dolomite, which resulted in larger pore throats and leached voids and an overall improvement in reservoir properties. The diagenetic events most likely responsible for porosity development in the medium crystalline dolomite are the same as those discussed for the finely crystalline dolomite above.

The Reservoir Model—Norphlet Formation

The contact between the Norphlet Formation and overlying Smackover Formation is a very distinct lithologic boundary that can easily be recognized in both cores and logs. The top of the Norphlet Formation was used in the reservoir model as the marker or datum from which the Norphlet stratigraphic framework was built.

The wells at HPU did not penetrate the entire Norphlet Formation because they encountered the gas/water contact well above its base. Thus, the total thickness of the Norphlet Formation at HPU is unknown, but from well penetrations we know it is in excess of 520 feet (159 meters). The total thickness of the Norphlet Formation above the gas/water contact reaches a maximum of approximately 404 feet (123 meters) in the HPU 10-2#1 well.

Figure 44 displays the ratio of net pay thickness to total Norphlet Formation thickness above the gas/water contact. It dramatically illustrates that, contrary to the overlying Smackover Formation,

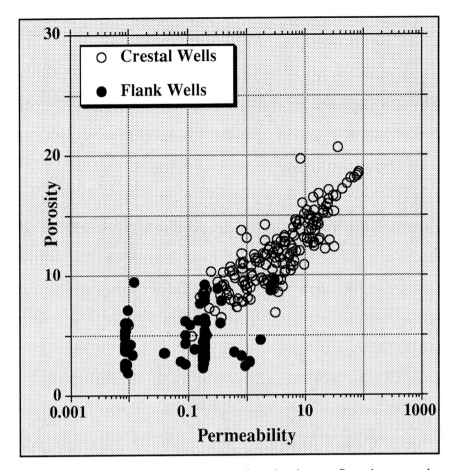

Figure 42. Porosity–permeability crossplot for the lower Smackover unit, which is composed entirely of the medium crystalline lithofacies. The two distinct populations in the crossplot indicate that reservoir quality is directly related to position on the structure (crest vs. flank). The data in this plot are only from four crestal wells and two flank wells in order to minimize overposting and cluttering of the data. Figure 43, shows the average porosity relative to structure within the lower Smackover Formation using data from all wells.

reservoir quality in the Norphlet Formation improves toward the structural flanks. This trend is related to the updip increase in intergranular quartz cement associated with a decrease in authigenic illite as discussed earlier (see *Diagenetic Illite* and Fig. 27).

Figure 45 is a net pay thickness map for the Norphlet Formation The structural contours for the top of the Norphlet Formation are posted on the map in order to emphasize the distribution of net pay with respect to structure. Figure 45 shows that: 1) although the *net-to-gross ratio* in the crestal Norphlet Formation is lower (see Fig. 44), and reservoir quality in the Norphlet Formation increases offstructure, total pay thickness is generally higher along the crest because the crest has the highest total interval thickness above the gas/water contact; and 2) there is a localized region in the central part of the crest where the reservoir quality is significantly reduced. Partitioning the Norphlet Formation into an upper and lower section helps to better understand the distribution of reservoir.

Although reservoir quality improves flankward, the upper Norphlet net pay thickness (Fig.46A) is highest along the crest of the structure. This is because the majority of the upper Norphlet exceeds the porosity and permeability cutoffs. Therefore, the thicker the gross interval above the gas/water contact, the thicker the resulting net pay. This is illustrated in Figure 46B, an average

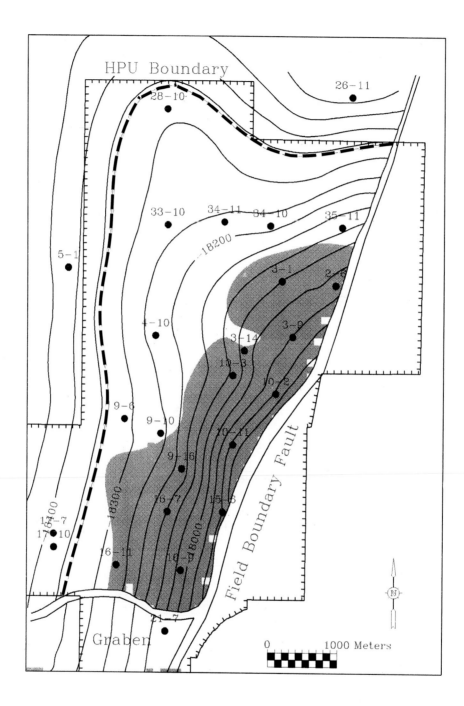

Figure 43. Combined average total porosity (pay *and* nonpay) and structure map of the lower Smackover lithofacies package as modeled in SGM. The map centers on the main fault block because it contains the majority of well data. The solid black contours represent the structural top of the lower Smackover Formation and the contour interval is 50 feet. The black dashed contour is the gas/water contact. The gray shaded area is that portion of the lower Smackover Formation that has an average porosity greater than 8 percent. Similar to the middle Smackover Formation (see Figure 40), reservoir quality in the lower Smackover Formation is best on the structural crest and diminishes rapidly toward the flanks.

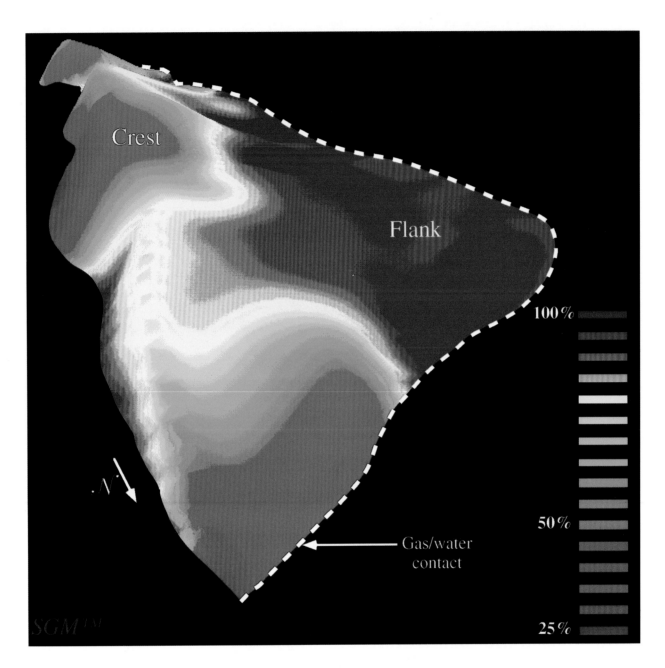

Figure 44. Image captured from SGM™ showing the ratio of net pay thickness in the Norphlet Formation relative to its gross interval thickness *above the gas/water contact*. The net–to–gross ratio has been draped over the Norphlet structure to show the relationship between reservoir quality and structural position (crest vs. flank). Red regions show areas of the Norphlet Formation that are 95 to 100 percent pay. The light blue regions along the crest represent a ratio of about 35 to 40 percent net–to–gross. The view is from above, looking toward the south, and the crestal area in the distance (upper left corner of the display) is the south fault block, which shows a relatively low net–to–gross ratio as well.

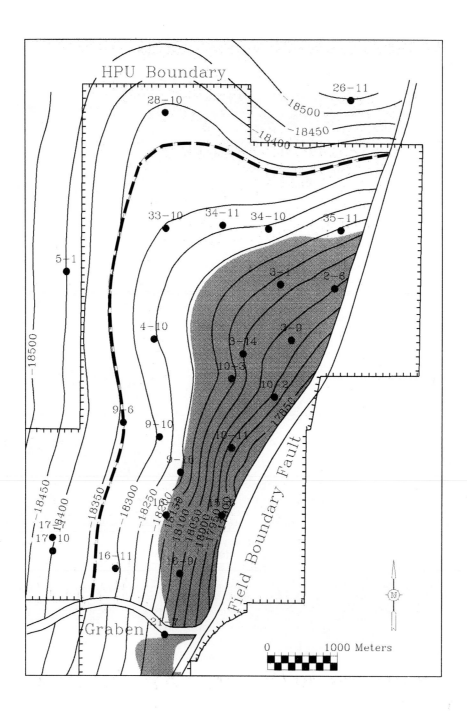

Figure 45. Combined net pay thickness and structure map of the Norphlet Formation as modeled in SGM. The map centers on the main fault block because it contains the majority of well data. The solid black contours represent the structural top of the Norphlet Formation and the contour interval is 50 feet. The black dashed contour is the gas/water contact. The gray shaded area represents net pay thickness of over 150 feet for the Norphlet Formation. Although the thickest net pay is focused along the structural crest, this is more attributable to greater overall thickness above the gas/water contact along the crest than it is better reservoir quality (see Figures 46-49).

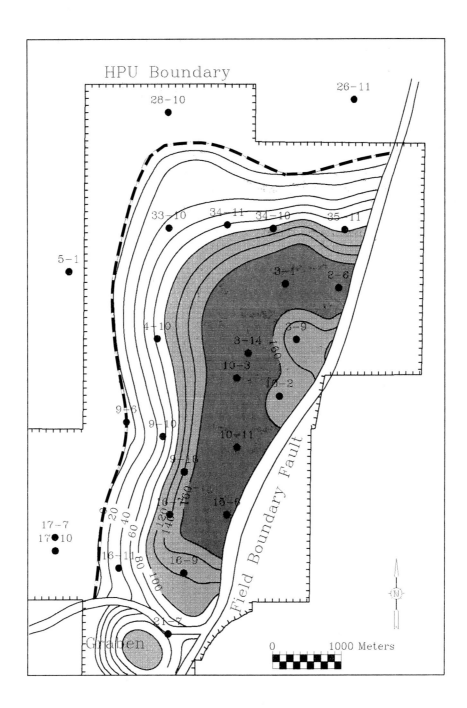

Figure 46A. Net pay thickness map of the upper part of the Norphlet Formation as modeled in SGM. The map centers on the main fault block because it contains the majority of well data. The solid black contours represent the net pay thickness in the upper Norphlet Formation and the contour interval is 20 feet. The black dashed contour is the gas/water contact. The light gray color represents net pay thickness between 100 and 160 feet, and the dark gray everything above 160 feet thick. The different shading emphasizes the localized (relative) "thin" in the upper Norphlet Formation around the HPU 3-9#1 and 10-2#1 wells. Figure 46B shows an average porosity map for the same reservoir interval.

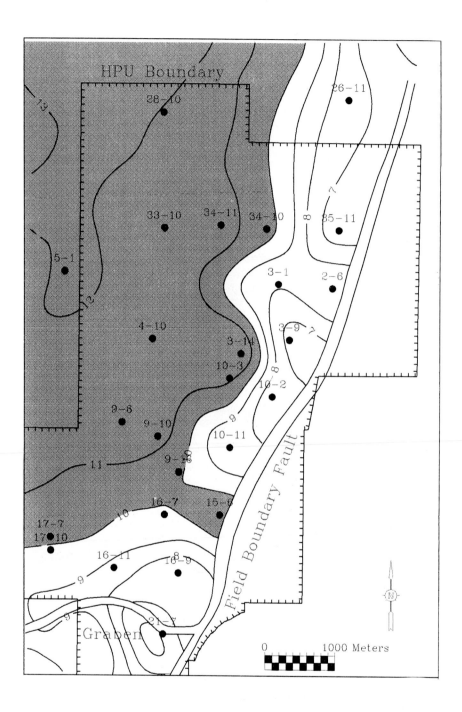

Figure 46B. Average porosity map of the upper part of the Norphlet Formation, including both pay and nonpay, as modeled in SGM. The map centers on the main fault block because it contains the majority of well data. The solid black contours represent average porosity in the upper Norphlet Formation at a one–percent contour interval. The light gray is the region where average porosity is greater than 10 percent. This map clearly shows that reservoir quality in the upper Norphlet Formation increases flankward.

porosity map for the entire upper Norphlet (i.e., pay *and* nonpay). This figure shows that porosity in the upper Norphlet increases flankward, but porosities along the crest average about 6 to 7 percent, which is above the porosity cutoff. The higher porosities in the upper Norphlet Formation, relative to the underlying section, are interpreted to have formed through dissolution of a pre–existing mineral cement, a cement which arrested mechanical compaction and the associated porosity reduction (see *Pressure Solution and Evidence for Secondary Porosity* and Fig. 31).

The distribution of reservoir in the lower Norphlet Formation (Figs. 46C and D, Fig. 47) differs significantly from the distribution in the overlying, upper Norphlet Formation. In the central part of the crest, the majority of the lower Norphlet is non–reservoir. Figure 46 C shows that net pay thickness in the region of the HPU 10-2#1 well is less than 10 feet (3 meters). However, just to the southwest in the region of HPU 16-9#1 well, net pay thickness is approximately 180 feet (55 meters). These regions are approximately along strike with respect to each other but there is tremendous disparity in their reservoir quality! Figure 46 D is an average porosity map for the entire lower Norphlet (i.e., pay *and* nonpay) and it shows why so much of the lower Norphlet is nonpay—the average porosity in the region of HPU 10-2#1 is approximately 3.5 percent.

Norphlet Formation simulation layer geometry

The Norphlet Formation appears to have been deposited on an erg uninfluenced by basement structure. A stratigraphic section hung on top of the Norphlet (see Fig. 22) shows several zones that correlate horizontally across the unit. The uppermost marker is also associated with the zone of secondary porosity development in the HPU 16-9#1, and it appears that the precursor cement deposited in this zone was stratigraphically influenced. It should be noted that the conventional core showed distinct dune boundary surfaces at the top and base of that particular zone, strongly suggesting that Norphlet Formation diagenesis followed a sedimentary template.

Having established that the Norphlet Formation correlates laterally with little stratigraphic variation in thickness, uniform cell thicknesses of two feet were used in SGM to construct the stratigraphic framework layers for the Norphlet Formation. Numerous stratigraphic cross–sections datumed on the top of the Norphlet Formation were constructed in SGM for the purposes of correlating zones of similar reservoir properties. SGM layer numbers were recorded for the correlated zones, and these layers ultimately were combined to form the reservoir simulation layers. Figures 48 and 49 present a structural cross–section along the crest of the Norphlet Formation, showing the distribution of pay and nonpay with respect to the simulation layers. As can be seen in the displays, an effort was made to identify zones of "extreme" reservoir properties, such that flow "conduits" and barriers within the "fine–scale", geologically–based model were preserved when scaled up in Western Atlas' GeoLink® for reservoir simulation.

ENGINEERING STUDIES—RESERVOIR SIMULATION

Reservoir simulation studies are commonly used to evaluate both new and mature fields and to determine the most efficient reservoir management schemes. The main business incentive for performing such studies is that of economics; that is, to increase profitability through better reservoir management. A realistic reservoir model can be a very effective tool for determining hydrocarbon reserves, for developing plans for new fields, for estimating facility needs, and for evaluating alternative plans to improve field productivity.

A numerical reservoir simulator is simply a collection of computer software that has the ability to solve the complex set of equations that govern fluid flow in porous media. It can thus calculate the flow of oil, gas, and water in underground petroleum reservoirs under a variety of well and field constraints. In order to do this, though, the region of interest must be completely defined;

Figure 46C. Net pay thickness map of the lower Norphlet Formation as modeled in SGM. The map centers on the main fault block because it contains the majority of well data. The solid black contours represent the net pay thickness in the lower Norphlet Formation and the contour interval is 20 feet. The black dashed contour is the gas/water contact. The gray areas are regions where net pay thickness is greater than 100 feet. The distribution of net pay in the lower Norphlet dramatically illustrates that the crest of the structure is *not* necessarily the thickest and best part of the reservoir. The HPU 10-2#1, 10-11#3 and 15-6#1 wells, all of which are along the structural crest, have net pay thicknesses of approximately 20, 80, and 90 feet, respectively. The gross interval thickness of the lower Norphlet above the gas/water contact for these wells ranges from about 225 to 250 feet, meaning that the HPU 10-2#1 well has a net pay–to–gross thickness ratio of less than 10 percent! Figure 47 presents a rather dramatic image of this region.

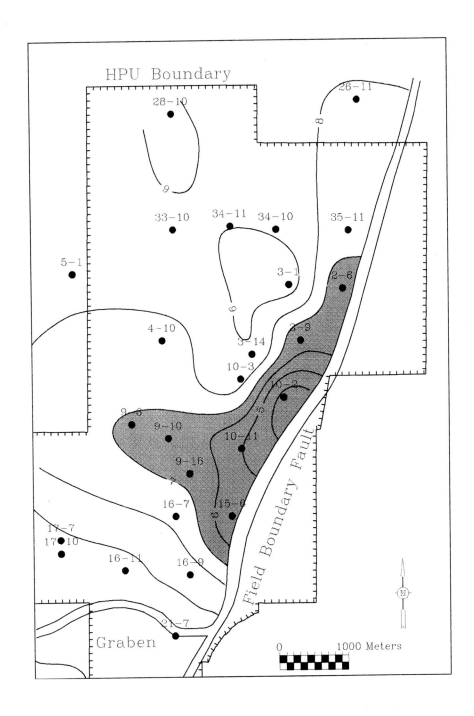

Figure 46D. Average porosity map of the lower Norphlet Formation, including both pay and nonpay, as modeled in SGM. The map centers on the main fault block because it contains the majority of well data. The solid black contours represent average porosity in the upper Norphlet Formation at a one–percent contour interval. The light gray is region where average total porosity is *less than* seven percent. This map emphasizes the reason for low net pay thicknesses along the crest of the structure (see Figs. 46C and 47).

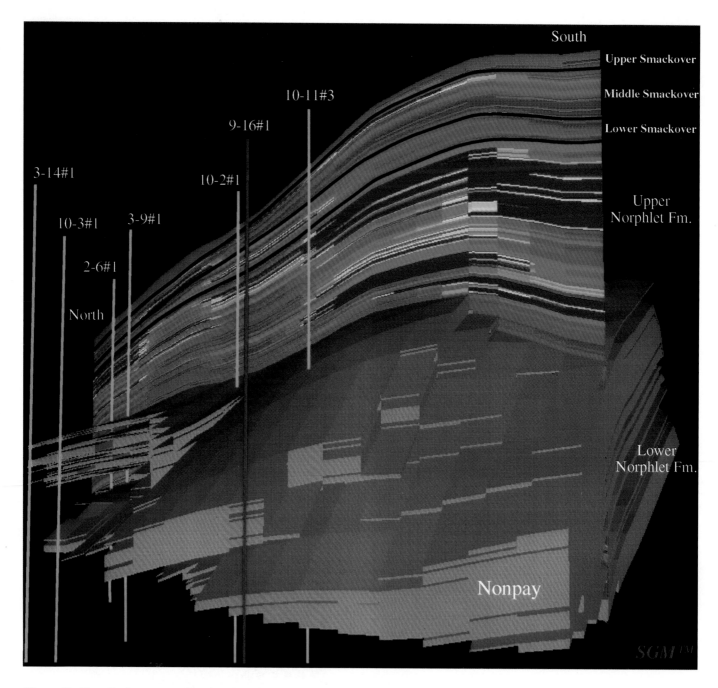

Figure 47. Two displays are combined to show the distribution of net pay and nonpay in the Smackover and Norphlet Formations along the crest of the main area. The view is looking toward the northeast at a cross-section panel that vertically stretches from the top of the Smackover Formation to the gas/water contact within the Norphlet Formation. The colors show permeability from less than 0.01 md (blue) to greater than 20 md (red), with the nonpay regions in gray. The tops of the lithofacies packages are traced on the cross-section as black lines. The second part of the display shows, in gray, the three-dimensional distribution of the large nonpay region within the lower Norphlet Formation discussed in the text. *Except for the cross-section panel, pay is "transparent"* **in the lower part of the Norphlet Formation.** Thus, places where the cross-section is visible along the lower left and right side of the display represent "hollow" zones of reservoir quality sandstone pay. Notice that the nonpay in the lower Norphlet Formation is relatively massive around the HPU 10-2#1 and 10-11#3 wells (center), but it thins and becomes discontinuous toward the north (left) around the HPU 3-9#1 well. There are several isolated pay compartments illustrated along the right (south) side of the display. These lower Norphlet Formation compartments are isolated from the upper Norphlet by a thick and continuous barrier. Furthermore, they are sealed updip (on the east) by the field-bounding fault and downdip (west) as the formation dips beneath the gas/water contact. The green wells are active producers, the red well is an active gas injector, and the blue well is inactive. The thickness of the lower Norphlet nonpay zone along the right margin of the display is approximately 200 feet (61 meters).

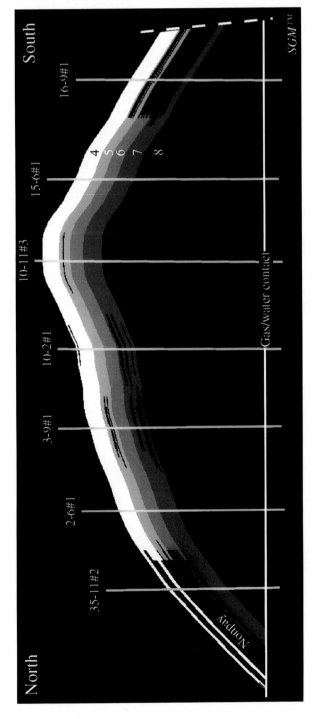

Figure 48A. North-south cross-section panel across the main fault block showing the relationship between reservoir quality and simulation layers for the upper Norphlet Formation. All nonpay *and everything below simulation layer no. 8* has been filtered out and appears black on the panel. The overlying Smackover Formation has been stripped off as well. The upper Norphlet Formation is represented by simulation layers 4 through 8, where reservoir quality is excellent except along the flanks (e.g., HPU 35-11#2 and 16-9#1 wells). The north flank graben fault is shown by a white dashed line along the right margin of the display. Total thickness of this interval is approximately 170 feet (52 meters). See Figure 5 for location of the wells.

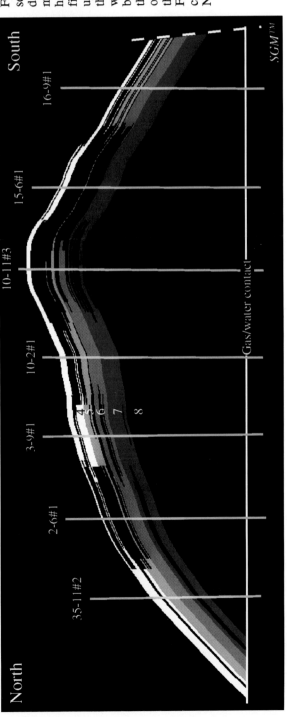

Figure 48B. Same north-south cross-section panel as Figure 48A, but in this display, all reservoir above 5.0 millidarcies permeability, a relatively high permeability at HPU, has been filtered out and appears black. The upper Norphlet is especially good along the crest (e.g., 10-11#3 and 15-6#1 wells). Notice the correspondence between high-permeability zones and the simulation layers. This was done in order to minimize the affects of scaling the model up for reservoir simulation. Figure 49 shows this same set of cross-section panels for the lower Norphlet Formation.

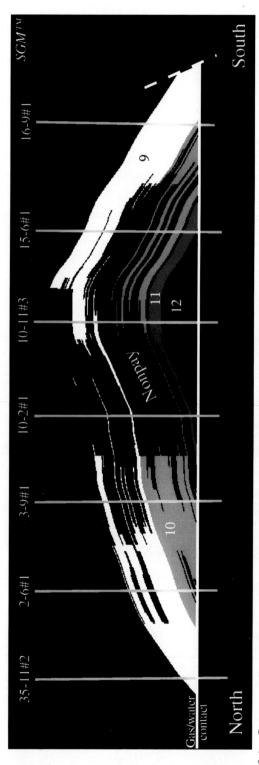

Figure 49A. Same north-south cross-section panel as Figure 48, but showing the relationship between reservoir quality and simulation layers for the *lower* Norphlet Formation, represented by simulation layers 9 through 12. A great deal of the lower Norphlet Formation (simulation layers 9 and 10) is non–reservoir along the crest. This is the reason for the poor net-to-gross ratio along the crest (see Figure 44) and corresponds to the large nonpay zone illustrated in Figure 47. The maximum total thickness of this interval is approximately 265 feet (81 meters). See Figure 5 for location of the wells.

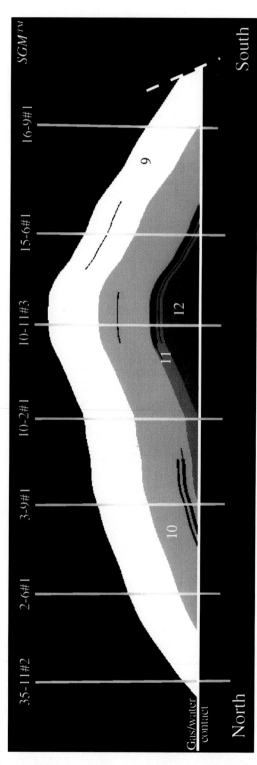

Figure 49B. This figure shows the distribution of reservoir with permeabilities above 5.0 millidarcies in the lower Norphlet Formation in the same north-south cross-section panel as Figures 48 and 49A. The lower Norphlet Formation has very few "high permeability" zones and most of the gas injection occurs in the upper Norphlet Formation, which is why the simulation layers are much thicker here, relative to the upper Norphlet Formation.

including reservoir structure and topography, reservoir rock properties, fluid properties, and initial distribution of the fluids within the reservoir. Well locations and their historical flow rates also need to be known.

The present Hatter's Pond Study was conducted using Western Atlas' VIP™ Compositional Model. The system can simulate gas condensate flow behavior, taking into account the fact that fluid properties and phase behavior varies strongly with fluid temperature, pressure, and composition. Fluid properties and phase equilibrium are governed by a generalized cubic equation of state in which reservoir fluid is treated as a mixture containing an arbitrary number of hydrocarbon and non–hydrocarbon components.

Accuracy and usefulness of any reservoir model is heavily dependent on the quality of the underlying geologic model and on the retention of the character of the model on scaling–up to the chosen simulation grid by whatever means selected. Development of a geological data base via Stratamodel's SGM, described above, formed a solid basis for the Hatter's Pond reservoir simulation model.

Geologic Model Conversion to Reservoir Simulation Scale

Complex three–dimensional geologic models developed using state–of–the–art workstation software can embody a large amount of detail. For the Hatter's Pond model, the SGM representation contained on the order of three million individual grid elements defined on a vertical scale of one to four feet. Once this model was complete, it had to be converted to a format and scale acceptable for reservoir simulation. Reduction of the number of grid cells by orders of magnitude is necessary because of the limitations of computer software and hardware to solve the fluid flow equations in a reasonable time span.

Western Atlas' GeoLink is a software product that can be used for three–dimensional visualization and model conversion to simulation scale. It provides a direct interface with SGM for input, and with the VIP–Executive reservoir simulator for output. GeoLink also provides a simple programming language for creating additional geological properties from previously defined measured properties. Using SGM and GeoLink, the geologist and the reservoir engineer participated jointly to implement the conversion of the detailed geologically–based reservoir model to a model based on hydraulic flow units. Here, a representative set of permeability cross–sections were studied (see Fig. 48) and the layers originally defined in SGM were grouped in accordance with the ability of each grouping to act collectively as a single flow unit with regard to potential fluid movement. In making these selections, the emphasis was on hydraulic continuity and not so much on stratigraphic and lithologic definition. The Hatter's Pond geologic model contained more than 300 SGM layers, which were exported to GeoLink where the scale–up process resulted in 13 simulation model layers or flow units, 3 in the Smackover and 10 in the Norphlet. The Norphlet layers were a further subdivision of four layers used in previous simulations which were found to be inadequate flow definition (Stoudt and others, 1992). The reservoir properties as defined on the geogrid were averaged or summed appropriately to obtain porosity, permeability, gross thickness, net thickness, and water saturation values for the grossed–up grid. Arithmetic averages were used for porosity, geometric averages for horizontal permeability, and harmonic averages for vertical permeability.

The final step was to design an aerial simulation grid. Theoretically, a separate grid can be defined for each layer to accommodate vertical definition of the graben sloping faults. However, development problems with the mesh–building software at the time of model construction led us to use of one grid projected onto each simulation layer, with vertical faults defined by the fault traces on the top of the Smackover. The final mesh was 17x37x13 for a total of 8177 cells, 629 per layer (Fig. 50). The zero–porosity blocks were considered inactive, leaving 3272 active cells.

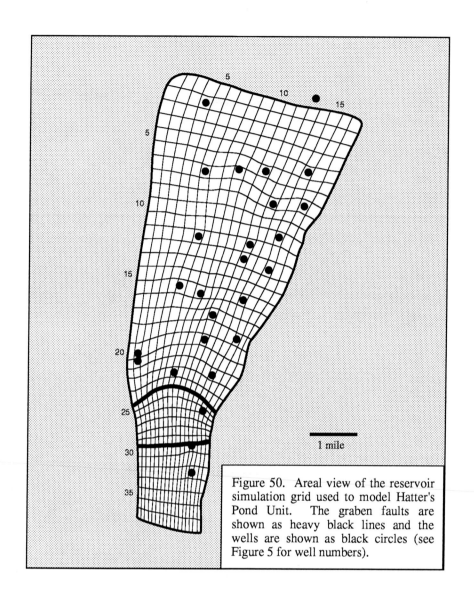

Figure 50. Areal view of the reservoir simulation grid used to model Hatter's Pond Unit. The graben faults are shown as heavy black lines and the wells are shown as black circles (see Figure 5 for well numbers).

Averaging of reservoir properties from the 100m x 100m geocells (2 acres) provided the needed properties defined on the course grid (about 20 acres). The grid–construction algorithm also centered wells within the appropriate cell for each defined well location. This resulted in some minor localized distortion of the grid.

Reservoir Fluid Description

Special fluid analysis tests conducted in 1975–76 indicate that the HPU reservoir fluid is a rich gas condensate. It was shown to exist totally as a gas at the original reservoir pressure of 9150 psi and reservoir temperature of 325 degrees Fahrenheit. As pressure is reduced below the original value, the fluid remains a gas until the dewpoint pressure is reached at 3050 psi. At that point, hydrocarbon liquids begin to condense. Further reductions in pressure result in increasing liquid saturations, reaching a maximum of about 40.0 percent of the hydrocarbon pore volume at 2965 psi. Note that this change is very rapid occurring with only an 85 psi pressure drop. Still further reductions result in liquid revaporization and decreasing condensate saturations in the rock. This behavior is illustrated by the constant composition expansion experiment shown in Figure 51.

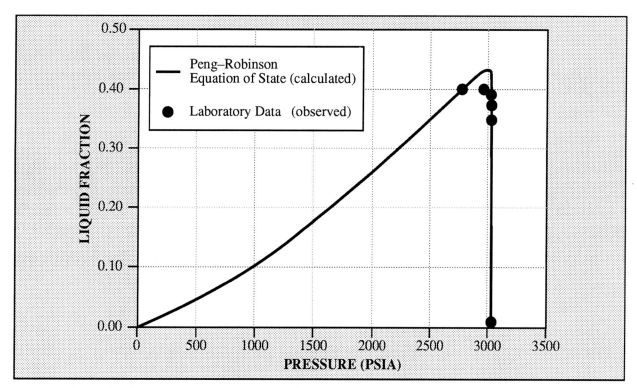

Figure 51. HPU 34-11#1 constant composition expansion.

Initial reservoir hydrocarbon fluid was assumed to be the same as the recombination of separator samples taken during initial testing of the Peter Klein 3-14#1 discovery well in 1975, and also from the Creola Minerals 34-11#1 later that same year (note: names such as Peter Klein and Creola Minerals refer to their original lease names; since unitization in 1988, all wells are now preceded by HPU). The sample compositions were nearly identical. Additional fluid sampling also indicated that the same fluid is common to both the Smackover and Norphlet formations. The Peter Klein 3-14#1 composition is given in Table 1 along with composition of the separator products (800 psig and 114° F). Note that the intermediate and heavy hydrocarbons are present in fairly high concentrations. As a result, the Hatter's Pond fluid is very rich in liquid content compared to most reservoir condensates. It typically yields 255 barrels of liquid condensate (C_{5+}), 65 barrels of natural gas liquids (C_2–C_4), and 687 thousand cubic feet of residue gas (C_1,N_2,CO_2) for each million cubic feet of full well stream when processed through the current surface facilities.

Compositional simulation requires fluid characterization through an equation of state which matches laboratory-measured PVT experiments. Specifically, the Peng-Robinson equation was chosen for this work. Western Atlas' DESKTOP-PVT™ program was used to simulate an extensive body of experimental data and to systematically vary specific parameters in the equation until a best match of the data was achieved. Nonlinear least squares regression was used to obtain a fine-tuned equation that was used in the simulator to calculate phase behavior and phase densities as a function of temperature, pressure, and composition. Twelve components were employed, namely: N_2, CO_2, C_1, C_2, C_3, C_4, C_5, C_6, and four pseudo-components representing the C_{7+} fraction.

TABLE 1.—COMPOSITION OF HATTER'S POND UNIT FLUIDS

Component	CO_2	N_2	C_1	C_2	C_3	C_4	C_5	C_6	C_{7+}
Reservoir Fluid	.0613	.0283	.4625	.0958	.0666	.0644	.0453	.0342	.1416
Separator Gas	.0804	.0476	.6970	.1008	.0428	.0220	.0065	.0020	.0009
Separator Liquid	.0364	.0033	.1575	.0892	.0976	.1195	.0957	.0761	.3247

Field Operations

From 1976 to 1988, the field was produced by simple pressure depletion; that is, gas was produced but since no gas reinjection occurred, pressure declined to about 5000 psi. The field was unitized in 1985 for the purpose of gas cycling, and reinjection of dry residue gas from the gas processing facilities began in 1988. It is anticipated that gas cycling will continue to 1998 for a total injection period of 10 years.

The purpose of gas cycling is to slow the decline of reservoir pressure, delaying the onset of liquid condensation in the reservoir at pressures below the dewpoint. Hydrocarbon liquid in the rock is undesirable because it interferes with the ability of gas to flow to the wellbore and thus can reduce well productivity. Also, the liquid itself lacks sufficient saturation to flow through the reservoir to a well and be produced. In fact, it can sometimes cause liquid to build–up in the region surrounding a well further restricting gas flow, possibly even killing the well. All of these are relative permeability effects. Gas cycling also introduces dry gas into the reservoir which helps sweep the original fluid to nearby wells. Dry gas mixed with the original wet gas results in a gas of diluted liquid content. The diluted gas has a higher dewpoint than the original fluid but does not condense near as much liquid below its dewpoint. A 50–50 mixture of dry gas with the original Hatter's Pond fluid will result in a maximum liquid dropout of 6.5 percent hydrocarbon pore volume. This is not enough to affect well productivity.

Produced gas is initially passed through equilibrium separators. The separator liquid is treated by a stabilization (distillation) process to remove light hydrocarbons leaving a stabilized condensate revenue stream. The separator gas and stabilizer gas are merged and sent to a gas processing plant where cryogenic processing separates the inlet stream into natural gas liquids and a dry residue gas with sulfur, CO_2 and water removed. The residue is either sold, or reinjected, or it is split and part is sold and part is reinjected (Fig. 52). Currently, the unit is operating with 100 percent gas reinjection.

The unit has 10 active producers and 3 injectors (September 1994). All of these wells are on the east (crestal) side of the unit where the Norphlet is most prolific. The western downdip half has never contributed significantly to the overall success of the operation. Full well stream production is 52 million cu ft/D with about 12,000 bbl/D of associated liquids. Gas injection rate is 32 million cu ft/D. To date the unit has produced 241 billion cu ft of full well stream which has yielded 58 million barrels of condensate, 15 million barrels of natural gas liquids, and 168 billion cu ft of dry gas, 47 billion cu ft of which has been reinjected. Plans are being made for additional drilling in 1995.

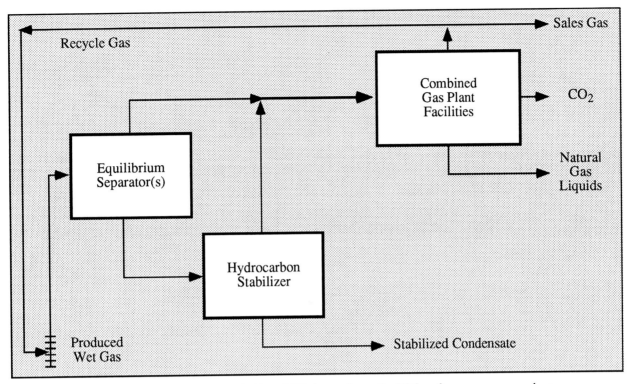

Figure 52. Schematic of the Hatter's Pond Unit surface facilities for gas processing.

History Matching of Historical Data

Various aspects of the simulation model definition have been described above. These included the reservoir layering, rock property definition for each grid cell, relative permeability and capillary pressure data specification, fluid component specification, fluid property definition by means of an equation of state, and initial equilibrium pressure and saturation conditions. Additionally, well locations and completion intervals had to be provided along with tables that relate tubinghead pressure and flowing bottomhole pressure with well flow rate and separator gas/condensate ratio, the latter being a measure of produced fluid composition. Once the model has been completed, history matching can begin.

The historical gas production rates and injection rates for each well were used as model input. The primary field measured data used for matching calculated simulator response were the reservoir pressures and condensate yields by individual well. Certain modification had to be made to the data to achieve reasonable matches, but this was not a major effort due to the excellent nature of the geologic model.

The shut–in pressure match was obtained by adjusting model pore volume, both by increasing the initial water saturation and by localized modifications to porosity. Known barriers to flow between certain flow units were also added, i.e., between the Smackover and Norphlet in the south fault block, and between Smackover Units 3 and 4 in the northeast portion of the main reservoir. Both of the graben faults were modeled as sealing although it is now known that the northern one is in pressure communication with the main section. In previous versions of the model, it was necessary to compartmentalize some reservoir regions to achieve observed rapid pressure gradients in the associated well. However, the upper Smackover was recently remodeled in SGM to more accurately represent the isolated and distinct pods of high porosity and permeability. This allowed removal of artificially inserted barriers and provided a more satisfying reservoir description which, in turn, resulted in satisfactory pressure matches.

Matching of breakthrough time and subsequent liquid yield reduction with time required modest permeability increase in several flow units in the regions between the injection well HPU 3-1#1 and producers HPU 2-6#1 and between injector HPU 9-16#1 and producer HPU 10-11#3. Gas breakthrough to HPU 3-9#1 from the HPU 3-1#1 required a larger increase in permeability to match field results but a reasonably close match was ultimately achieved. Earlier model versions that used only four Norphlet layers for simulation were unable to match breakthrough times and liquid yield variations. But by using additional layers to more accurately represent the actual permeability variations, much better results were obtained.

Pressure history matches for four wells, HPU 2-6#1, HPU 3-14#1, HPU 10-11#3, and HPU 15-6#1 are shown in Figure 53. These are a representative group of wells from the main fault block. Each sub–figure shows the separator gas production rate along the bottom and the calculated reservoir shut–in pressure above. The solid lines are the calculated pressures from the simulation model while the discrete data points are measured pressures from field shut–in pressure build–up tests. These all show generally satisfactory agreement, although some improvement might be made in the late–time behavior of well HPU 10-11#3 (1988 and later). The discovery well HPU 3-14#1 was lost in 1985 due to tubing collapse and other mechanical failure.

Reservoir Simulation Production Forecasting

Field performance forecasting follows the history matching process. It is this step that provides information that can assist management in planning for continued exploitation of the reservoir. Predictions give the user a means to evaluate the future performance of a reservoir or well under different operating strategies. He can examine a variety of scenarios and select a strategy that likely will result in the most desirable performance, usually based on economic comparisons.

The first case run is often a base case involving only continued unit operations with no major changes. This provides a standard by which other cases can be compared. Modifications to the base–case situation can then be considered. As an example, production declines might suggest that a new well is needed to keep the gas processing plant operating at full capacity. The geologists and engineers would pick promising locations and a series of runs would be made to test the effect of a new well in each of the prospective new locations. Spreadsheets would be constructed from the results listing annualized full well stream production, injection volumes, and revenue streams for each of the locations. Economic analysis and comparison can determine if any or all of the new wells could be justified and then rank them on the basis of increased profitability. This methodology provides management with a basis for making informed decisions.

The procedure described above has been carried out on numerous occasions for the HPU. The results have led to changes in injection well locations, drilling of additional production wells, opening of additional perforations, and increasing the amount of gas being reinjected. It has also provided justification for drilling replacement wells for wells that have been lost for mechanical reasons. Other studies include: evaluation of an injection project in the south fault block, evaluation of field performance sensitivity to length of the gas injection period (10 years vs. 12 or 15 years), evaluation of the effect of increasing the capacity of the gas processing plant, evaluation of the effect of using full pressure maintenance vs. partial pressure maintenance, evaluation of the effect of lowering surface operating pressure, and evaluation of gas lift potential.

The impact of operational changes must be assessed to determine the overall economic viability of a *combination* of changes. Thus, it is useful to execute a 'most–likely' operational scenario

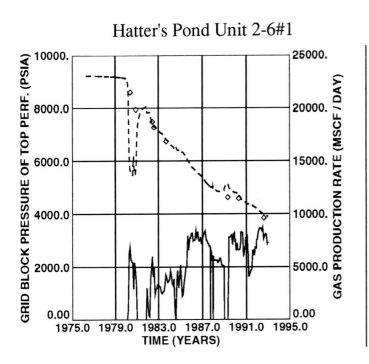
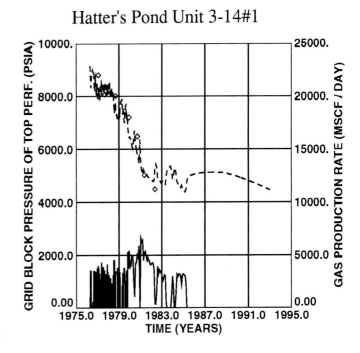
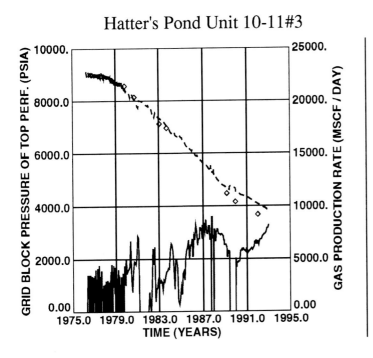
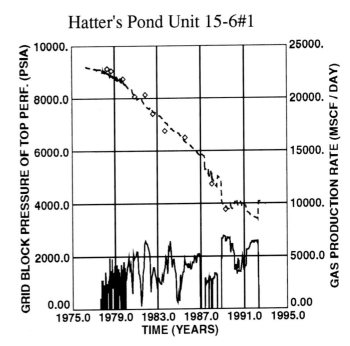

Figure 53. Pressure history match plots for four representative wells from the main area of Hatter's Pond Unit. Gas production rate is shown as a solid line, observed pressures are shown as diamonds, and simulation history match of pressures is shown as a dashed line.

whereby all anticipated changes are implemented to quantify expected recovery. This case should be updated periodically in order to remain current.

Finally, a primary depletion case was made with one producer per section. Here the reservoir gas was produced with no reinjection at any time. This case helped to establish primary recoverable reserves. The results provide support to other reservoir engineering techniques to establish figures used for official reserve bookings. It also provides a base condition for comparisons with gas cycling cases to help quantify incremental secondary reserves.

Simulation results are presented in the form of tabular output, time plots of the primary variables, and/or interactive three-dimensional visual displays on a workstation. The latter provides a means to view changes of any reservoir attribute over time. Thus, physical changes, such as frontal movements or pressure changes can be evaluated which ultimately improves understanding of the reservoir and the processes being applied to it.

A graphical image from one such display was captured and is shown in Figure 54. This was taken from a production forecast testing the feasibility of moving the current injection location from HPU 3-1#1 to the northeast to HPU 35-11#2. The figure shows methane concentration in the reservoir at 1/1/1998, near the end of the 10-year injection period. Methane content of the original fluid is 46.25 percent (red in the figure) while the injected residue gas contains 80.7 percent (dark blue). The top five simulation layers are stripped off in the main fault block and graben for a better look at the projected gas distribution at that time. All of the layers are present in the south fault block and the figure shows high methane concentrations in the Smackover which contained the injection perforations. None of the gas migrated to the Norphlet since the two formations are separated by a flow barrier. Notice that dry injection gas appears to have broken through in several producers, HPU 2-6#1, HPU 3-9#1, and HPU 10-11#3. These are the unit wells that have already experienced significant reductions in liquid yield.

It is clear from the success of reservoir simulation applied to the HPU that the geologic model constructed from a detailed knowledge of the rocks enabled us to model the unit with reasonable accuracy and that this ability will continue to have a significant influence on the reservoir management decisions to be made over its remaining life.

Future Work

Reservoir description is usually the area of greatest uncertainty for all reservoir simulation models. Therefore, to maintain their reliability and usefulness, models should be updated as new data become available. This could be the result of core data from new wells in regions with sparse well control, updated geophysical interpretations, updated geological interpretations based on reservoir performance and/or, attainment of new data from other sources. Thus, there should be a continuing evolution of the model right up to field abandonment. Accordingly, plans are in effect to improve the HPU model in the near future by implementing the following changes:

- Add core and log data from new wells to the SGM 3D model.

- Create new fault representations to match the current geophysically-based structural interpretation.

- Replace the gas-water contact, currently flat in the model, with a variable surface that more closely represents the actual gas-water contact as defined by core, log and capillary pressure data.

- Regrid with more simulation layers to provide better vertical definition, perhaps 18

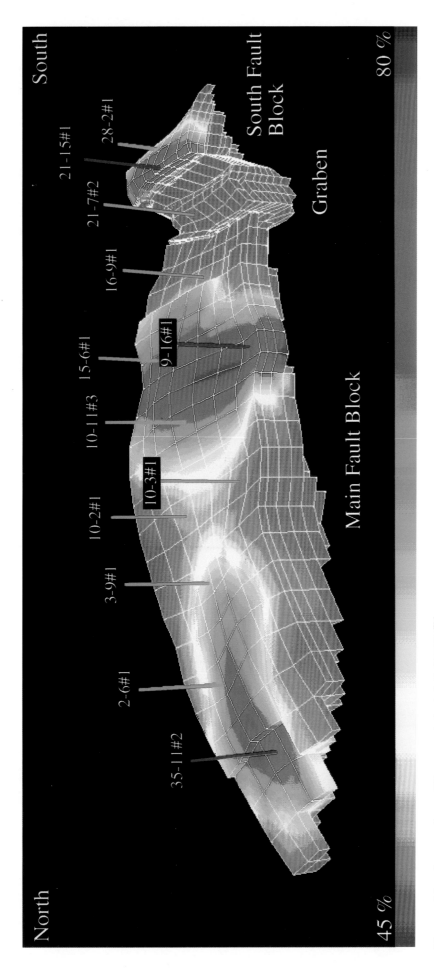

Figure 54. Image from Western Atlas' 3-DVIEW™ showing forecasted distribution of methane gas as of 1/1/98 as predicted through reservoir simulation. The color bar along the bottom indicates methane gas concentration. The original condensate contains about 46% methane (see Table 1), so red areas in this simulation have been essentially unaffected by gas injection. Gas injectors are shown in red and active producers are green. The Smackover Formation and upper part of the Norphlet have been stripped off of the top in the main fault block and in the graben so that the middle and lower Norphlet are shown down to the gas/water contact. The south fault block shows the reservoir from the top of the Smackover to the gas/water contact. The western flank of the field has been sliced away as well in order to show the gas distribution in the middle Norphlet. Note significant gas breakthrough is predicted (indicated by high gas content in green and blue) in the producing wells.

layers instead of the present 13. These will be added to the upper Norphlet where the flow of injected gas is most prevalent..

- Allow partial communication between the main part of the field and the graben fault block.

- Incorporate local grid refinement into grid cells containing wells.

- Consider the possibility of applying geostatistics to the well data instead of the present interpolation schemes used within SGM. Having several equally probable reservoir descriptions would provide a means to evaluate uncertainty in the performance forecast results.

- Update the historical production/injection data and repeat the history match process.

ACKNOWLEDGMENTS

The authors would like to thank the following individuals and organizations for making this publication happen:

- Four Star Oil and Gas Co., Texaco E & P Inc., New Orleans Onshore Division, and Texaco E & P Technology Department for permission to publish,
- Mike O'Friel and Kent Tribble, Texaco E & P. Inc., New Orleans Onshore Division, for critical technical advice and review,
- Leon Williams and Dave Byrd for log analysis work,
- Amy Blackwell for drafting of many of the figures,
- Bill Lawrence and Greg Vardilos for the core photography, and
- Christie Callendar for collecting EDS spectra of the breunnerite samples.

REFERENCES

BARIA, L. R., STOUDT, D. L., HARRIS, P. M., AND CREVELLO, P. D., 1982, Upper Jurassic reefs of Smackover Formation, United States Gulf Coast: American Association of Petroleum Geologists Bulletin, v. 66, no. 10, p. 1449-1482.

BRAUNSTEIN, J., HUDDLESTUN, P., AND BIEL, R., 1988, Gulf Coast Region COSUNA chart: American Association of Petroleum Geologists Correlation of Stratigraphic Units of North America (COSUNA) Project.

DIXON, S. A., SUMMERS, D. M., AND SURDAM, R. C., 1989, Diagenesis and preservation of porosity in Norphlet Formation (Upper Jurassic), southern Alabama: American Association of Petroleum Geologists Bulletin 73, p. 707-728.

HOUSEKNECHT, D. W., 1984, Influence of grain size and temperature on intergranular pressure solution, quartz cementation, and porosity in a quartzose sandstone: Journal of Sedimentary Petrology 54, p. 348-361.

KUGLER, R. L., AND McHUGH, A., 1990, Regional diagenetic variation in Norphlet Sandstone: implications for reservoir quality and the origin of porosity: Gulf Coast Association of Geological Societies Transaction 40, p. 411-423.

MARZANO, M. S., PENSE, G. M., AND ANDRONACO, P., 1988, A comparison of the Jurassic Norphlet Formation in Mary Ann Field, Mobile Bay, Alabama, to onshore regional Norphlet Trends: Gulf Coast Association of Geological Societies Transaction 38, p. 85-100.

McBRIDE, E. F., LAND, L. S., AND MACK, L. E., 1987, Diagenesis of eolian and fluvial feldspathic sandstones, Norphlet Formation (Upper Jurassic), Rankin County, Mississippi, and Mobile County, Alabama: American Association of Petroleum Geologists Bulletin 71, p. 1019-1034.

PRATHER, B. E., 1992, Origin of dolostone reservoir rocks, Smackover Formation (Oxfordian), northeastern Gulf Coast, United States of America: American Association of Petroleum Geologists Bulletin, v. 76, no. 2, p. 133-163.

SAIGAL, G. C., BJORLYKKE, K., AND LARTER, S., 1992, The effects of oil emplacement on diagenetic processes: examples from the Fulmar Reservoir sandstones, central North Sea: American Association of Petroleum Geologists Bulletin 76, p. 1024-1033.

SIBLEY, D. S., AND BLATT, H., 1976, Intergranular pressure solution and cementation of the Tuscarora orthoquartzite: Journal of Sedimentary Petrology, v. 46, p. 881-896.

SIEBERT, R. M., 1985, The origin of hydrogen sulfide, elemental sulfur, carbon dioxide, and nitrogen in reservoirs, in Timing of Siliciclastic Diagenesis: Relationship to Hydrocarbon Migration: Sixth Annual Research Conference of Gulf Coast Section of SEPM Abstracts, p. 30-31.

STOUDT, E. L., THOMAS, A. R., GINGER, E. P., AND VINOPAL, R. J., 1992, Geologic reservoir characterization for engineering simulation, Hatter's Pond Field, Mobile County, Alabama: Society of Petroleum Engineers Paper 24713, presented at the 67th Annual Technical Conference of the Society of Petroleum Engineers, Washington, D.C., p. 521-534.

SZABO, J. O., AND PAXTON, S. T., 1991, Intergranular volume (IGV) decline curves for evaluating and predicting compaction and porosity loss in sandstones: American Association of Petroleum Geologists Meeting Abstract, 1991 Annual Convention Program, p. 214.

TEW, B. H., MINK, R. M., MANN, S. D., BEARDEN, B. L., AND MANCINI, E. A., 1991, Geologic framework of Norphlet and pre-Norphlet strata of the onshore and offshore eastern Gulf of Mexico area: Gulf Coast Association of Geological Societies Transaction 41, p. 590-600.

THOMAS, A. R., DAHL, W. M., HALL, C. M., AND YORK, D., 1993, $^{40}Ar/^{39}Ar$ analyses of authigenic muscovite, timing of stylolitization, and implications for pressure solution mechanisms: Jurassic Norphlet Formation, offshore Alabama: Clays and Clay Minerals, v. 41, no. 3, p. 269-279.

THOMSON, A., 1979, Preservation of porosity in the deep Woodbine/Tuscaloosa trend, Louisiana: Gulf Coast Association of Geological Societies Transactions, v. 29, p. 1156-1162.

VINET, M. J., 1984, Geochemistry and origin of Smackover and Buckner dolomites (Upper Jurassic), Jay Field area, Alabama-Florida, in Ventress, W. P. S., Bebout, D. G., Perkins, B. F., and Moore, C. H., eds., The Jurassic of the Gulf Rim: Proceedings of the 3rd Annual Research Conference, Gulf Coast Section of the Society of Economic Paleontologists and Mineralogists Foundation (GCS/SEPM), p. 365-374.

WALDERHAUG, O., 1990, A fluid inclusion study of quartz-cemented sandstones from offshore mid-Norway - possible evidence for continued quartz cementation during oil emplacement: Journal of Sedimentary Petrology 60, p. 203-210.

WELLER, J. M., 1959, Compaction of sediments: American Association of Petroleum Geologists Bulletin, v. 43, p. 273-310.

APPENDIX A.—X–RAY DIFFRACTION AND PETROPHYSICAL DATA, HPU 16-9#1, HATTER'S POND UNIT, MOBILE COUNTY, ALABAMA

phi	k	depth	illite	quartz	kspar	plag	calc	dolo	pyrite	halite	breunnerite
12.3	23	18129	2	87	6	6	0	0	0	0	0
11.5	11	18130	2	80	9	9	0	0	-99	0	0
12	33	18133	2	85	7	6	0	0	0	0	0
10.4	4.54	18134	3	80	9	8	0	0	0	0	0
10.1	9.23	18138	1	87	5	5	0	1	0	0	0
10.8	12	18140	1	87	6	6	0	0	-99	0	0
10	2.72	18141	2	83	7	6	0	2	-99	0	0
10.3	17	18142	2	88	5	5	0	0	-99	0	0
8.6	1.09	18143	4	82	6	7	0	2	0	0	0
10.7	9.28	18144	4	82	6	7	0	1	0	0	0
9.6	4.01	18146	1	86	6	7	0	0	0	0	0
7	0.37	18149	4	82	6	7	0	1	-99	0	0
11.8	29	18154	2	86	6	5	0	1	0	0	0
5.1	0.25	18160	2	86	5	5	0	2	0	0	0
7.8	0.83	18161	2	82	7	6	0	3	0	0	0
4.4	0.22	18165	4	83	6	7	0	0	0	0	0
5.1	0.28	18166	1	84	7	8	0	0	0	0	0
7.7	1	18183	2	81	7	8	0	2	0	0	0
5.3	0.15	18188	2	84	5	7	0	2	0	0	0
5.2	0.06	18192	4	77	8	8	0	2	0	0	0
4.8	0.07	18194	2	77	7	10	0	4	0	0	0
4	0.09	18197	7	72	10	10	0	2	0	0	0
5.3	0.23	18200	5	80	6	5	0	3	0	0	0
4.3	0.1	18209	4	81	9	6	0	0	0	0	0
4.2	0.14	18213	4	82	5	5	0	3	0	0	0
4.1	0.03	18215	3	81	7	6	0	3	0	0	0
4.3	0.11	18218	8	77	8	5	0	2	0	0	0
4.1	0.11	18222	2	84	6	4	0	3	0	0	0
3.8	0.05	18224	1	85	6	5	0	3	0	0	0
3.4	0.06	18227	3	85	5	5	0	2	0	0	0
6.8	0.28	18233	8	68	9	8	0	7	0	0	0
4.7	0.12	18236	2	76	9	7	0	6	0	0	0
6.3	0.31	18239	6	72	6	9	0	6	0	0	0
4.8	0.12	18241	3	79	7	8	0	3	0	0	0
6.6	1.27	18244	3	82	6	5	0	3	0	0	0
9.1	2.8	18246	6	77	6	6	0	4	0	0	0
10.5	9.47	18253	5	79	7	5	0	4	0	0	0
10.1	6.64	18255	4	80	8	6	0	2	0	0	0
12.2	10	18260	4	79	7	7	0	2	0	0	0
10	3.95	18263	3	83	5	5	0	4	0	0	0
10.2	4.31	18269	9	78	6	5	0	2	0	0	0
10.9	22	18272	3	77	6	7	0	7	0	0	0
10.9	6.46	18275	3	75	8	7	0	6	0	0	0
12.9	21	18277	5	79	8	6	0	3	0	0	0
13.3	64	18280	5	81	6	6	0	2	0	0	0
6.2	0.4	18283	7	78	7	6	0	2	0	0	0
8.7	1.18	18286	6	85	3	3	0	2	0	0	0
8	0.7	18289	9	74	6	8	0	2	0	0	0
8.1	0.79	18292	5	79	6	6	0	5	0	0	0
8	0.7	18294	6	83	5	5	0	1	0	0	0
8.9	0.71	18297	5	80	6	5	0	3	0	0	0
10.6	1.42	18302	7	76	6	8	0	2	0	0	2

APPENDIX A (CONT).—X–RAY DIFFRACTION AND PETROPHYSICAL DATA, HPU 16-9#1, HATTER'S POND UNIT, MOBILE COUNTY, ALABAMA

phi	k	depth	illite	quartz	kspar	plag	calc	dolo	pyrite	halite	breunnerite
9	0.66	18306	12	72	7	6	0	1	0	0	3
10.3	0.76	18310	9	76	6	5	0	2	0	0	2
11.3	2.25	18314	8	75	6	6	0	3	0	0	2
7.5	0.79	18322	11	77	4	4	0	1	0	0	2
9.6	2.31	18326	5	78	4	6	0	3	0	0	4
10.8	3.65	18330	10	77	5	4	0	2	0	0	2
11.7	6.33	18333	7	75	5	3	0	9	0	0	0
9.3	1.66	18337	8	79	4	6	0	1	0	0	1
12.8	3.01	18343	9	78	5	2	0	4	0	0	2
11.9	1.79	18349	12	78	5	1	0	4	0	0	0
11.4	1.48	18353	12	77	4	2	0	2	0	0	3
11.9	1.54	18357	12	77	4	1	0	5	0	0	1
11.1	1.31	18361	12	77	4	2	0	3	0	0	2
12.4	1.38	18366	12	78	5	0	0	3	0	0	2
12.2	1.34	18370	14	80	4	0	0	2	0	0	0
8.2	1.37	18381	7	77	5	6	0	3	0	0	3
10.3	2.09	18383	7	78	5	3	0	3	0	0	3
8.4	0.62	18388	10	72	5	4	2	4	0	0	4
9.5	2.24	18393	9	80	4	4	0	3	0	0	0
8.6	2.51	18399	10	76	5	4	0	4	0	0	2
8.8	0.97	18401	11	76	5	3	0	3	0	0	2
8.6	3.27	18403	9	73	4	5	0	5	0	0	4
9	2.89	18408	11	77	5	5	0	2	0	0	0
5.7	0.21	18412	12	71	5	4	0	2	0	0	4
5.2	0.96	18416	12	67	5	7	0	3	0	0	6
8	0.51	18423	15	73	4	4	0	2	0	0	3
7.9	0.33	18426	13	70	5	2	0	1	0	0	10
9.2	0.29	18427	14	79	5	1	0	1	0	0	0
8.8	0.44	18428	14	74	4	2	0	3	0	0	3
7.6	0.31	18429	17	73	5	3	0	2	0	-99	0
10.2	1.69	18430	11	78	5	1	0	2	0	2	0
8.6	1.12	18431	19	71	4	1	0	1	0	1	3

(-99 equals trace quantity)

THE IMPACT OF GEOLOGIC RESERVOIR CHARACTERIZATION ON THE FLOW UNIT MODELING AT THE KERN RIVER FIELD, CALIFORNIA, USA

ELLIOTT P. GINGER,[1] WILLIAM R. ALMON,[1] SUSAN A. LONGACRE,[1] AND CYNTHIA A. HUGGINS[2]

[1] Texaco E & P Technology Department, 3901 Briarpark, Houston, Texas 77042; and
[2] Texaco U.S.A., Bakersfield Division, 1546 China Grade Loop, Bakersfield, California 93308-9700

ABSTRACT

A multidisciplinary Kern River Geology Team was charged with developing geologically-based reservoir models that improve understanding of the steam flood process and the encroachment of cool water in the Kern River field. The focus of the effort was on investigating reservoir layers, continuity and boundary conditions, as they relate to movement of steam, oil and water through the system. This paper presents the results of a pilot study of steam movement in a localized area in Kern River field known as Project D-159.

Sands of the Kern River Formation that serve as reservoir to hydrocarbons in the field were deposited in a set of braided river channels that repeatedly crossed the field area from late Miocene through Pleistocene time. Channel sand deposition within the study area was not continuous through time, but rather was episodic. At times the depositional system changed significantly, such that the channels shifted to another area and the only deposition in the study area was a succession of floodplain silts. Where these silt units are not breached by subsequent rejuvenated river channel systems, they are barriers between successive large sand reservoirs.

A spatial "description" of the sand and silt components in the reservoir system was developed using both conventional geologic methods and geostatistical methods. The conventional geologic methods include stratigraphic cross-sections, developing a layering pattern within the reservoir, and expressing that layering pattern on cross-sections and maps. The layering system was based on a set of "rules" that focused on the unconformable bases of the major sand episodes and that included the overlying silts with the sand body with which it was genetically linked.

Using GRIDSTAT – proprietary software developed at Texaco – a series of rapidly constructed geostatistical cross-sections were made that closely approximated the conventional cross-sections. A rigorous statistical comparison of the conventional and geostatistical cross-sections indicates that mean rates of differences in lithology assignment on the GRIDSTAT sections, relative to the conventional cross-sections, are less than 3 percent. The average difference in lithology assignment between GRIDSTAT and traditional techniques is approximately the same as would be present between cross-sections generated by different geologists. However, on GRIDSTAT cross-sections approximately 5 percent of the wells have lithology assignment error rates that are in excess of 10 percent. Such difference rates have the potential to produce significant errors in determining sand thickness and are unacceptably high; geological intervention is required.

Average total correlation difference between GRIDSTAT sections and traditional cross-sections is 5.5 percent. There is a tendency for GRIDSTAT to fail to find all geologic correlations; this tendency is significant at the 0.95 confidence level.

Stratamodel's program SGM™ was used to model the 3-D inter–well distribution of reservoir properties in Project D-159. A special layering style was used in order to best represent the internal stratigraphic character of the reservoir. A cell thickness of five feet was chosen within each sequence in order to minimize the vertical averaging of well data while keeping the size of the model within reason. Two types of well data have been modeled at D-159 to date—short–normal resistivity (from fifty–four wells) and temperature measurements (monthly profiles from eleven temperature observation (TO) wells for January through November, 1992). Geobodies and model operations are used in evaluating and tracking the rate of heat growth or dissipation. If temperature geobodies are built in a time sequence, growth in the volume of successive geobodies based on specific temperatures would allow calculation of the heat added to the reservoir; while shrinkage would indicate heat dissipation.

INTRODUCTION

In order for a steam flood to be both efficient and effective, geologists and engineers must have knowledge about the location of those elements in the reservoir that will confine the injected steam to the desired flow units, and they must know about those elements that will permit steam to escape. Of primary concern is the lateral continuity of major confining barriers and the lesser baffles within reservoir horizons. A secondary concern that often assumes major proportions is the presence of breaches in the confining barriers that result in sand-on-sand contacts of local to nearly field-wide extent.

This paper presents the results of a multidisciplinary team study of the shallow Tertiary to Quaternary Kern River Formation as it hosts heavy oil in the Kern River field. Conventional geologic methodologies were used in tandem with geostatistical methods to develop a geological reservoir model for ultimate use by the production geologists and engineers as well as the simulation engineers. The geological reservoir model was rendered in several dimensions: one- and two-dimensions from cores, well logs, outcrops, cross-sections and maps. Three-dimensional renderings of the reservoir and selected thermal data were designed to be used in tracking the dynamic flow of steam within selected reservoir flow units and in monitoring unconfined steam that encountered sand-on-sand contacts or "holes" between reservoir layers.

The field data were examined on two scales: a field-wide scale (5x6 miles or 8x9.7 km) that helped define the major flow units and general patterns of sand-silt distribution; and a fine scale (a 3000 ft^2 or 915m^2 area that included the Project D-159) that permitted a close look at lithologies, local layers and detailed thermal data taken over time. Additional studies, both planned and in progress, are attempting to use a geological reservoir model and modern computer technology to ultimately improve heat management by way of optimizing recovery, helping to manage water production and improve reliability of reserves estimation.

GEOLOGIC SETTING

Location

The Kern River field occurs in the southern part of California's Great Valley, in a segment known as the San Joaquin basin (Fig. 1). Although this region of the state is famous for its tectonic and seismic activity, the field is located on the relatively quiescent southeastern side of the basin, at the foot of the Sierra Nevada Mountains. Situated just north of the Kern County city of Bakersfield, the field occupies the better part of fourteen square miles (36.3 km^2) of the much larger, upper Cenozoic braidplain built by the Kern River as it exited from a steep-walled canyon in the Sierras.

Stratigraphic Framework

The Kern River Formation is an unconsolidated succession of sand and silt with minor, local claystone and calcareous caliche intervals. Although the basal part of the Kern River accumulated during the late Miocene, most of the formation is of Pliocene and Pleistocene age (Figs. 2 & 3). The formation is capped by gravelly, unsorted Holocene alluvium and colluvium of variable thickness.

Deposition of the Kern River Formation was largely within the braidplain of the ancestral Kern River (Fig. 4). That river flowed into a marine embayment during the late Miocene, with the marine Etchegoin Formation being a lateral equivalent. During the Pliocene that marine environment was progressively displaced to the west by high rates of sedimentation on

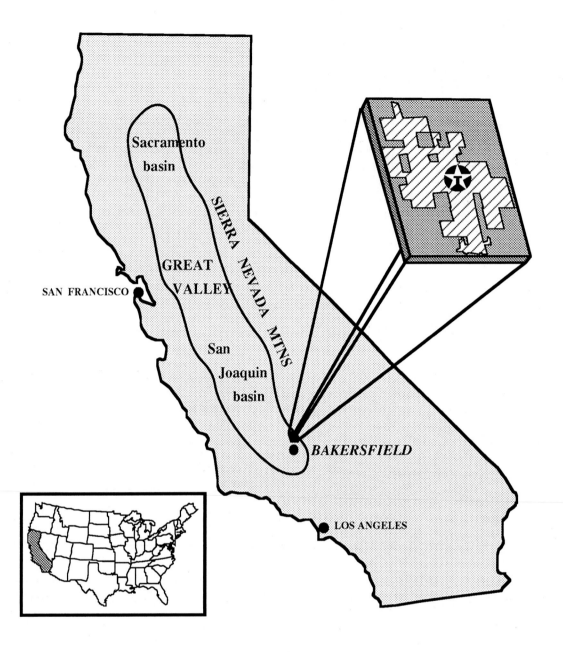

Figure 1. The Kern River field is located in the southern San Joaquin basin portion of California's Great Valley, just to the west of the Sierra Nevada Mountains. The map of the field shows that portion of the field operated by Texaco.

the shelf and slope (Bartow, 1987, 1991). During the Pliocene, the marine connection into the basin was closed (Bartow, 1987, 1991) and brackish to fresh water lakes occupied the center of the basin. Thus the Pliocene and younger deposition in the basin was influenced by local base levels and not simply changes in eustatic sea level. Cyclic deposition within the Kern River Formation appears to be variably controlled 1) by climate cycles in the Sierra Nevada Mountains, particularly during the late Pliocene and Pleistocene, 2) by progressive changes in the drainage area of the Kern River due to stream capture processes in the mountains, 3) by changes in relative base level, and 4) by normal avulsion processes.

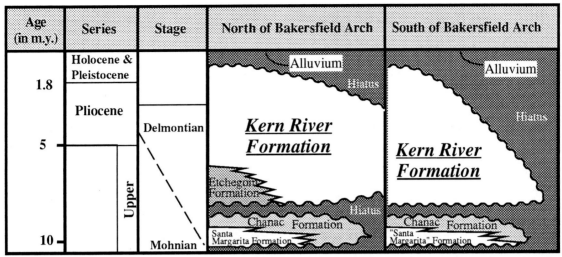

Figure 2. Correlation chart of the youngest Tertiary and Quaternary formations in the southeastern San Joaquin Valley (Bartow & McDougall, 1986).

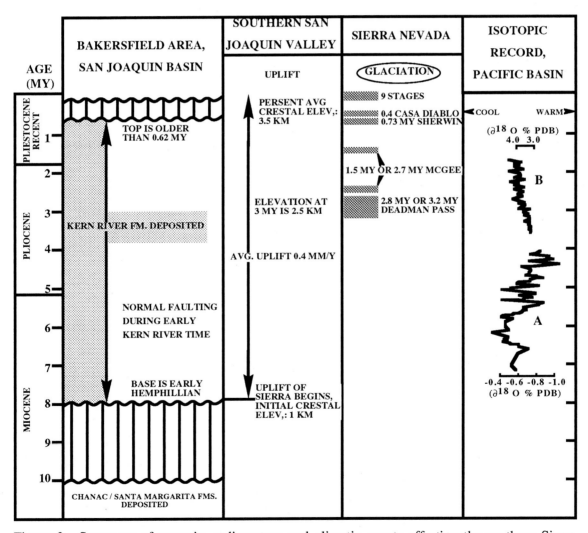

Figure 3. Summary of tectonic, sedimentary and climatic events affecting the southern Sierra Nevada Mountains and the San Joaquin basin during the last 10my (Graham et al., 1988).

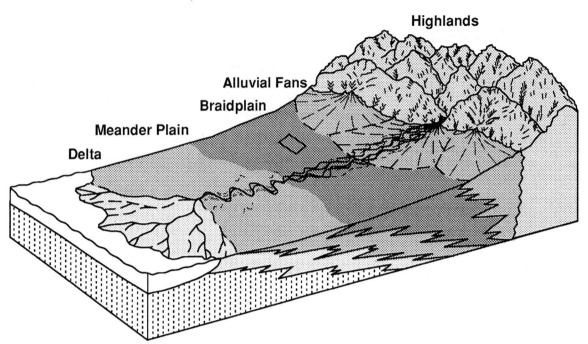

Figure 4. Schematic block diagram of the larger depositional system for the Kern River Formation. Field location is shown by the box above (north) of the channel fairway.

The general stratigraphy of sand and silt layers within the producing part of the formation shows a preponderance of sand over silt layers, and many of the sand layers appear to be amalgamations of several smaller-scaled sand units. A long history of stratigraphic correlations between very closely spaced wells resulted in the identification of 11 major reservoir horizons or flow units, nine of which are noted by the alpha-numeric labels on the type log of Fig. 5. In general, most of the produced oil came from the R and K sand series, and that is where most of the steam displacement takes place today. As a consequence, most of this paper will focus on those deeper horizons.

Structure

A schematic field-wide cross-section (Fig. 6) illustrates the generally west-southwest homoclinal dip of 4-6°. Bartow and McDougall (1984) and Bartow (1987 and 1991) note that the area of the Kern River field is located on the relatively quiescent north flank of the Bakersfield Arch, a basement feature that Sheehan (1986) related to tectonic evolution of the southern Sierras and adjacent granite-cored transpressed mountain systems.

An active normal fault on the west side of the Kern River field separates it from the Kern Front field (Bartow, 1991); virtually no production within Kern River field comes from wells immediately adjacent to that down-to-the-west fault. A major down-to-the-south normal fault system bounds the field along its southern margin, and some producing wells occur right up to that fault. Production from south of that fault system is isolated from the Kern River field. There is neither structural nor stratigraphic closure on the up-dip eastern and northern areas of the field. Localization of the hydrocarbons is thought to be due to some combination of hydrodynamic trapping on the top of the regional groundwater table (the "floating oil pool" idea of Kodl et al. (1990)) or biodegradation and deasphalting that occurs where the oil sands extend close or into surface outcrops.

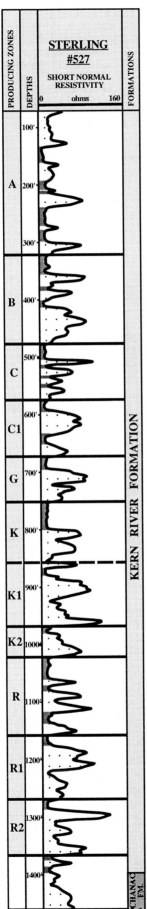

Figure 5. Type well for the D-159 Project area and an outcrop photo from one of the K sand bodies. A) The most useful type of well log is the short normal resistivity which responds to the sediments and the saturating oil and fresh water. Because of the lack of resistivity contrast between the fluids, the log is an excellent lithology discriminator for the silts and sands. To the left of the log are the alphanumeric labels for the major reservoir layers recognized in the field. Note that the layer boundaries are placed at the base of major sand packages and that the bounding silts are combined with the sand with which they are most genetically related. Note the "hole" or sand-on-sand contact where the K Layer rests directly on the K1 Layer, without a bounding silt. B) Outcrop photograph of some of the amalgamated channel sands of one of the K layers.

TABLE 1.—SUMMARY INFORMATION ON THE KERN RIVER FIELD

Trap Type:	Stratigraphic / Hydrodynamic
Reservoir Rocks:	
Age	late Miocene through Pleistocene
Stratigraphic Unit	Kern River Formation
Lithology	Unconsolidated conglomeratic medium- to very-coarse grained feldspathic litharenites
Depositional Environment	Fluvial, braided stream
Productive Lithofacies	Channel sandstones
Entrapping Lithofacies	Floodplain silts and muds
Petrophysical Properties	
Pore Type	Intergranular
Average Porosity	31%
Average Permeability	3000 md
Reservoir Temperature:	
Ambient	90° F
Steam Chest	250° F
Reservoir Pressure:	60 psi
Fluid Data:	
API Gravity	13.5
Oil Viscosity:	
Ambient	800 cp
Steam Chest	15 cp
Field Size:	
Total	9,660 productive acres
Texaco	5,888 productive acres (61%)
Wells:	
Total Active Wells	Approximately 10,000
Texaco's Active Producers	>4,400
Texaco's Active Injectors	>1,500
Injection and Production Statistics:	
Original Oil-in-Place	±3.8 billion barrels
Texaco's OOIP	±2.4 billion barrels
Recoverable Reserves	±1.9 billion barrels
Texaco's Recoverable Reserves	±1.2 billion barrels
Daily Water Injected	±250,000 barrels as 80% quality steam
Daily Oil Produced	±80,000 barrels
Daily Water Produced	±700,000 barrels

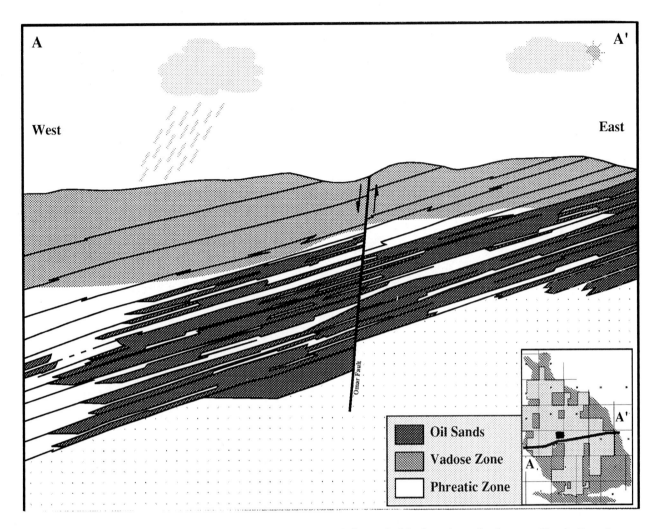

Figure 6. Schematic cross-section of the Kern River field showing the homoclinal dip, the vadose and phreatic zones within the field, and the distribution of hydrocarbons within the major sand and silt layers. Modified from Kodl, et al., 1990.

Within the field, three relatively small normal faults are recognized (Fig. 7): the Green & Whittier, Canfield, and Sterling-Omar faults. The latter one is the longest, has throw of up to 64', and has significant impact on the steam flood and related hydrocarbon production (see later section). The Sterling fault cuts across the northeast corner of the Project D-159 area, the detailed study area of this paper. Also shown on Fig. 7 are the locations of ten field-wide cross-sections that will not be included in this paper but that will be part of the core workshop.

FIELD SUMMARY

Table 1 summarizes most of the pertinent general statistics concerning the Kern River field, particularly the 61 percent operated by Texaco. The most impressive numbers concern the wells themselves: with nearly 10,000 wells in roughly 14 square miles (36.3 square kilometers), that's an *average* of 714 wells per square mile or definitely more than one per acre (Fig 8). In the heart of the field, within the quarter section that contains the D-159 Project, nearly 400 wells have been drilled, including injectors, producers, disposal wells, and temperature observation (TO) wells.

Although most wells are spaced 150-200' (46-61 m) apart, at the extreme end of the spatial distribution are wells so close that it is difficult to get a field vehicle between them!

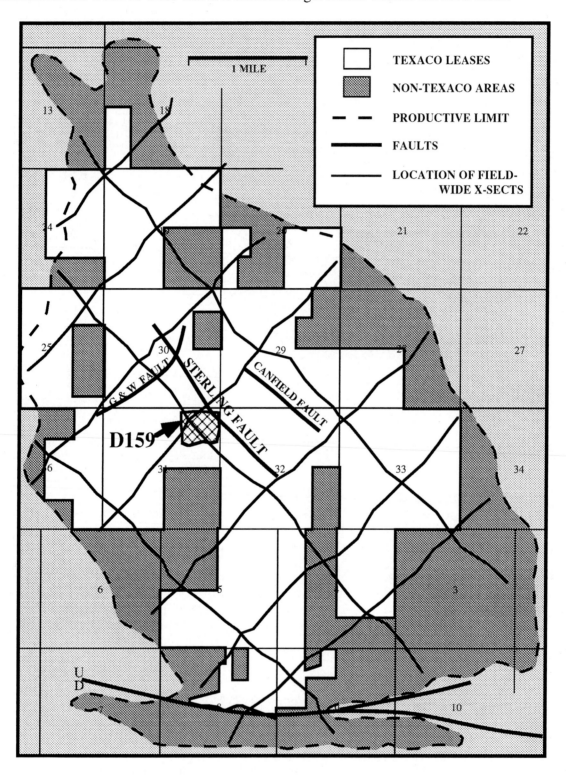

Figure 7. Map of the field showing the outline of the productive part of the field, the location of the D-159 Project area, major and lesser faults that impact the distribution of hydrocarbons in the field, and the location of the set of field-wide cross-sections. Shown in white are the Texaco-operated steam displacement properties.

Figure 8. Photograph of part of the Kern River field showing the high density of wells and abundant production facilities required to support a steam displacement operation. Note pickup trucks at middle-left for scale.

ONE- AND TWO-DIMENSIONAL MODELING

Lithofacies and Depositional Facies

Outcrop data set.—
Although various interpretations of the depositional system had been proposed previously, examination of outcrops in the field and on nearby roadcuts confirmed that the sediments of the upper Kern River Formation that contain hydrocarbons in the field are the product of fluvial deposition (Figs. 9 and 10). Moreover, the river systems delivering sediment to the area were almost exclusively braided rivers formed on the floodplain or braidplain downslope from an alluvial fan at the mouth of Kern Canyon, as shown schematically in Fig. 4.

One characteristic of braided river systems is a multiplicity of channels that repeatedly split and join. Since the braided system is characteristically overloaded with sediment, the abandonment of a choked channel and the formation of a new one is a common process. During high flow or floods, the separate channels may merge into a single wider, deeper channel that often removes at least some of the underlying, pre-existing channel and overbank deposits. A long history of floods can produce a thick, widespread layer of mostly silt-sized floodplain material flanking a fairway of interconnected channel sands.

A braided river system produces two main types of deposits: channel deposits, which are coarse-grained; and sheets of overbank or floodplain deposits, which are generally fine-grained. The sands and silts in the Kern River reservoir are interpreted, respectively, to be channel and overbank deposits from an ancestral braided Kern River. The roadcut in Figure 10 exposes the dark floodplain silts and the lighter-colored amalgamated channel sand layers.

Figure 9. Five-foot (1.5 m) high outcrop of several thin cycles of channel sands from the R series of sands. Note the coarser layer or lag at the base of the cycles, overlain by better sorted and planar to trough cross-bedded channel-fill gravelly sands.

Figure 10. Four-foot (1.2 m) high outcrop showing a different texture and bedding style from Figure 9. Note the overall more uniform coarse grain size with some embedded gravel fragments and the preservation of both stoss and lee slopes in the avalanche cross-beds.

Study of the outcrops provided us with "snapshots" of selected intervals that were useful in several ways. Both geostatisticians and geologists got a sense for the lateral continuity of silt layers, of individual channels, and of the channel systems. The outcrops further taught us that there were **three** lithologic fabrics in the system, not just the two that had previously been recognized:

* <u>in-place sand units</u>, that are the product of deposition in scour-based flooded river channels as high-volume flow decreases (Figs. 9, 10, 11 and 12A, lighter-colored intervals);

* <u>in-place silt layers</u>, that are the widely-distributed floodplain silt-sized sediment deposited by overbank flood waters (Figs. 11 & 12A, darker-appearing intervals; and

* the newly-seen <u>transported silt lumps or mud balls,</u> eroded from the floodplain silts at channel margins, rolled around a bit, and finally deposited along with gravel and sand in the channels (Fig. 12B).

Figure 11. Highway roadcut east of the field showing in the middle and left distance the light-colored amalgamated sands of reservoir-equivalent sediments, overlain by the darker layer of overbank or floodplain silts. The continuity and integrity of those silt units is of major concern when underlying sands are being flooded with steam.

Figure 12A. An excavated location within the field that shows about 10+ feet (3 m) of dark alluvium (Holocene) on top of the sands and silts of the Kern River Formation. Height of the face is about 50 ft (15.3 m). The fairly dark silt that is continuous across the left/top of the photo is thinned to nothing on the right/middle by scouring of the subsequent channel that deposited the overlying light-colored sands. We're looking at the edge of a "hole" in the bounding silt, a hole through which steam could pass.

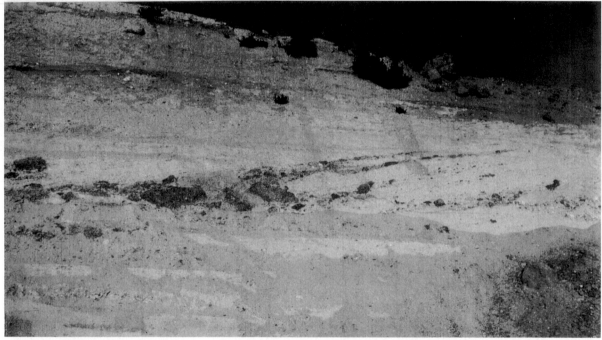

Figure 12B. This 50 ft (15.3 m) high excavation shows some large clasts of floodplain silt (the dark masses at center and left) eroded from a channel margin. Although a resistivity log run through one of those clasts might respond as if it were continuous bedded silt, steam will find a way around these objects. Low resistivity log responses must be carefully examined before drawing conclusions about silt continuity.

Core data set.—

The general conceptual model from the outcrop was enhanced and extended through examination of more than 50 cores, mostly via their photographs. For examples of the facies seen in core, please refer to Figures 13 to 16 and their extensive captions.

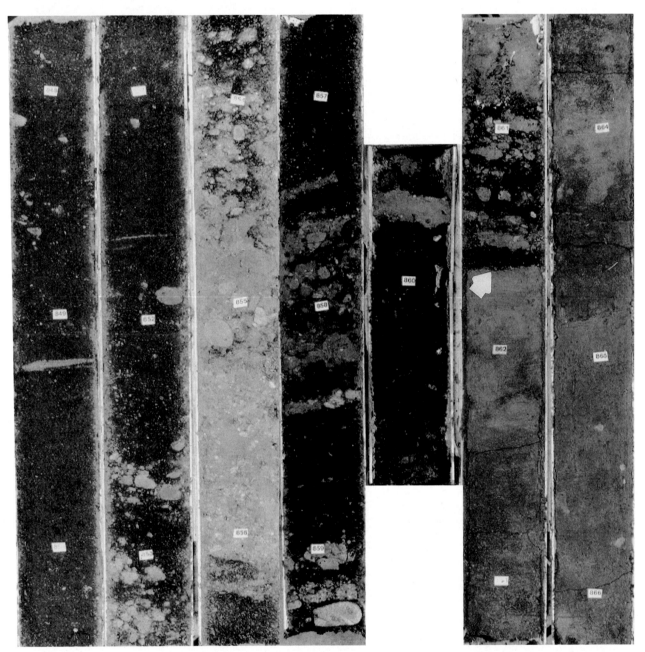

Figure 13. Continuous core, 848-866 ft (259-264 m), from the Cordes TO #2 well, just to the southeast of D-159. The oil saturated sand (dark) is from the base of the K2 Layer, with the boundary of the underlying, bedded, topmost R silt occurring at 861.6' (arrow). Note the sharpness of the contact and the abundance of rip-up clasts of overbank or floodplain silt or mud balls (light-colored) in the K2 sands, particularly in the interval 853-858'. The clast-rich intervals are interpreted to be the initial filling of scoured channels, with the multiplicity of such intervals reflecting repeatedly reactivated channeling and stacking of sands. The core is almost 5" (12.7 cm) wide.

Figure 14. Two intervals of core from the Cordes TO #2 well, 900-908' and 913-921' (274.5-277 m and 278.5-281 m, respectively). The dark, oil saturated sands are from the uppermost channel sand in the R Layer. Note the sharp basal contact with the bedded silt just above 919' (arrow), which is interpreted to be a scoured contact. The contact between the sand and the overlying, bounding silt of the R Layer appears to be rather abrupt, suggesting infilling by overbank sediments after the main channel fairway had shifted elsewhere. The core is almost 5" (12.7 cm) wide.

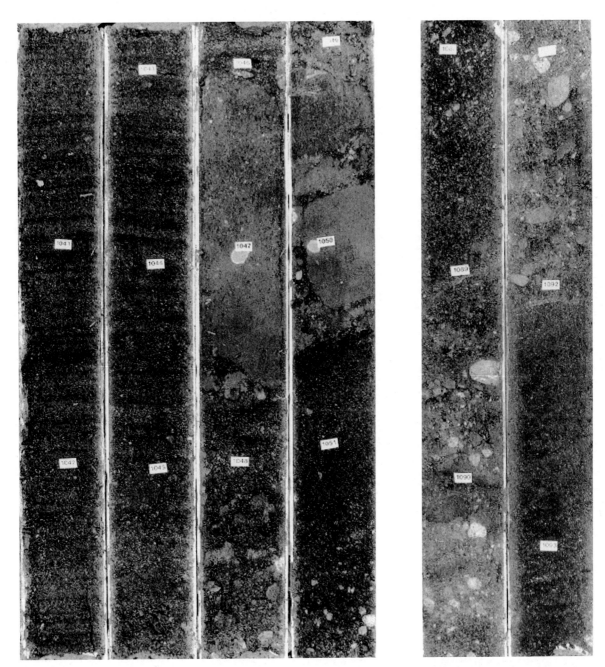

Figure 15. Two intervals of core from the Cordes TO #2 well, 1040-1052 and 1088-1093' (317-321 m and 332-333.4 m, respectively). Both sand intervals are from within the R1 Layer, with the four lengths at the left coming from the middle of the R1 and the two lengths on the right from the lower part of the R1 Layer. Both segments contain abundant mud clasts. The mud balls in the lower R1 are coarse-sand to gravel sized and are easily seen. The mud balls in the middle R1 range from greater than 5" (12.7 cm) in diameter (at 1050') to millimeter sized in the laminated interval in the leftmost two lengths of core. The short normal resistivity log response to these clasts is reduced resistance, in the case of the clast at 1050', almost back to the "shale base line". Although intervals such as this do have reduced resistivity, steam will move through a clast-bearing interval by going around the clasts, whatever their size. The core is almost 5" (12.7 cm) wide.

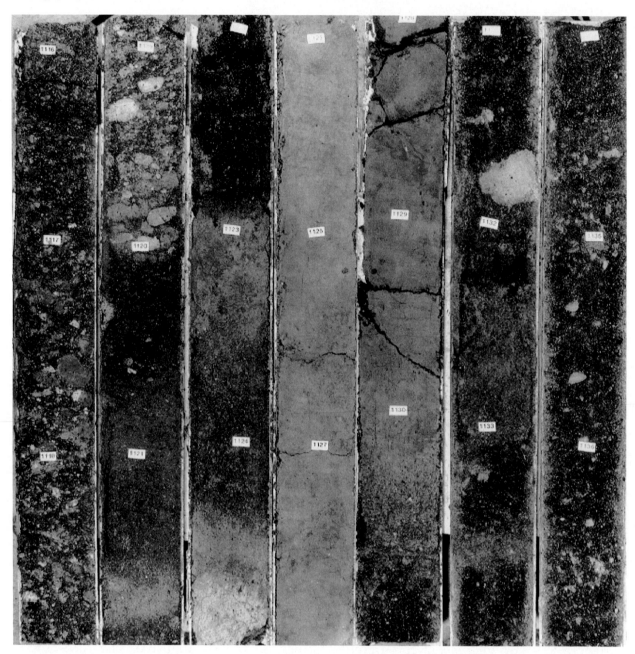

Figure 16. Continuous core from the Cordes TO #2, from 1116-1137 ft (340-346.7 m). Starting at the top of this interval is the sand at the base of the R1 Layer, the bounding silt at the top of the R2 Layer, and the sand in the uppermost R2. Note the gradational contact at the top of the R2 sand, in foot number 1130, suggesting a gradual waning of energy prior to deposition of seven-plus feet (2 plus meters) of overbank or floodplain silt. A full "shale base-line" response is developed by the resistivity curve through the bedded, bounding silt. The basal R1 sand is relatively well sorted for the first few feet, has a relatively sharp base, and contains mud-clast-rich intervals above 1120 ft that will produce a reduced resistivity response but that will also permit the transmission of steam. The core is almost 5" (12.7 cm)wide.

Well log data set.—

Since the formation water in the field is basically fresh and thereby indistinguishable from oil, the SP log is virtually useless and the resistivity logs became lithology logs useful for distinguishing sands from silts. Core-calibrated shallow resistivity logs (short normal) are the fundamental data type used in this study; nonetheless, Fig. 17 shows a full suite of logs. Also shown on Fig 17 is a lithology column that includes the three lithologic fabrics seen in cores and on outcrop. Note the differences on the resistivity log between the sands and the bedded silts, and between the bedded silts and the mud-ball-bearing sands.

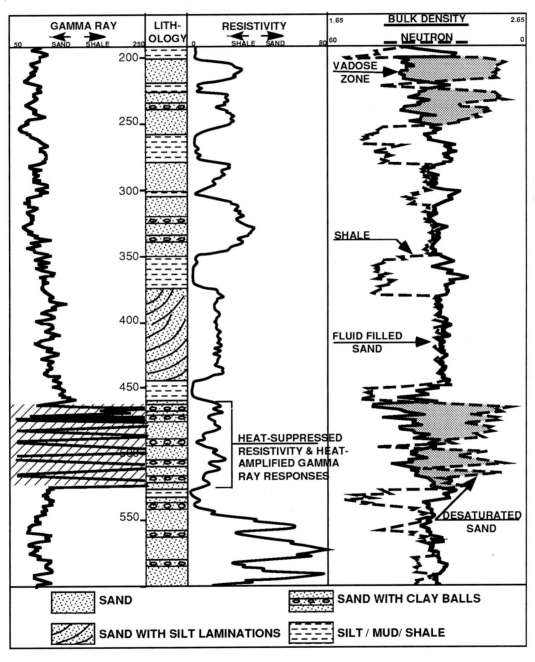

Figure 17. Typical logs for the Kern River field. Combined with the electric log curves is a lithology column that illustrates the resistivity response to various combinations of sand and silt. Note the off-scale response of the gamma curve to the steam chest in the lower part of the log. The porosity curves demonstrate the differing responses to silts, to fluid-filled sands and to sands that are filled with either air or steam.

The newly-seen-and-appreciated silt lumps or mud balls range in diameter from millimeters to at least eight feet (Fig. 12 and from the cores, Figs. 13, 15 & 16). The large clasts will certainly create an impact on resistivity logs run either through or near to them. From outcrops it was clear that although well logs would respond to large lumps as if they were continuous silt layers, steam would go right around them. Thus it is important to distinguish on logs the presence of silt lumps in sand layers from laterally continuous, layered silt.

Diagenesis in the field is almost non-existent. The sands are unconsolidated everywhere except for minor calcium carbonate caliche soil horizons or concretions. The presence of less than 5 percent clay minerals, even in the fine-grained units, the silts, attests to a dominance of mechanical over chemical weathering in both the provenance and the depositional environments. The low concentration of clay minerals and the approximately equal amounts of potassium-bearing feldspathic and micaceous minerals in both the sands and silts makes it impossible to distinguish silts on the gamma ray log, as is commonly done elsewhere.

The Pore System

Pores in the sands and silts of the Kern River Formation are virtually all interparticle, with only minor amounts of secondary dissolution porosity within some of the less chemically stable minerals. The amount of porosity in the sands and silts is about the same, averaging 29-32 percent. On the other hand, differences in permeability are at least one order of magnitude, with the silts in the range of 100-300 md and sands in the range of 1000 to 10,000 md. The high density, interfacial tension and viscosity of the oil, as well as low buoyant forces in the reservoir, combine to keep the oil confined to the sands.

Correlations

Our approach to the relationship between sands and silts in adjacent wells was governed by the concept of channel systems scouring down into and across pre-existing overbank silt under the influence of a significantly lowered base level, or a strong shift in the climate cycle. Our focus was on the bases of the major sand packages, not their tops. Our concern was with the ability of the channel system to either plane off only part of the underlying silt, leaving a barrier between adjacent sands, or to remove all the barrier-producing silts, leaving a sand-on-sand contact. Many of the sand units have fairly flat, dipping basal surfaces, planed off by the rejuvenated channel system. The events that resulted in the migration of the channel system to another area, yielding only overbank silts to accumulate in the area of interest, often continued to provide limited sands via smaller channels, which in turn made for an irregular top to the overall sand package. We did not try to flatten the tops of any sand units; the natural sedimentary system would not produce such a geometry.

Correlations between wells were made using the shallow resistivity log readings, with some supporting interpretations from SP and the few neutron-density logs available. A qualitative shale base-line was developed for each resistivity log, facilitating the discrimination of sands, silts, and intervals with mud balls. The conceptual geologic model guided the correlation process, even to the point of shaping silt/sand interfaces to reflect channel-margin geometries.

Cross-Sections, Grids and Maps

Until quite recently, geological maps and cross-sections were the principal ways in which a "geologic description" was conveyed to others, and they were essentially end products. These geological graphics now are viewed as a series of fundamental steps in rendering the description in three dimensions. The steps still have to be taken: to confirm correlations, to read the larger "story", to evaluate continuity of features, to quality check the data and interpretations, and to have a template against which subsequent computer interpretations can be tested.

The 151 wells in the D-159 study were arranged in seven NW-SE cross-sections and seven NE-SW cross-sections; these orientations are close to alignment with strike and dip, respectively. Three additional sections bound the west, south, and east margin of the study area. These cross-sections used every injector well along the line and all the TO wells in the area. Two of the D-159 cross-sections, DB and SC, coincide with parts of the network of field-wide sections shown in Fig. 7. Those field-wide geologic cross-sections were constructed using every fourth injector, a spacing that is sufficient to develop equivalent geological correlations but that is not sufficient for the concurrent geostatistical analysis.

The correlations on the cross-sections were used to define the reservoir layers or major flow units that are the subject of the next section of the paper. Once the layer bases had been defined, the tabulated elevations of those "picks" were gridded and contoured using Z-MAP *Plus* ™. The resulting structural contour maps on the bases of 11 major sand packages, from the base of A to the base of R2, are the fundamental maps for this study.

Given the 10-foot contouring interval, the maps all show a gentle and regular dip to the SW or WSW. The contours on the base of the K1 Layer (Fig. 18) show this gentle dip, as was also seen in the schematic cross-section of Figure 6. Irregularities on the surfaces reflect topography on the top of the underlying floodplain silt, produced by the erosive channel base and by resistant, or at least non-eroded, remnants of the underlying silts. The most irregular surfaces are those at the base of the B, C and K sands; the extreme irregularities are due to the relative paucity of sand in these layers and due to defining the layers on the base of the lowest sand interval in the layer.

The geology in D-159 is somewhat complicated by the presence of faults, as noted on the map of Figure 18, where the Sterling fault cuts NW-SE across the northeast corner of the area of interest (AOI). The main strand of the Sterling fault was included in the SGM ™ model of the study area.

Reservoir Layering

To accommodate the requirements of at least two different numerical models, the oil-bearing part of the D-159 reservoir was divided into a series of Layers or flow units that would reflect major inter-connected intervals with similar flow features and initially similar bounding silts. Ideal thickness for such a Layer would be 100 feet or more.

Layering in D-159 essentially parallels the large-scale packages of traditionally-identified sands and silts, but our layering differs because different "rules" were used. Fig. 19 presents the southeastern end of cross-section DC (see Fig. 18 for location of the line) and illustrates the results of our applying the following "rules":

* The Layers of this study are defined by the **base** of the major sand package from which the layer takes its name.

* The major bounding silts are bundled into the Layer with their **underlying** package of stacked channel sands, reflecting their genetic link with those sand units.

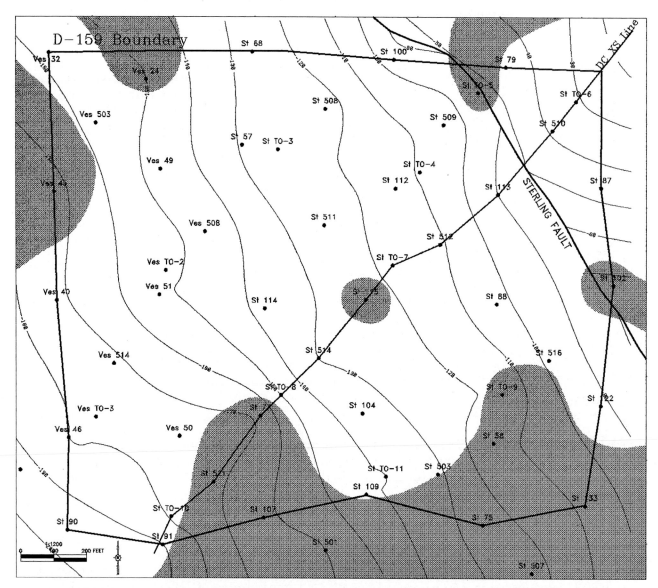

Figure 18. Structural contour map on the base of the K1 sand, which coincides with the base of the K1 Layer or major flow unit. Note the gentle dip to the SW. The dark gray pattern covers those areas of D-159 in which there is sand-on-sand contact with the underlying R sands at the top of the R Layer. In the unpatterned area, the bounding silt between the two layers should be intact.

> * When correlations indicate that all of a pre-existing bounding silt was removed from between major sand packages, the layer boundary on the cross-section was dashed across a sand-on-sand interface; these are the "holes".

Generally, application of the first two rules puts the higher-energy sands at the base of the gross interval or Layer and the low energy silts at the top. An expectation that follows from using the above "rules" is that the greatest lateral continuity in the sands should occur in the lower part of a Layer, with increasing <u>potential</u> for discontinuity at the tops of sands and highest <u>potential</u> for sand discontinuity in the overlying silt. The inverse situation

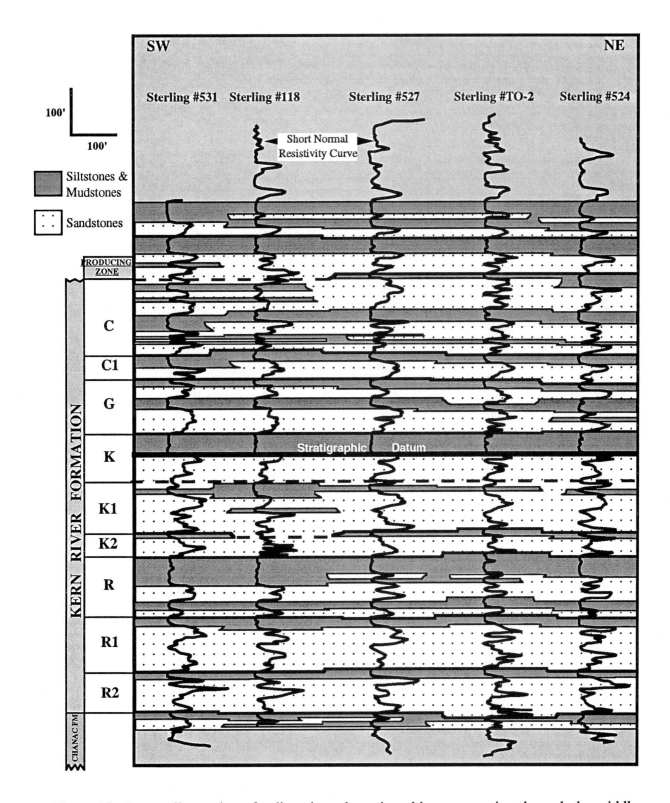

Figure 19. Down-dip portion of a dip-oriented stratigraphic cross-section through the middle of Project D-159. Dashed contacts between major reservoir flow units highlight those units that are in sand-on-sand contact and through which steam will pass. Note also that the boundaries between flow units are defined on the **bases** of the major sand packages.

pertains to the silts: their lowest _potential_ for continuity should be in the lower parts of the sand packages, it should increase toward the top of the sand intervals, and the greatest continuity in the silt beds should occur at the top of the overall Layer or couplet or flow unit.

However, when base level is next lowered or the climate cycle shifts and puts significantly more water into the drainage system or for whatever reason the river energy escalates, the capping silts of the underlying Layer are vulnerable to erosion, to either partial removal or total elimination, depending upon the depth to which the subsequent channel systems can erode.

In a sand-dominated channel system or in a channel system that has the power and the time to scour a flat base across an area, we should expect that the sand unit has a flat base and that the basal sand is everywhere part of the same channel system; they will be "time-equivalent" sands. This situation exists at the bases of the G Layer or flow unit, most of the C and K2 Layers and all of the R series Layers. If, however, the unit is silt-dominated or if the channel system doesn't have the power or the time to create that planar surface, we should suspect that the basal sands in the flow unit may not all be part of the same family of channels and may be of different ages; time lines from the sands could lap-out on silt beds serving as channel margins.

A final expectation is that although some sand units will have rather flat bases, few or none of them will have flat tops across the entire area.

Returning to Figure 19, note that the bases of Layers or flow units are marked by heavy solid lines where sand rests on silt, and by heavy dashed lines where the basal sand scoured down into sands at the top of the underlying unit. Note that in the K2 and the underlying R Layers, the basal sands *on this particular part of this cross-section* show sands resting on continuous silts; we should not expect there to be vertical cross-flow in these units in this area. In contrast, the sand-on-sand contact at the bases of the K and K1 Layers show the "holes" between units and point to avenues for vertical migration of fluids.

"Holes" in the separating or bounding silts in the D-159 area may be defined by as few as one or as many as 60 wells. Since the identification of these holes between major sand packages will have a strong impact on field operations, a series of subcrop maps that locate the holes were made for the base of each layer; Figure 18 is one of these maps. This is a structural contour map on the base of the K1 Layer, with shaded areas showing the sand-on-sand contacts or "holes" between major Layers through which fluids might flow. The occurrence of the basal sand of the K1 Layer on the bedded, continuous silt lithology of the _underlying_ unit is shown by the lack of pattern. From the full set of similar illustrations of this type (since these are not all included in the paper, refer back to Figure 19), it appears that the silt at the top of the R Layer is the only one of the major silts that is continuous across all of D-159, *so far as these data indicate*. Production, thermal and/or RFT data can be used to validate the absence of holes between the injectors used in this study. When the entire field is viewed at this same horizon, as much as 30 percent of the field area has sand-on-sand contacts.

Integrating data from cores and logs in the D-159 area and illustrating them on cross-sections taught us that the river system responsible for the reservoir Layers or flow units did not always have the same character. The R Layers were deposited by a sand-and-gravel dominated river system that swept broadly across the area, with many sand-on-sand channel contacts within the major sand horizons. Many sand channel bases were lined with abundant eroded and transported mud balls that complicate the log correlations but that don't have that much impact on the movement of steam. Between the major sand intervals in the R series, while the main channel system was elsewhere on the braidplain, floodplain silts deposited in the D-159 area were generally continuous and, with few exceptions, were not pierced by the erosive downcutting of the subsequent channel system that swept across the area; that is, at D-159 there are few holes between the major R series of flow units.

The sands in the K series of flow units differ from those in the R series by having reduced resistivity responses, probably produced by the increased inclusion of small mud chips or mud balls as grains in the sands. In contrast to the channel systems of the K and K2 Layers, the K1 channel systems were very efficient at removing overbank silts, which enhanced sand continuity, produced the silt chips and mud balls that lowered the resistivity response, amalgamated the sands to the point that definition of pulses of channel development is almost impossible, and produced inter-connected reservoir units.

In contrast, the K, G, and C1 Layers are so silt-rich that the pulses or episodes of major channel development can be discriminated. Apparently the character of the river system and some of its controls were quite different from those of the lower K and R series.

Geostatistical Modeling of Lithology

A major objective of the detailed geologic study of Project D-159 was a rigorous comparison of stratigraphic cross-sections generated by traditional geologic methods and a comparable set that can be rapidly generated by Texaco's proprietary geostatistical software, GRIDSTAT. The program uses normalized shallow resistivity log data to simultaneously correlate groups of well logs. These well logs can represent 2-D cross-sections or they can be a 3-D volume. Once a 3-D volume is correlated, cross-sections can be extracted.

Resistivity log response in the sands varies from Layer to Layer, probably as the result of lithologic factors, because the formation water is fresh and resistivity does not appear to vary as a function of water saturation.. Characteristically the R series sands (R, R-1, and R-2 Layers) are the most resistive sands in the Kern River Formation. They are generally more than twice as resistive as the shallower C, and G sands (C, C-1 and G Layers). The R sandstones are the coarsest-grained, highest-energy sandstones. Thus, the higher resistivity in the R sandstones may result from reduced clay/mica content and lower surface area/volume ratios in the pore system relative to the finer-grained, lower-energy C and G sandstones. The K series sandstones (K, K-1 and K-2 Layers) are the least resistive. Their characteristic response is approximately twenty percent lower than that of the C and G Layers. The lower resistivity response of the K series of sands is most likely the result of numerous, thin silt laminae that are extremely common in the K sands. *In-situ* bedded silts are represented by low resistivity values and low spontaneous potential values. Resistivity values of the bedded floodplain silts do not appear to vary from one flow unit or Layer to another. However, both resistivity and spontaneous potential gradually increase as the silts became sandier.

The probability distribution function for the resistivity of a typical Kern River well shows that the resistivity distribution is bimodal, with one population occurring below 6 ohm-meters and another, more resistive than 8 ohm-meters. Based on a comparison of shallow resistivity log response with the lithology distribution present in core from six wells in and around the D-159 area, the low resistivity population (below 6 ohm-meters) represents silts, while the high resistivity population (above 8 ohm-meters) represents clean sands. The intermediate values represent sandy silts. For lithology assignment, sandy silts and "clean" silts were lumped together, and the 8 ohm-meter value was used to separate essentially clean reservoir sandstones from non-reservoir silts and sandy muds.

A statistical analysis examined the significance of differences in lithology distribution and well-to-well correlations generated by the two techniques. First, variations in the distribution of sand and silt/mud lithologies within individual wells were compared, <u>on an individual well basis</u>. Then, differences in the individual sand and silt/mud unit correlations between <u>pairs of wells</u> were evaluated, focusing on the continuity attributed to each lithologic unit by each technique. Differences in the lithologic distribution and the well-to-well correlation of the individual units were cataloged separately for faulted and non-faulted wells.

A rigorous statistical comparison of the conventional and geostatistical cross-sections indicates that mean rates of differences in lithology assignment on the GRIDSTAT sections, relative to the conventional cross-sections, are less than 4 percent for all but perimeter cross-sections. Average total correlation difference between GRIDSTAT sections and traditional cross-sections is 6.18 percent. The greatest variations in lithology assignment occur on dip-oriented cross-sections as a result of GRIDSTAT's tendency to generate additional correlations that are not present on geologically-derived cross-sections. This tendency causes reservoir intervals to appear capped by continuous silts when they are not.

The increased wellbore temperature associated with steam flood operations tends to reduce formation resistivity by a factor of 2 or 3 (Mansure et al., 1990, and Ranganayaki, et al., 1992) relative to the pre-steam flood resistivity. Thus, the resistivity data in each well need to be carefully examined, either removing any wells with suppressed resistivity data before using GRIDSTAT or normalizing the data in such a way that the suppressed resistivity data is increased to match the normal data. If wells with suppressed resistivity are deleted from the data set, the rate of differences in lithology assignment drops to a mean of 2.92 percent and directional characteristics (strike vs. dip) disappear. Using data sets from which wells with anomalous resistivity have been deleted, the average difference in lithology assignment between GRIDSTAT and traditional techniques is approximately the same as would be present between cross-sections generated by different geologists.

Even though the average difference rate for lithology assignment is acceptable, approximately 5 percent of the wells on GRIDSTAT cross-sections have lithology assignment error rates that are in excess of 10 percent. Such difference rates have the potential to produce significant errors in determining sand thickness and are unacceptably high. To obtain acceptable lithology assignment in these wells requires the intervention of a geologist.

When wells with suppressed resistivity are deleted from the data set, the average total correlation difference between GRIDSTAT sections and traditional cross-sections improves to 5.53 percent (down from 6.18 percent) and directional differences in total correlation differences cease to be statistically significant. However, there is a tendency for GRIDSTAT to produce cross-sections that are less continuous in the strike direction and more continuous in the dip direction, relative to traditional geologic cross-sections. This tendency is significant at the 0.95 confidence level.

Evaluation of the total correlation difference for faulted and non-faulted wells with normal resistivity indicates that faulted wells contain more total correlation differences than non-faulted wells. Most of the differences arise from additional correlations generated by GRIDSTAT when correlating a pair of wells, one of which is faulted. However, the mean total correlation difference variation is not statistically significant because the faulted sample population is small and its standard deviation is large.

Comparing the percentage manual correlation difference (pMCD) and percentage GRIDSTAT correlation difference (pGCD), the individual components of the total correlation difference of faulted wells indicate that GRIDSTAT-correlated wells (mean pGCD = 8.56 percent) are more likely to contain additional correlations that are not present on the traditional geological cross-sections than to contain errors produced by GRIDSTAT failing to find all geological correlations (mean pMCD = 3.10 percent). The difference is significant at the 0.975 level. Thus, faulted wells on GRIDSTAT-correlated cross-sections are likely to contain silts that appear to be continuous between wells but which will not be present on hand-correlated cross-sections.

GRIDSTAT is able to successfully locate the majority of faults on cross-sections that approached the fault plane at angles greater than 45°. However, the location of the fault cut in 20 to 30 percent of faulted wells is mis-located to such an extent that producer-injector pairs may be "located" in different fault blocks. Traditional cross-sectioning techniques identified a number of faults on cross-sections that approached the fault at a low angle (<30°). This is a geometry that the GRIDSTAT software is unable to deal with because the fault throw produces minimal changes in the nominal dip. GRIDSTAT was unable to locate any faults on cross-sections that approached the fault plane at angles of less than 30°. Nonetheless, when properly-oriented cross-sections are used, GRIDSTAT is an excellent screening tool for finding the general location of faults.

In the hands of a competent geologist, GRIDSTAT is a powerful tool for distributing lithology and performing well-to-well correlations, so long as the data base has been cleared of wells with suppressed resistivity responses. The advantage of increased speed of cross-section generation more than compensates for the small differences in lithology distribution. The tendency for GRIDSTAT to produce more-continuous correlations or cross-sections than would be expected from traditional methods can probably be reduced by modifying the variogram used for kriging.

THREE-DIMENSIONAL MODELING

The primary purpose(s) of building quantitative, three-dimensional (3-D) models such as the one for Project D-159 may be one (or more) of several. At Kern River, the main focus is on building reservoir models that directly aid in the planning, monitoring, and improving steam flood projects in order to maximize hydrocarbon recovery. For instance, these models will be used to impact decisions on perforation, injection, and production strategies.

In other cases, while optimum reservoir management is the ultimate goal, geologically-based reservoir models serve as the numerical input to reservoir simulation and forecasting studies. Tremendous care must be given to scaling up the geologic reservoir model to a format and scale acceptable for reservoir simulation.

The software program SGM™ (**S**tratigraphic **G**eocellular **M**odeling from Landmark/Stratamodel) was used to model the inter-well distribution of reservoir properties in Project D-159. SGM™ uses stratigraphic patterns and correlations to generate a 3-D framework of cells within layers. These layers define stratigraphically equivalent well data (or intervals of well data) and ensure geologic-based interpolations that are constrained by key reservoir features such as lithofacies changes, faults, and unconformities.

An internal architecture (fine-scale layering) built for D-159 was based on the stratigraphic relationships established by visits to field outcrops and correlation work on traditionally constructed geologic cross–sections. As discussed above, each surface represents the unconformable (erosional) *base* of a major channel–sand package or L̲ayer as used in the earlier parts of this paper; these sequence boundaries in SGM™ guide the

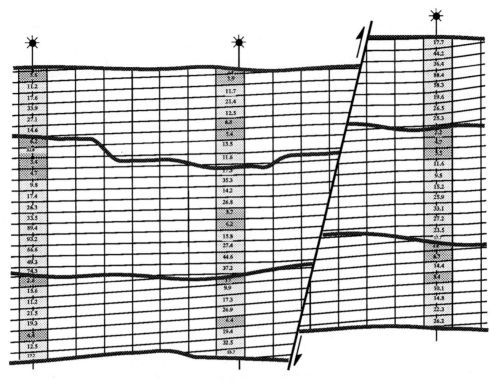

Figure 20. Schematic cross-section illustrating how the "optional depositional pattern" arranges sequence boundaries and internal fine-scale layering for Project D-159. Sequence boundaries (the heavy patterned lines) were loaded using grids from the major flow units, with different grids on the two sides of the fault. The shapes of the sequence boundaries are determined by the shapes of the surfaces at the bases of major reservoir flow units. The layers are 5 ft (1.5 m) thick, and the well data are averaged across that interval.

skeletal framework for the model. In order to accurately model the intersections between the basal surfaces and the Sterling fault, as well as to properly build the fine-scale layers within the sequences, *two* separate grids across the entire AOI were made for each layer's basal surface—one grid for the upthrown surface and one grid for the downthrown surface. Thus, the 12 basal surfaces constructed for the D-159 model are represented by 24 projected or "naive" grids. The projected parts of the surface grids were truncated by the Sterling fault during building of the stratigraphic framework. In addition to the basal surface grids, a grid of the Sterling fault plane was created in Z-MAP *Plus*™.

In the D-159 model, the grids were loaded in the following sequence :

- The naive surface grids from the upthrown fault block were loaded first, from bottom to top

- The Sterling fault grid was input next, truncating all of the "naive" (or projected) parts of the upthrown surfaces—that is, the parts of the upthrown surface grids that extended to the southwest of the Sterling fault. The truncated, southwestern part of the grid was discarded.

- The downthrown surfaces were loaded last, from bottom to top. Similar to the upthrown surfaces, the downthrown grids were truncated at the Sterling fault and the "naive" portions (the portions of the surfaces that extended beyond the fault to the northeast) were discarded.

Building the Stratigraphic Framework—Layering Style and Thickness

The type of fine-scale layering used to build a stratigraphic framework in SGM™ is dependent upon the geological character of the reservoir. The layering style will significantly influence the well-to-well correlation of attributes. In Project D-159, a special unconformable layering style was used in order to best represent the internal stratigraphic character of the reservoir (Fig.20). This layering style involves the use of an "optional depositional pattern", a pattern (i.e., surface grid) that conforms to neither the upper nor lower sequence boundary. As shown schematically in Figure 20, layers are built within the sequence that onlap (filling the lows) the base of the sequence and are truncated (scoured) by the overlying sequence boundary–a layering style that most closely represents the character of a braided river system.

When an optional depositional pattern is used in SGM™, layers of user-specified thickness are built parallel to the depositional pattern. In the D-159 project, a thickness of five feet was used for the layers in each sequence in order to minimize the vertical averaging of well data while keeping the size of the model within reason.

From Well Data to the 3-D Attribute Model

Once the framework was created, thermally-adjusted and normalized resistivity data from 54 wells in Project D-159 were used to populate the 3-D attribute model. Additionally, temperature profiles from eleven TO wells were used to model the monthly change in temperature distribution from January through November of 1992. Table 2 summarizes the parameters that were used to build the attribute model.

Model Analysis and Visualization

The authors believe the potential impact of SGM™ as a reservoir management tool is virtually limitless, especially given the wealth of data available from Kern River field! The most important features of SGM™ include the ability to:

- geologically bias attribute calculations,

- quantitatively analyze the arrangement of attributes, and

- visualize the distribution of attributes in two– or three–dimensions.

A qualitative comparison of the traditional geologic model to the SGM™ model shows that they are very similar with regard to the larger features. Figures 18 and 21 compare very favorably in their location of holes between two major sand intervals, even though Figure 18 is a structural contour map with holes located using the geological cross-sections and Figure 21 is an SGM™ isopach of all the silts within the sands and silts of the K2 Layer. There are, however, differences between the two models with respect to smaller features such as thin, discontinuous sandstones and siltstones. This is not surprising and is most likely a result of the averaging of well data when integrated into the stratigraphic framework. The discrepancies can presumably be minimized by using thin layers in the stratigraphic

TABLE 2.—PARAMETERS AND VALUES USED TO BUILD AN ATTRIBUTE MODEL FOR PROJECT D-159.

OVERALL RUN PARAMETERS*	
Search Radius	1500'
Minimum Well Distance	0**
Number of Search Sectors	16
Allowable Adjacent Empty Sectors	15
Wells per Sector	10
Secondary Search Limit	0**
Well Distance Override	0**
CALCULATION PARAMETERS+	
Calculation Code	Deterministic
Power Function	2
Attribute Dependency	Not used
Vertical Averaging	Not used
DIRECTIONAL BIAS	
Major Axis	80°
Bias Weight	1.7

* parameters specified once for all attributes; cannot vary from one attribute to another

** a zero is SGM™'s designation to ignore this parameter

+ parameters specified separately for *each* attribute; parameters can vary from one attribute to another but parameters were the same for modeling resistivity and temperature data

TABLE 3.—STEAM INJECTION HISTORY FOR THE D-159 AREA.

START DATE	SAND INJECTED	COMMENTS
Aug. 1969	K1	A small region of D-159 area included as part of Project D-013
Jan. 1972	K1	The rest of D-159 area included as part of Project D-021
Jan. 1978	R1	The updip portion of D-159 area recompleted in the R1 sand as part of Project D-079
Jan. 1978	R2	The downdip portion of D-159 area recompleted in the R2 sand as part of Project D-073
Aug. 1983	R	Initiation of Project D-159, all wells recompleted in the R sand & steam injected through 1992
Mar. 1993	K	The updip portion of D-159 recompleted in the K sand as part of Project D-243
Mar. 1993	R1	The downdip portion of D-159 recompleted in the R1 sand as part of Project D-239

framework, therefore decreasing the number of well data points averaged together in each cell. Nevertheless, caution must be exercised when choosing a layer thickness—the effect of decreasing layer thickness in order to minimize averaging of well data is to increase the overall size of the model.

Over the years, the D-159 area has been part of several steam flood projects in a number of different producing zones. Table 3 summarizes the rather complex steam injection history in

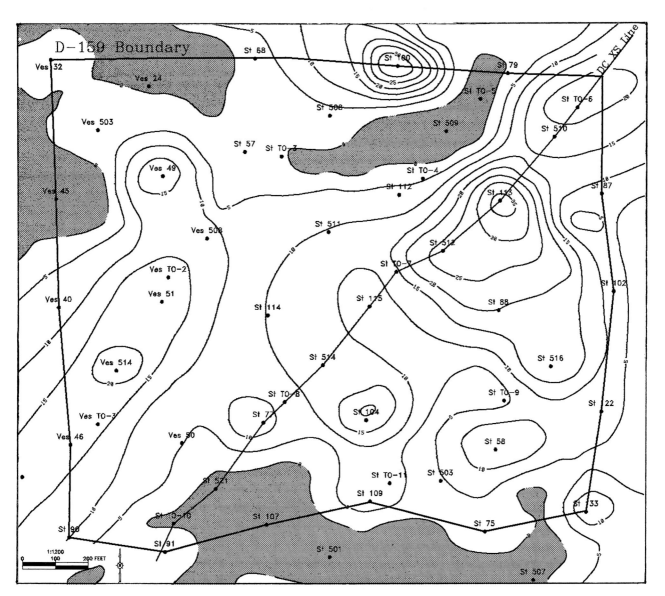

Figure 21. Isopach map of the silt thickness in the K2 Layer of Project D-159. The contour interval is 5 feet (1.5 m); regions of zero thickness (i.e., no silt) are shaded gray. The grid for this map was constructed in SGM™ by summing up the thickness of all cells within the K2 flow unit that contain a resistivity value of less than 10 ohms. The grid was then exported from SGM™ and imported into Z-MAP *Plus*™ for contouring and labeling. Notice the similarity between the shaded ("no silt") regions in this map and the "holes" illustrated on Figure 18, which were derived from traditional geologic cross-sections. The cross-section line annotated **DC XS Line** (labeled in the upper right corner) is shown in Figures 22 and 23.

and around the D-159 area, much of which occurred as part of various other steam flood projects. Steam injection in the sands of the R Layer at Project D-159 began in 1983 and continued into early 1993, but the affects of the previous (and subsequent) events can be seen even today, as will be shown in subsequent figures.

Figure 22 shows a dip–oriented cross–section that cuts through the central part of Project D-159. The colors represent temperatures in the sands as measured in January, 1992. The diverse steam injection history in and around Project D-159 is reflected in the temperature distribution on the cross–section. For instance, steam has not been injected into the K1 sand since January of 1978; however more than fourteen years later temperatures in the K1 (and K2) sand(s) were still relatively high with respect to the overlying zones. The elevated temperatures in the K1 and K2 sands are most likely a combined effect of residual heat from the K1 steam flood as well as heat conducted from the underlying R sand steam flood. The relatively hot temperatures in the R sand, of course, are a result of that zone being an active steam flood at the time of the temperature survey.

There are indications from the temperature distribution in Figure 22 that the Sterling fault acts as an effective barrier to updip steam migration, including:

- the cooler zones on the updip side of the fault in the K1 and K2 sands, and

- the thickened high temperature zone against the fault the extends from the top of the R down into the upper part of the R1 sand, suggesting downward growth and expansion of the steam zone.

Some of the heat against the Sterling fault in the upper R1 sand may be residual from the 1978–to–1983 R1 steam injection (as part of Project D-079—see Table 3). However, there is little difference in temperature between the heated R1 sand and the overlying R sand (typically less than 10° F), suggesting the two zones are being heated at the same time. In addition, the discontinuous character of the silt between the R and R1 sands (or the bounding silt at the top of the R1 Layer) against the Sterling fault would allow communication of fluids between the two zones.

Model operations.--

The use of "model operations" is one of the most robust means of analyzing data in SGM™. Model operations are a way of calculating and/or summarizing data by allowing the user to define and execute mathematical operations, including Boolean (*if, and, or* statements) logic. The operations are applied cell–by–cell and can be used to calculate new attributes, sum or average existing attributes, perform volumetrics, and/or scale–up reservoir properties into two-dimensional grids.

The D-159 data set includes nearly a year's worth of monthly temperature data gathered from 11 TO wells. The change in temperature from January to November 1992, for Project D-159 was calculated using model operations (Fig.23). This was done simply by subtracting one attribute from another—the November 1992, temperature distribution from the January 1992, temperature distribution.

The change in temperature through time was very useful in determining where the injected steam had and had not gone. Several observations about Project D-159 were made from the 1992 change in temperature:

- Temperatures in the R sand, which was actively steamed during 1992, *decreased* by an average of about 4 to 5° F. The reduced temperatures were caused by a decrease in reservoir pressure due to production *and* by the gradual curtailment over the course of the year in the amount of steam injected into the R sand (1992 was the last year of steam injection into the R sand; systematic reduction in the amount of steam injected in the last year of a project is common field operating procedure).

- Temperatures significantly increased in the downdip part of the upper R1 sand during 1992. The steam originated in an adjacent steam flood project (Project D-212) that began injecting steam into the R1 sand in January 1992. The steam migrated updip into Project D-159 along the base of the R1 siltstone until it encountered a hole. The steam then moved up into the R sand, creating a localized "hot spot" in a zone that otherwise cooled down during 1992.

- Temperatures significantly increased during 1992 in several places where there was no steam intentionally introduced. Temperatures increased during 1992 by as much as 50° F in localized regions of the C and C1 sand. The temperature increases were most likely a product of casing leaks in injectors or producers and/or open perforations in producers that allowed heated fluids from below to escape into the formation.

Geobodies.--

The creation of geobodies puts a powerful analytical and visual tool into the hands of the reservoir worker. Geobodies are 3–D volumes of <u>connected</u> attributes having a common property (or properties) and/or satisfying certain criteria. Visualizing geobodies can be *extremely* useful for depicting the continuity, or lack thereof, of various reservoir attributes.

Silt geobodies.-- Paramount to the proper management of a steam flood is the spatial distribution of the bounding silts at the tops of the reservoir flow units. Figure 24A and B illustrate the 3-D distribution of silts for several of the producing zones in Project D-159. In these figures only the silts are shown—the sands are transparent (black). It is clear from images such as these that some silts, for example the ones within and between the K2, R1 and R2 sands (Fig. 24A), will, at best, act as baffles to fluid migration. Steam injected in these layers will pass around the silts or through the holes and will ultimately rise into, and heat the overlying sands. Yet other silts, such as those at the top of the K and R Layers (Fig. 24B), are relatively thick and continuous and will very effectively confine fluids.

Temperature/Sand geobodies.-- Figure 25 illustrates the distribution of *sands* (i.e., the silts are transparent) that had a temperature above 225° F in November 1992. They include:

- nearly all of the R sand, but this is no surprise since steam was being injected into this zone when this temperature data were collected;

- a significant portion of the downdip K1 and K2 sands even though no steam has been directly injected into the K series since 1978;

Figure 22 (top). Dip–oriented cross–section line across Project D-159 (see Fig. 21 for location of line) showing the distribution of silts and sands as well as the temperature distribution for January 1992. The silts have been "filtered out" of the display and are transparent (black) on the cross–section, which gives a sense for the thickness and continuity (or lack thereof) of various units. For scale, the **R Layer** (which includes the R1 silt) is approximately 130 feet thick (39.7 m), and the entire length of the cross–section is about 3000 feet (915 m). The producing sands have been annotated along the margins of the displays and their bases partially marked with white dashed lines. The colors represent temperatures—red areas are $\geq 225°$ F; dark blue regions are $\leq 100°$ F. Notice that the **K1 and K2** sands are 50° to 60° cooler on the upthrown side of the Sterling fault. In addition, note the downward thickening of the hot zone (black arrows) from the **R** sand into the upper **R1** sand against the Sterling fault. These factors suggest that the Sterling fault provides updip confinement within Project D-159, acting as a barrier to the migration of steam.

Figure 23 (bottom). The same cross–section as above with the same horizontal and vertical scales, but the *colors* now represent the **change in temperature** from January, 1992 to November, 1992. Red represents the sands that have *increased* by more than 15° F; the dark blue areas have *decreased* by more than 10° F. The light blue that covers most of the **R** sand (i.e., the active steam zone during 1992) represents a decrease of about 4° to 5° F. The heated (red) zone in the upper **R1** sand was caused by the updip migration of steam along the base of the R1 silt from the adjacent D-212 steam flood, an **R1** project that began in January 1992. The steam was confined to the upper **R1** sand until it neared the Sterling fault, where it found a hole in the silt separating it from the overlying **R** sand, a hole that allowed the steam to escape upward into the upper **R** sand causing a localized "hot spot" (white oval). A much larger "hot spot" can be seen in the **C and C1** sands (white square). This temperature anomaly is most likely the result of a casing leak in an injector or producer, and/or perhaps open perforations in a producer that have allowed some of the hotter produced fluids from below to escape into the cooler formation.

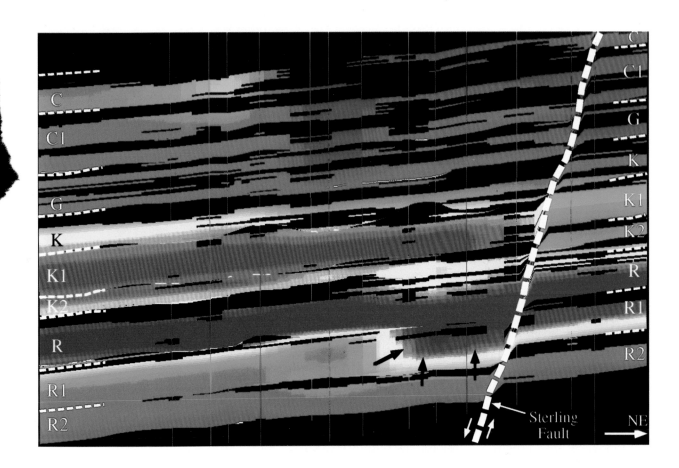

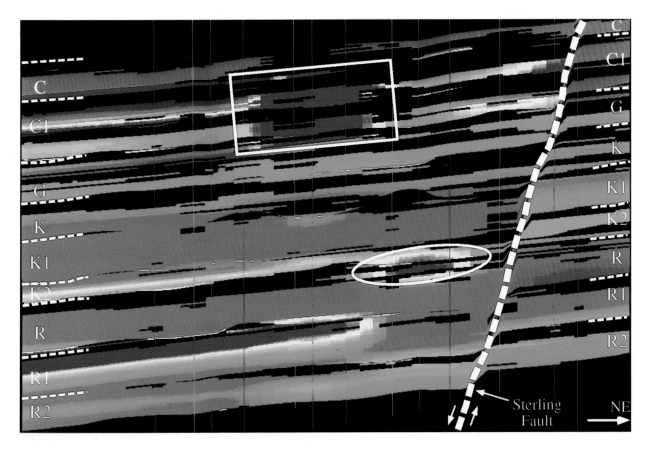

Figure 24A (top). Three–dimensional distribution of silts within the **K2** (green), **R1** (lighter blue) and **R2** (slightly darker blue) Layers *only in the downthrown side of the Sterling fault* The intervening sands (the **K2, R and R1** sands) and silts (the R silt), and all of the upthrown fault block are "transparent" and do not show. The view is looking updip, toward the Sterling fault. For scale, the southern edge of the **R1** silts (i.e., along the right margin of the display) is approximately 2000 feet (610 m) across; its thickness is noted in the display. Obviously, steam injected into sands within any of these silts will not be confined for very far or for very long, particularly within the **K2** Layer. The holes in the bounding **R1** silt near the Sterling fault (yellow arrows) are the same as those noted in the cross–section display (Fig. 23) at the top of the **R1** Layer. These holes have allowed steam and heat to mix and move between the **R and R1** Layers within Project D-159.

Figure 24B (bottom). Three–dimensional distribution of silts within the **K** (red), and **R** (yellow) zones *in the downthrown block only*. The intervening sands (the **K, K1 and K2** sands) and silts (the **K1 and K2** silts), and all of the upthrown fault block, are "transparent" and do not show. The view is looking updip, toward the Sterling fault. For scale, the southern edge of the **R1** silts (i.e., along the right margin of the display) is approximately 2000 feet (610 m) across; its thickness is noted in the display. These silts, at least within the D-159 project area, are relatively thick (in excess of 60 feet or 18.3 m) and continuous and provide true barriers to fluid migration. In other regions of Kern River field, however, these same silt units are thin and discontinuous and are no more than baffles.